Classical Mechanics in Geophysical Fluid Dynamics

This new edition of *Classical Mechanics in Geophysical Fluid Dynamics* describes the motions of rigid bodies and shows how classical mechanics has important applications to geophysics, as in the precessions of the earth, oceanic tides, and the retreat of the moon from the earth owing to the tidal friction. Unlike the more general mechanics textbooks this gives a unique presentation of these applications. The coverage of geophysical fluid dynamics has been revised, with a new chapter on various kinds of gravity waves, a new section on geostrophic turbulence, and new material on the Euler angles, the precession and nutation of a Lagrange top, Rayleigh–Bénard convection, and the Ekman flow.

This textbook for senior undergraduate and graduate students outlines and provides links between classical mechanics and geophysical fluid dynamics. It is particularly suitable for geophysics, meteorology, and oceanography students on mechanics and fluid dynamics courses, as well as serving as a general textbook for a course on geophysical fluid dynamics.

Classical Mechanics in Geophysical Fluid Dynamics
Second Edition

Osamu Morita

CRC Press
Taylor & Francis Group
Boca Raton London New York

CRC Press is an imprint of the
Taylor & Francis Group, an **informa** business

Second edition published 2023
by CRC Press
6000 Broken Sound Parkway NW, Suite 300, Boca Raton, FL 33487-2742

and by CRC Press
4 Park Square, Milton Park, Abingdon, Oxon, OX14 4RN

CRC Press is an imprint of Taylor & Francis Group, LLC

First edition published by CRC Press 2019

ISBN: 978-1-032-31503-4 (hbk)
ISBN: 978-1-032-31504-1 (pbk)
ISBN: 978-1-003-31006-8 (ebk)

DOI: 10.1201/9781003310068

Typeset in CMR10 font
by KnowledgeWorks Global Ltd.

Publisher's note: This book has been prepared from camera-ready copy provided by the authors.

Contents

Preface

The first edition of this book was published at August, 2019. The purpose of the book was to be a link between the classical mechanics and the geophysical fluid dynamics. I think that the purpose was fairly accomplished. After two years passed from the publishing of the first edition, I began to revise the whole manuscript and add a few contents which I withheld in the first edition. They are the Euler angles, the precession and nutation of a Lagrange top, and Rayleigh–Bénard convection with both rigid boundaries and with rigid and free boundary. I enhanced the description of the Ekman flow taking into account of the Southern Hemispheric case. Further, I wrote a new chapter about various kinds of gravity waves and a new section of the geostrophic turbulence.

This book is written for the undergraduate course of two semesters (Chapter 1–Chapter 8, Chapter 9–Chapter 11) and the graduate course of one semester (Chapter 12–Chapter 15). The fundamentals of classical mechanics and general physics are presented in Chapter 1, Introduction. In Chapter 2, Kinematics, the relation between position vector, displacement vector, velocity, and acceleration is described. Chapter 3, Force and Motion, is devoted to Newton's three laws of motion. In this chapter, Newton's second law (the equation of motion) is applied to the idealized object, a particle or a point mass. In Chapter 4, Inertial Forces, inertial forces (apparent forces) in non-inertial reference frames are discussed. Two inertial forces in a rotating system are derived and the problem of the Foucault pendulum is discussed precisely. In Chapter 5, Work and Energy, the new physical concepts in the process of the first transformation of the Newton's second law are introduced. A strict discussion is concerning conservative forces and potential energy, and it is revealed that mechanical energy of a system is conserved without external forces exerting on it. In Chapter 6, Oscillatory Motion, damped and forced oscillations are taken up. The method of solving second-order ordinary differential equations is shown precisely for the first-year undergraduate students who have not yet learned about ordinary differential equations in mathematics lectures. In Chapter 7, Mechanics of Rigid Bodies, the center of mass is defined and it is shown that the motion of the center of mass of a many-particle system obeys Newton's second law. Chapter 8, Momentum and Impulse, presents the second transformation of the Newton's second law. The transformed equation is called the momentum equation, which shows that momentum is conserved in the absence of external forces. Chapter 9, Angular Momentum Equation, is devoted to the angular momentum equation, which is the third transformation of the Newton's second law and relates the changing rate of the angular momentum to torque. The angular momentum of a system is conserved in the absence of torque exerting on the system. In Chapter 10, Motion of Rigid Bodies, the angular momentum equation is transformed to a concise form, in which the new concept, i.e., the moments of inertia, is defined. In this chapter, the calculations of the moment of inertia of various shaped rigid bodies are shown. The Euler angles, which relate the reference frame on the inertial system and that on the rotating rigid body, are introduced. The Euler's equations, which are equations of motion of a rigid body on the reference frame bound to the rigid body, are derived. As a representative example, a discussion is made of the precession and nutation of a Lagrange top. It is presented that the precession of the Earth is one of the causes of climate change, which was proposed by the Serbian geophysicist, Milutin Milanković. In Chapter 11, Orbital Motion of Planets, Newton's law of universal gravitation, and the process Newton discovered it are reviewed. Kepler's three laws are strictly proved. And oceanic tides are discussed as an illustrative example of the law of universal gravitation. An approximate calculation is made for the speed of the retreat of the Moon from the Earth owing to tidal friction. Chapters 1 to 11 are compulsory for undergraduate students. In Chapter 12, Introduction Geophysical Fluid Dynamics, the fundamental basis for geophysical fluid dynamics is presented. We will use the Boussinesq approximation, by which we can treat the phenomena of Earth's atmosphere, oceans, and laboratory systems comprehensively. In Chapter 13, Phenomena in Geophysical Fluids: Part I, we discuss the Taylor–Proudman theorem, Ekman boundary layer, Kelvin–Helmholtz instability, Rayleigh–Bénard convection and Taylor vortices. In Chapter 14, Phenomena in Geophysical Fluids: Part II, shallow water gravity waves, internal gravity waves, and inertio-gravity waves are presented. Chapter 15, Phenomena in Geophysical Fluids: Part III, inertial oscillations, Rossby waves, the barotropic instability, the baroclinic instability, and geostrophic turbulence are discussed. These four chapters are written for graduate students studying geophysics, meteorology, and oceanography.

Interspersed throughout the text are brief biographies of ten noteworthy physicists, astronomers, and geophysicists. I believe that they will refresh readers fatigued with following equations and solving problems to learn about some of the remarkable persons responsible for the development of classical mechanics and geophysical fluid dynamics.

Acknowledgements

I heartily acknowledge Dr. Yasumasa Okochi, Emeritus Professor of National Institute of Technology, Kumamoto College, and Dr. Takahiro Iwayama, Professor of Fukuoka University, for their serious proofreading of the draft and giving much helpful advice. Thanks are due to Ms. Chiho Iseki, my sister, for making sophisticated figures in Chapter 10. I am grateful to the Master and Fellows of Trinity College, Cambridge, England for permission to use a photograph of Sir G. I. Taylor. I also thank the American Meteorological Society for permission to use figures in the Journal of the Atmospheric Sciences. Special thanks are due to Mr. Tony Moore, Senior Editor at Taylor & Francis Group Ltd. for his decision to publish this edition. Finally, I thank the four anonymous referees for their valuable advice and critical comments to polish this book.

March, 2022
Osamu Morita

Author bio

Osamu Morita is a Guest Professor at Fukuoka University with a background in meteorology, physics and of the earth and planetary sciences.

Chapter 1

Introduction

In this chapter, minimal prerequisites not only for classical mechanics but also for physics, in general, are introduced. Physical dimensions, SI base units, SI-derived units, and significant figures are discussed. Four coordinate systems, Cartesian coordinates, plane polar coordinates, cylindrical coordinates, and spherical coordinates are concisely shown. The coordinate transformation, which will be used in later chapters, is discussed. A Taylor series expansion is often used in physics, so that it is precisely explained.

1.1 Physical Dimensions and Units

In general, physical laws give the relationships between physical properties. They have physical dimensions expressed in terms of multiples and ratios of dimensionally independent seven properties: length, mass, time, thermodynamic temperature, amount of substance, electric current, and luminous intensity. Usually they are measured in SI[1] base units, as seen in Table 1.1. Other properties are measured in terms of the SI-derived units, which are formed by multiples and ratios of the SI base units. Some important derived units have special names and symbols, which are shown in Table 1.2.

In order to keep numerical values of physical properties within convenient limits, it is conventional to use decimal multiples and submultiples of SI units. Prefixes used to indicate them are listed in Table 1.3. They are affixed to any of the SI base units and the SI-derived units except the kilogram (kg). Because the kilogram is already a prefixed unit, prefixes for mass should be affixed to the gram (g) not to the kilogram.

---

[1]SI is the abbreviated designation of Le Système International d'Unit.

Table 1.1: SI base units.

Property	Name	Symbol
Length	meter	m
Mass	kilogram	kg
Time	second	s
Temperature	Kelvin	K
Amount of Substance	mole	mol
Electric Current	Ampere	A
Luminous Intensity	Candela	cd

Table 1.2: SI-derived units.

Property	Name	Symbol
Frequency	Hertz	$Hz\,(s^{-1})$
Force	Newton	$N\,(kg\,m\,s^{-2})$
Pressure	Pascal	$Pa\,(N\,m^{-2})$
Energy	Joule	$J\,(N\,m)$
Power	Watt	$W\,(J\,s^{-1})$
Electric Charge	Coulomb	$C\,(A\,s)$
Electromotive Force	Volt	$V\,(W\,A^{-1})$

Table 1.3: Prefixes for decimal multiples and submultiples of SI units.

Multiple	Prefix	Symbol	Submultiple	Prefix	Symbol
10^{24}	yotta	Y	10^{-24}	yocto	y
10^{21}	zetta	Z	10^{-21}	zepto	z
10^{18}	exa	E	10^{-18}	atto	a
10^{15}	peta	P	10^{-15}	femto	f
10^{12}	tela	T	10^{-12}	pico	p
10^{9}	giga	G	10^{-9}	nano	n
10^{6}	mega	M	10^{-6}	micro	μ
10^{3}	kilo	k	10^{-3}	milli	m
10^{2}	hecto	h	10^{-2}	centi	c
10^{1}	deca	da	10^{-1}	deci	d

1.2 Significant Numbers

There are two kinds of physical properties. The value of the first kind is directly measured by appropriate measuring instruments, e.g., length, mass, temperature, etc. The value of the second kind is calculated by products and ratios of directly measured values, e.g., area, volume, density, etc. The former is called the directly measured value and the latter is the indirectly measured value.

There are two types of measuring instruments, one is the analogue type and the other is the digital one. When one uses analogue measuring instruments, it is usual to read to one-tenth of their minimum graduation by estimation. The significant figures of directly measured values depend on the accuracy of the measuring instruments. Suppose that we measure the length of a 100 m dash course and describe it as 100 m. In this description, we can't know whether the value is significant to the rank of 1 m or 0.1 m. When it is significant to the rank of 0.1 m, it is clear to write as 1.000×10^2 m or 100.0 m.

The significant figure of an indirectly measured value is limited by the minimum of directly measured values used for calculation. To avoid meaningless calculations, we should keep the rank of directly measured values the same as the minimum one.

Problem 1. There is a cylinder made of copper. Measuring by a caliper, the diameter is 55.25 mm and the height is 102.35 mm. Calculate the volume of the cylinder in the unit of mm^3.

1.3 Coordinate Systems

In physics, several coordinate systems are used, corresponding to the subject to treat or the scale of phenomena. Although these coordinate systems are basically three-dimensional, two-dimensional, or one-dimensional coordinate systems are also used, corresponding to the degeneracy of the degrees of freedom. Coordinate axes are orthogonal to each other, while they are the combinations of straight lines and curved lines. The coordinate system composed only of straight lines is called a rectangular coordinate system or a Cartesian coordinate system. The coordinate systems composed of more than one curved line are called orthogonal curvilinear coordinate systems.

1.3.1 Cartesian Coordinates

A Cartesian coordinate system is the most familiar coordinate system. Three coordinate axes are usually represented as (x, y, z), which are straight lines and obey the right-hand rule (Fig. 1.1).

1.3.2 Plane Polar Coordinates and Cylindrical Coordinates

While any point on a two-dimensional plane is uniquely designated by (x, y) in Cartesian coordinates, another coordinate system is sometimes convenient using a radius r, the distance outward from the origin, and azimuth angle θ

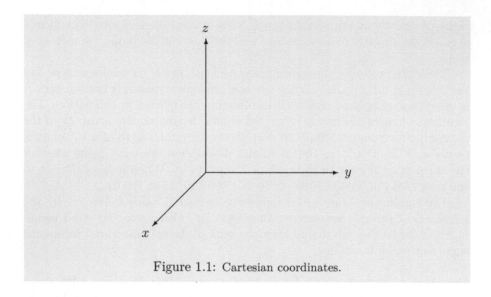

Figure 1.1: Cartesian coordinates.

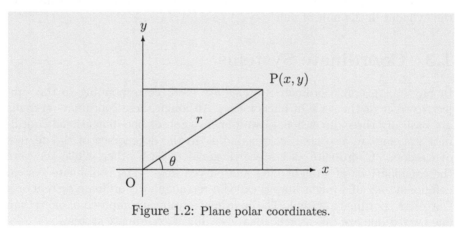

Figure 1.2: Plane polar coordinates.

measured counterclockwise from the x-axis. This coordinate system is referred to as a plane polar coordinate system (Fig. 1.2). The formulae of the coordinate transformation from plane polar coordinates to Cartesian coordinates are

$$x = r \cos \theta, \tag{1.1}$$
$$y = r \sin \theta. \tag{1.2}$$

And the inverse transformation formulae from Cartesian coordinates to plane polar coordinates are

$$r = \sqrt{x^2 + y^2}, \tag{1.3}$$

$$\theta = \tan^{-1}\left(\frac{y}{x}\right). \tag{1.4}$$

The area element of plane polar coordinates at (r, θ) is $dS = r d\theta dr$. The three-dimensional coordinates added the z-axis perpendicular to the r-θ plane of plane polar coordinates are called cylindrical coordinates. The volume element of cylindrical coordinates at (r, θ, z) is $dV = r d\theta dr dz$.

1.3.3 Spherical Coordinates

When we calculate physical properties of a spherical body or discuss the global scale motions, we should use spherical coordinates (Fig. 1.3). Any point in the three-dimensional space is designated by radius r, azimuth angle ϕ, and zenith angle θ. The transformation formulae from spherical coordinates to Cartesian coordinates are

$$x = r \cos\theta \cos\phi, \tag{1.5}$$
$$y = r \cos\theta \sin\phi, \tag{1.6}$$
$$z = r \sin\theta. \tag{1.7}$$

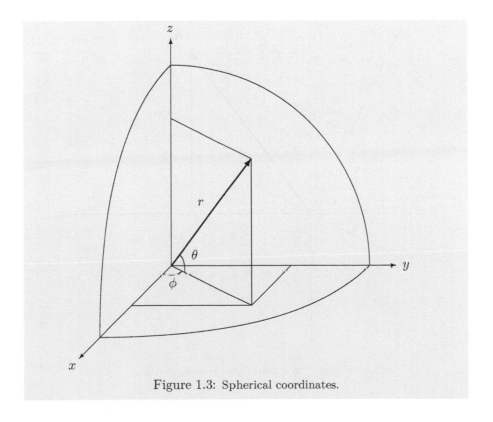

Figure 1.3: Spherical coordinates.

The area element at (ϕ, θ) on a sphere of the fixed radius r is $dS = r^2 \cos\theta d\phi d\theta$, and the volume element at (r, ϕ, θ) is $dV = r^2 \cos\theta dr d\phi d\theta$.

1.4   Coordinate Transformation

Suppose that two-dimensional Cartesian coordinates (x, y) are rotated θ counterclockwise about the origin O, and let the axes of the rotated coordinates be (x', y'). How is an arbitrary vector $\mathbf{A} = (A_x, A_y)$ on the original Cartesian coordinates expressed on the rotated coordinates? Supposing that x'- and y'-components of vector $\mathbf{A}$ are (A'_x, A'_y), we find from Fig. 1.4,

$$A'_x = A_x \cos\theta + A_y \sin\theta,$$
$$A'_y = -A_x \sin\theta + A_y \cos\theta.$$

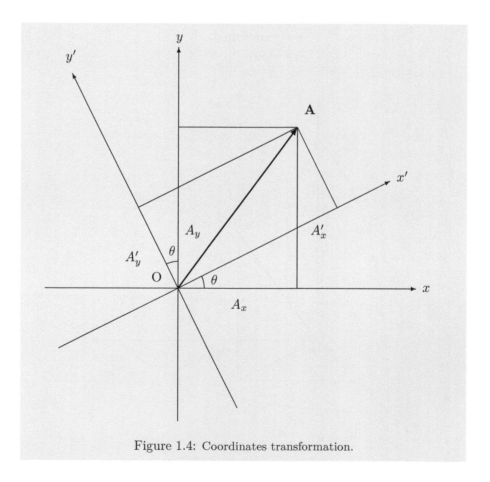

Figure 1.4: Coordinates transformation.

The coordinates transformation formulae are expressed using the matrix form as,

$$\begin{pmatrix} A'_x \\ A'_y \end{pmatrix} = \begin{pmatrix} \cos\theta & \sin\theta \\ -\sin\theta & \cos\theta \end{pmatrix} \begin{pmatrix} A_x \\ A_y \end{pmatrix}. \tag{1.8}$$

The inverse transformation formula from (A'_x, A'_y) to (A_x, A_y) is obtained by operating the inverse matrix on (1.8),

$$\begin{aligned} \begin{pmatrix} A_x \\ A_y \end{pmatrix} &= \begin{pmatrix} \cos\theta & \sin\theta \\ -\sin\theta & \cos\theta \end{pmatrix}^{-1} \begin{pmatrix} A'_x \\ A'_y \end{pmatrix} \\ &= \begin{pmatrix} \cos\theta & -\sin\theta \\ \sin\theta & \cos\theta \end{pmatrix} \begin{pmatrix} A'_x \\ A'_y \end{pmatrix}. \end{aligned} \tag{1.9}$$

The transformation formulae of the coordinate rotation (1.8) and (1.9) are used in the discussion of the Foucault pendulum in section 4.3 and the Euler angles in subsection 10.4.1.

1.5 Taylor Series

If a function $f(x)$ is continuously differentiable, it can be expanded to a power series of $(x - x_0)$ around $x = x_0$. The power series is referred to as a *Taylor series*. When the power series agrees with the function $f(x)$, it is said that $f(x)$ is able to be expanded to a Taylor series,

$$\begin{aligned} f(x) &= C_0 + C_1(x - x_0) + C_2(x - x_0)^2 + \cdots + C_n(x - x_0)^n + \cdots \\ &= \sum_{n=0}^{\infty} C_n(x - x_0)^n. \end{aligned} \tag{1.10}$$

Differentiating successively both sides of (1.10) with respect to x, we obtain

$$\begin{aligned} \frac{df(x)}{dx} &= C_1 + 2C_2(x - x_0) + \cdots + nC_n(x - x_0)^{n-1} + \cdots \\ &= \sum_{n=1}^{\infty} nC_n(x - x_0)^{n-1}, \end{aligned} \tag{1.11}$$

$$\begin{aligned} \frac{d^2 f(x)}{dx^2} &= 2!C_2 + 2 \cdot 3C_3(x - x_0) + \cdots + n(n-1)C_n(x - x_0)^{n-2} + \cdots \\ &= \sum_{n=2}^{\infty} n(n-1)C_n(x - x_0)^{n-2}, \end{aligned} \tag{1.12}$$

$$\vdots$$

$$\frac{d^m f(x)}{dx^m} = m!C_m + \frac{(m+1)!}{1!}C_{m+1}(x - x_0) + \cdots$$

$$+ \frac{n!}{(n-m)!} C_n (x-x_0)^{n-m} + \cdots$$

$$= \sum_{n=m}^{\infty} \frac{n!}{(n-m)!} C_n (x-x_0)^{n-m} . \tag{1.13}$$

$$\vdots$$

Substituting $x = x_0$ into (1.10), (1.11), (1.12), and (1.13), we get

$$C_0 = f(x_0), \tag{1.14}$$

$$C_1 = \frac{1}{1!} \frac{df(x_0)}{dx}, \tag{1.15}$$

$$C_2 = \frac{1}{2!} \frac{d^2 f(x_0)}{dx^2}, \tag{1.16}$$

$$\vdots$$

$$C_m = \frac{1}{m!} \frac{d^m f(x_0)}{dx^m}. \tag{1.17}$$

$$\vdots$$

Substituting from (1.14), (1.15), (1.16), and (1.17) into (1.10), we find

$$f(x) = \sum_{n=0}^{\infty} \frac{1}{n!} \frac{d^n f(x_0)}{dx^n} (x-x_0)^n. \tag{1.18}$$

When $x_0 = 0$, a Taylor series is called as a Maclaurin series.

Example 1. Expand $f(x) = \sin x$ to a Taylor series around $x = 0$.
Answer

$$\frac{df(x)}{dx} = \cos x \ , \quad \frac{d^2 f(x)}{dx^2} = -\sin x \ , \quad \frac{d^3 f(x)}{dx^3} = -\cos x \ , \quad \frac{d^4 f(x)}{dx^4} = \sin x.$$

Denoting the n-th derivative of $f(x)$ as $f^{(n)}(x)$,

$$f^{(n)}(0) = \begin{cases} (-1)^m \ , & \text{as} \quad n=2m+1 \\ 0 \ , & \text{as} \quad n=2m \end{cases} ,$$

$$\sin x = x - \frac{1}{3!} x^3 + \frac{1}{5!} x^5 - \frac{1}{7!} x^7 + \cdots$$

$$= \sum_{m=0}^{\infty} \frac{(-1)^m}{(2m+1)!} x^{2m+1}. \tag{1.19}$$

Example 2. Expand $f(x) = \cos x$ to a Taylor series around $x = 0$.
Answer

$$\frac{df(x)}{dx} = -\sin x \ , \quad \frac{d^2 f(x)}{dx^2} = -\cos x \ , \quad \frac{d^3 f(x)}{dx^3} = \sin x \ , \quad \frac{d^4 f(x)}{dx^4} = \cos x.$$

Therefore,

$$f^{(n)}(0) = \begin{cases} (-1)^m\,, & \text{as} \quad n=2m \\ 0\,, & \text{as} \quad n=2m+1 \end{cases}\,,$$

$$\cos x = 1 - \frac{1}{2!}x^2 + \frac{1}{4!}x^4 - \frac{1}{6!}x^6 + \cdots$$

$$= \sum_{m=0}^{\infty} \frac{(-1)^m}{(2m)!}x^{2m}. \tag{1.20}$$

Example 3. Expand $f(x) = \exp x$ to a Taylor series around $x = 0$.
Answer

$$\frac{d^m f(x)}{dx^m} = \exp x.$$

Therefore,

$$f^{(m)}(0) = 1,$$

$$\exp x = 1 + \frac{1}{1!}x + \frac{1}{2!}x^2 + \frac{1}{3!}x^3 + \cdots$$

$$= \sum_{m=0}^{\infty} \frac{1}{m!}x^m. \tag{1.21}$$

Example 4. Expand $f(x) = (1+x)^k$ to a Taylor series around $x = 0$.
Answer

$$f^{(1)}(x) = k(1+x)^{k-1},$$
$$f^{(2)}(x) = k(k-1)(1+x)^{k-2},$$
$$f^{(3)}(x) = k(k-1)(k-2)(1+x)^{k-3}.$$

$$\vdots$$

$$f^{(m)}(x) = \frac{k!}{(k-m)!}(1+x)^{k-m}.$$

$$\vdots$$

Substituting $x = 0$ into the above results, we find

$$f^{(m)}(0) = \frac{k!}{(k-m)!},$$

$$(1+x)^k = 1 + kx + \frac{k(k-1)}{2!}x^2 + \cdots + \frac{k!}{m!(k-m)!}x^m + \cdots$$

$$= \sum_{m=0}^{\infty} \frac{k!}{m!(k-m)!}x^m. \tag{1.22}$$

1.6   Problems

1. Using the area element $dS = r^2 \cos\theta d\phi d\theta$ in spherical coordinates, calculate the surface area of a sphere of radius a.

2. Using the volume element $dV = r^2 \cos\theta dr d\phi d\theta$ in spherical coordinates, calculate the volume of a sphere of radius a.

3. Prove the *Euler's formula* that

$$\exp\left(\tilde{\imath}x\right) = \cos x + \tilde{\imath}\sin x$$

   by making use of a Taylor series expansion of $\sin x$, $\cos x$, and $\exp x$, where $\tilde{\imath} = \sqrt{-1}$ is the imaginary unit.

4. Prove *de Moivre's theorem*

$$\cos n\theta + \tilde{\imath}\sin n\theta = \left(\cos\theta + \tilde{\imath}\sin\theta\right)^n,$$

   using the Euler's formula.

5. Estimate the round up error of a Taylor series expansion of $\sin x$, neglecting all terms of order x^2 and higher. Suppose that $x = 1.00 \times 10^{-2}$.

Chapter 2

Kinematics

The kernel of classical mechanics is Newton's second law in Chapter 3. It gives the relationship between mass of a body, its acceleration, and forces exerting on it. In this chapter, we will discuss the relationship between the position vector, displacement vector, velocity, and acceleration prior to discussing Newton's second law. Most of the above physical quantities are vectors, so we will study vector calculus, namely addition, subtraction, scalar product, and vector product. The acceleration for uniform circular motion, the centripetal acceleration, is derived. Further, the acceleration in a plane polar coordinate system is derived, in which the centripetal acceleration is included.

2.1   Vector Calculus

There are two kinds of physical quantities, i.e., scalar quantities and vector quantities. While the former has only a numerical value, the latter has a direction in addition to a numerical value. There are several ways to write vector quantities as $\mathbf{A}$, $\vec{A}$, $\tilde{A}$, A, etc. In this book the notation $\mathbf{A}$ is used. The basis vectors are denoted $\mathbf{e}$ adding a coordinate axis as a subscript, for instance $\mathbf{e}_x, \mathbf{e}_y, \mathbf{e}_z$ in Cartesian coordinates.

The magnitude of a vector $\mathbf{A}$ is expressed by $|\mathbf{A}|$ or A, which becomes in Cartesian coordinates as

$$|\mathbf{A}| = \sqrt{A_x{}^2 + A_y{}^2 + A_z{}^2}, \tag{2.1}$$

where A_x, A_y, A_z are the projections of $\mathbf{A}$ to the x, y, z-axes, which are called the x, y, z-components of $\mathbf{A}$ (Fig. 2.1).

2.1.1   Basis Vectors

Basis vectors are a set of linearly independent unit vectors, whose directions are in the coordinate directions. Any vector is expressed by a linear combina-

DOI: 10.1201/9781003310068-2

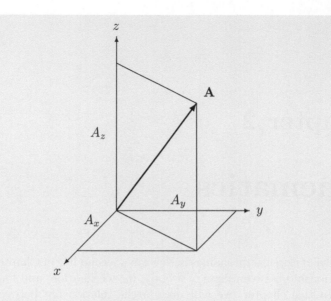

Figure 2.1: An arbitrary vector **A** and its Cartesian components.

tion of basis vectors. For instance, in the Cartesian coordinates, an arbitrary vector **A** is written as

$$\mathbf{A} = A_x \mathbf{e}_x + A_y \mathbf{e}_y + A_z \mathbf{e}_z, \tag{2.2}$$

where $(\mathbf{e}_x, \mathbf{e}_y, \mathbf{e}_z)$ are the basis vectors in a Cartesian coordinate system.

2.1.2   Addition of Vectors

The addition of two vectors $\mathbf{A}, \mathbf{B}$ is defined as the addition of components of **A** and **B** as

$$\begin{aligned}\mathbf{A} + \mathbf{B} &= (A_x \mathbf{e}_x + A_y \mathbf{e}_y + A_z \mathbf{e}_z) + (B_x \mathbf{e}_x + B_y \mathbf{e}_y + B_z \mathbf{e}_z) \\ &= (A_x + B_x)\mathbf{e}_x + (A_y + B_y)\mathbf{e}_y + (A_z + B_z)\mathbf{e}_z. \end{aligned} \tag{2.3}$$

Geometrically, an addition of two vectors is made by parallel translation of **B** putting the initial point of **B** to the final point of **A**, so the initial point and the final point of $\mathbf{A} + \mathbf{B}$ become the initial point of **A** and the final point of **B** (Fig. 2.2). This procedure is called the *triangle method*. Another method is called the *parallelogram method* (Fig. 2.3). First, we translate **B** in parallel, putting the initial point of **B** to the initial point of **A** and form a parallelogram whose adjacent sides are **A** and **B**. Then $\mathbf{A} + \mathbf{B}$ is given by the diagonal from the initial point to the opposite vertex of the parallelogram.

While the addition of vectors is called the vector composition, it is possible to decompose a vector into several vectors. This procedure is called the vector decomposition, and the decomposition of forces becomes important for making component equations from the equation of motion.

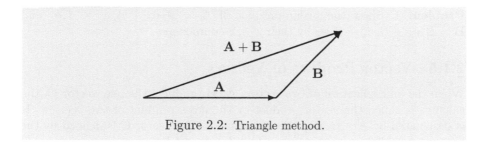

Figure 2.2: Triangle method.

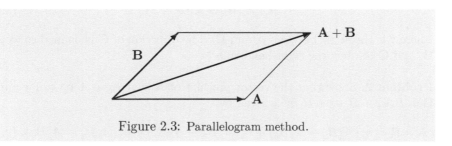

Figure 2.3: Parallelogram method.

2.1.3 Subtraction of Vectors

The subtraction of a vector **B** from a vector **A** is defined as the subtraction of components of two vectors. In Cartesian coordinates, it becomes

$$\mathbf{A} - \mathbf{B} = (A_x\mathbf{e}_x + A_y\mathbf{e}_y + A_z\mathbf{e}_z) - (B_x\mathbf{e}_x + B_y\mathbf{e}_y + B_z\mathbf{e}_z)$$
$$= (A_x - B_x)\mathbf{e}_x + (A_y - B_y)\mathbf{e}_y + (A_z - B_z)\mathbf{e}_z. \tag{2.4}$$

As $\mathbf{A} - \mathbf{B} = \mathbf{A} + (-\mathbf{B})$, the subtraction of two vectors is equivalent to the addition of two vectors **A** and $-\mathbf{B}$, which are done geometrically, making use of the triangle method or the parallelogram method.

2.1.4 Scalar Product of Vectors

When the multiplication of two vectors **A**, **B** produces a scalar, the calculus is called the scalar product or the dot product of two vectors. It is designated by $\mathbf{A} \cdot \mathbf{B}$, and is defined by

$$\mathbf{A} \cdot \mathbf{B} = |\mathbf{A}||\mathbf{B}| \cos\theta, \tag{2.5}$$

where θ is the angle between **A** and **B**.

The scalar products of the basis vectors in Cartesian coordinates are

$$\mathbf{e}_x \cdot \mathbf{e}_x = 1, \quad \mathbf{e}_y \cdot \mathbf{e}_y = 1, \quad \mathbf{e}_z \cdot \mathbf{e}_z = 1,$$
$$\mathbf{e}_x \cdot \mathbf{e}_y = 0, \quad \mathbf{e}_y \cdot \mathbf{e}_z = 0, \quad \mathbf{e}_z \cdot \mathbf{e}_x = 0.$$

Problem 1. Show the scalar product of $\mathbf{A} = A_x\mathbf{e}_x + A_y\mathbf{e}_y + A_z\mathbf{e}_z$ and $\mathbf{B} = B_x\mathbf{e}_x + B_y\mathbf{e}_y + B_z\mathbf{e}_z$ by their x, y, z-components.

2.1.5   Vector Product of Vectors

When the multiplication of two vectors $\mathbf{A}$, $\mathbf{B}$ produces the new vector $\mathbf{C}$, the calculus is called the vector product or the cross product of two vectors. It is designated by $\mathbf{A} \times \mathbf{B}$, and the magnitude of the vector $\mathbf{C}$ is defined by the area of a parallelogram with adjacent sides $\mathbf{A}$ and $\mathbf{B}$,

$$|\mathbf{C}| = |\mathbf{A}||\mathbf{B}| \sin\theta, \tag{2.6}$$

where θ is the angle between $\mathbf{A}$ and $\mathbf{B}$. The direction of $\mathbf{C}$ is defined so as $\mathbf{A}$, $\mathbf{B}$, and $\mathbf{C}$ to obey the right-hand rule.

Problem 2. Show that the vector product of $\mathbf{A} = A_x\mathbf{e}_x + A_y\mathbf{e}_y + A_z\mathbf{e}_z$ and $\mathbf{B} = B_x\mathbf{e}_x + B_y\mathbf{e}_y + B_z\mathbf{e}_z$ is given by

$$\mathbf{A} \times \mathbf{B} = \mathbf{e}_x(A_yB_z - A_zB_y) + \mathbf{e}_y(A_zB_x - A_xB_z) + \mathbf{e}_z(A_xB_y - A_yB_x). \tag{2.7}$$

$\mathbf{A} \times \mathbf{B}$ is written in the determinant form as

$$\mathbf{A} \times \mathbf{B} = \begin{vmatrix} \mathbf{e}_x & \mathbf{e}_y & \mathbf{e}_z \\ A_x & A_y & A_z \\ B_x & B_y & B_z \end{vmatrix}, \tag{2.8}$$

which is a convenient expression and easy to remember.

Problem 3. The magnitude of a vector product of $\mathbf{A}$ and $\mathbf{B}$ in the x-y plane is $(A_xB_y - A_yB_x)$. Prove that the area of a parallelogram OPQR is equal to $(A_xB_y - A_yB_x)$ in Fig. 2.4.

Problem 4. Prove the vector identity

$$\mathbf{A} \times (\mathbf{B} \times \mathbf{C}) = (\mathbf{A} \cdot \mathbf{C})\mathbf{B} - (\mathbf{A} \cdot \mathbf{B})\mathbf{C}.$$

2.2   Displacement and Velocity

Suppose that the position vector of an object at time t is $\mathbf{r}(t) = (x(t), y(t), z(t))$, and in a small time increment δt it moves to $\mathbf{r}(t + \delta t) = (x(t + \delta t), y(t +$

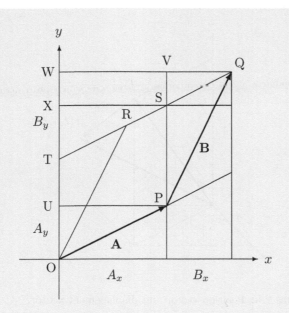

Figure 2.4: Vector product in the x-y plane.

δt), $z(t+\delta t)$) (Fig. 2.5). The difference between $\mathbf{r}(t+\delta t)$ and $\mathbf{r}(t)$ is called the displacement of the object, $\delta \mathbf{r}$. Dividing $\delta \mathbf{r}$ by δt, we obtain the mean velocity of the object $\bar{\mathbf{v}}$ between t and $t+\delta t$,

$$
\begin{aligned}
\bar{\mathbf{v}} &= \frac{\mathbf{r}(t+\delta t) - \mathbf{r}(t)}{\delta t} \\
&= \mathbf{e}_x \frac{x(t+\delta t) - x(t)}{\delta t} + \mathbf{e}_y \frac{y(t+\delta t) - y(t)}{\delta t} + \mathbf{e}_z \frac{z(t+\delta t) - z(t)}{\delta t}. \quad (2.9)
\end{aligned}
$$

Taking the limit $\delta t \to 0$, we get the instantaneous velocity of the object at time t,

$$
\begin{aligned}
\mathbf{v}(t) &= \lim_{\delta t \to 0} \frac{\mathbf{r}(t+\delta t) - \mathbf{r}(t)}{\delta t} = \frac{d\mathbf{r}}{dt} \\
&= \lim_{\delta t \to 0} \left\{ \mathbf{e}_x \frac{x(t+\delta t) - x(t)}{\delta t} + \mathbf{e}_y \frac{y(t+\delta t) - y(t)}{\delta t} \right. \\
&\quad \left. + \mathbf{e}_z \frac{z(t+\delta t) - z(t)}{\delta t} \right\} \\
&= \mathbf{e}_x \frac{dx}{dt} + \mathbf{e}_y \frac{dy}{dt} + \mathbf{e}_z \frac{dz}{dt} = \mathbf{e}_x u + \mathbf{e}_y v + \mathbf{e}_z w, \quad (2.10)
\end{aligned}
$$

where u, v, w are the x, y, z-components of velocity $\mathbf{v}$. The magnitude of the velocity is called speed and written as $|\mathbf{v}|$ or v. While the velocity is a vector,

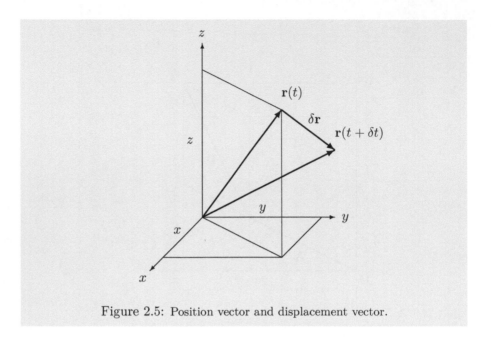

Figure 2.5: Position vector and displacement vector.

the speed is a scalar. Their units are

$$[v] = [\mathrm{m\,s^{-1}}].$$

Hereafter, units will be written with square brackets.

2.3 Velocity and Acceleration

Suppose that the velocity of an object at time t is $\mathbf{v}(t)$ in Cartesian coordinates. Let the velocity in a small time increment δt be $\mathbf{v}(t + \delta t)$ (Fig. 2.6). Dividing the velocity change $\delta \mathbf{v} = \mathbf{v}(t + \delta t) - \mathbf{v}(t)$ by δt, we get the mean acceleration of the object $\bar{\boldsymbol{\alpha}}$ between t and $t + \delta t$. Namely,

$$\bar{\boldsymbol{\alpha}} = \frac{\mathbf{v}(t + \delta t) - \mathbf{v}(t)}{\delta t}$$

$$= \mathbf{e}_x \frac{u(t + \delta t) - u(t)}{\delta t} + \mathbf{e}_y \frac{v(t + \delta t) - v(t)}{\delta t} + \mathbf{e}_z \frac{w(t + \delta t) - w(t)}{\delta t}. \quad (2.11)$$

Taking the limit $\delta t \to 0$, we obtain the instantaneous acceleration of the object at time t,

$$\boldsymbol{\alpha}(t) = \lim_{\delta t \to 0} \frac{\mathbf{v}(t + \delta t) - \mathbf{v}(t)}{\delta t} = \frac{d\mathbf{v}}{dt}$$

$$= \lim_{\delta t \to 0} \left\{ \mathbf{e}_x \frac{u(t + \delta t) - u(t)}{\delta t} + \mathbf{e}_y \frac{v(t + \delta t) - v(t)}{\delta t} \right.$$

$$+ \mathbf{e}_z \frac{w(t + \delta t) - w(t)}{\delta t} \Bigg\}$$

$$= \mathbf{e}_x \frac{du}{dt} + \mathbf{e}_y \frac{dv}{dt} + \mathbf{e}_z \frac{dw}{dt} = \mathbf{e}_x \alpha_x + \mathbf{e}_y \alpha_y + \mathbf{e}_z \alpha_z, \tag{2.12}$$

where $\alpha_x, \alpha_y, \alpha_z$ are the x, y, z-components of the acceleration $\boldsymbol{\alpha}$. From (2.10) and (2.12), we find

$$\boldsymbol{\alpha}(t) = \frac{d\mathbf{v}}{dt} = \frac{d^2\mathbf{r}}{dt^2}. \tag{2.13}$$

The magnitude of the acceleration is written as $|\boldsymbol{\alpha}|$ or α. And the unit of acceleration is

$$[\alpha] = [\mathrm{m\,s^{-2}}].$$

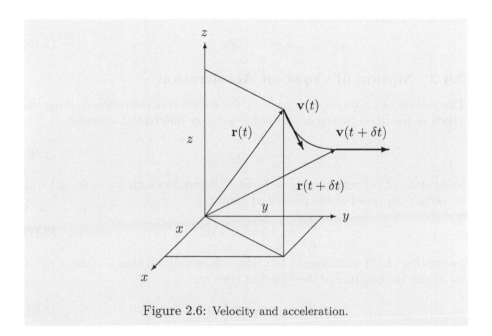

Figure 2.6: Velocity and acceleration.

2.4   One-Dimensional Motion

For simplicity, we will discuss one-dimensional motion in this section. One-dimensional motion is not a special motion, because when we solve two-dimensional or three-dimensional motion using the equation of motion, we decompose it into component equations. For instance, if we want to obtain the trajectory of the two-dimensional motion of an object, we will describe the motion by the equations of motion in the component form and solve them to get the position of the object as the functions of time like $x(t)$ and $y(t)$. The trajectory is obtained by eliminating a parameter t from $x(t)$ and $y(t)$.

2.4.1   Motion of Constant Velocity

The motion of an object moving with a constant speed u along a straight line is called the *linear motion of constant speed* or the *motion of constant velocity*, which is the basic state of motion. Taking the moving direction as the positive x-axis, the relationship between the position x and the speed u is given by the first-order ordinary differential equation,

$$\frac{dx}{dt} = u. \tag{2.14}$$

The position of the object at time t is obtained by integrating (2.14) with respect to t under the condition that $x = x_0$ at $t = 0$,

$$x = x_0 + ut. \tag{2.15}$$

2.4.2   Motion of Constant Acceleration

The motion of an object moving with the constant acceleration α along the x-axis is described by the second-order ordinary differential equation as,

$$\frac{d^2 x}{dt^2} = \alpha. \tag{2.16}$$

Integrating (2.16) with respect to t under the condition that $u = u_0$ at $t = 0$, we obtain the speed of the particle at time t,

$$\frac{dx}{dt} = u(t) = u_0 + \alpha t. \tag{2.17}$$

Integrating (2.17) with respect to t under the condition that $x = x_0$ at $t = 0$, we obtain the position of the object at time t,

$$x = x_0 + u_0 t + \frac{1}{2}\alpha t^2. \tag{2.18}$$

Eliminating time t using (2.17) and (2.18), we find

$$2\alpha(x - x_0) = u(t)^2 - u_0{}^2. \tag{2.19}$$

Equation (2.19) gives the relationship between the acceleration, displacement, initial speed, and final speed, without including time t explicitly.

Problem 5. Suppose that a person releases a stone at rest from a tower of 200.0 m in height at time t=0 s. The stone falls 44.1 m, 78.4 m, and 122.5 m at 3.00 s, 4.00 s, and 5.00 s, respectively. Answer the following questions.

1. What is the mean speed between $3.00\,\mathrm{s}$ and $4.00\,\mathrm{s}$?

2. What is the mean speed between $4.00\,\mathrm{s}$ and $5.00\,\mathrm{s}$?

3. What is the mean acceleration between $3.50\,\mathrm{s}$ and $4.50\,\mathrm{s}$?

Problem 6. An object at rest starts to move with constant acceleration $\alpha = 5.00\,\mathrm{m\,s^{-2}}$ along a straight line at time $t = 0\,\mathrm{s}$. What is the speed of the object at $t = 10.0\,\mathrm{s}$?

2.5 Two-Dimensional Motion

In this section, we will discuss the two-dimensional motion in the x-y plane. A trajectory of an object is obtained by eliminating time t from the position of the object, $x(t)$ and $y(t)$. Two examples, a parabolic trajectory and an elliptic trajectory, are shown.

2.5.1 Elliptic and Parabolic Trajectory

Example 5. Show the trajectory of an object moving in the x-y plane whose x- and y-positions are given by

$$x = a\cos\omega t, \tag{2.20}$$

$$y = b\sin\omega t. \tag{2.21}$$

Answer
Using the trigonometric formula $\sin^2(\omega t) + \cos^2(\omega t) = 1$, we find

$$\frac{x^2}{a^2} + \frac{y^2}{b^2} = 1,$$

which is the elliptic trajectory.

Example 6. Show the trajectory of an object moving in the x-y plane whose x- and y-positions are given by,

$$x = u_0 t\,, \tag{2.22}$$

$$y = \frac{1}{2}\alpha t^2. \tag{2.23}$$

Answer
From (2.22),

$$t = \frac{x}{u_0}. \tag{2.24}$$

Substituting from (2.24) into (2.23), we obtain

$$y = \frac{\alpha}{2u_0{}^2}x^2,$$

which is the parabolic trajectory.

2.5.2   Uniform Circular Motion

The circular motion of an object at constant tangential speed is the constant acceleration motion, because the magnitude of the velocity is constant but the direction of it is continuously changing. An acceleration of circular motion is obtained as follows by using Fig. 2.7.

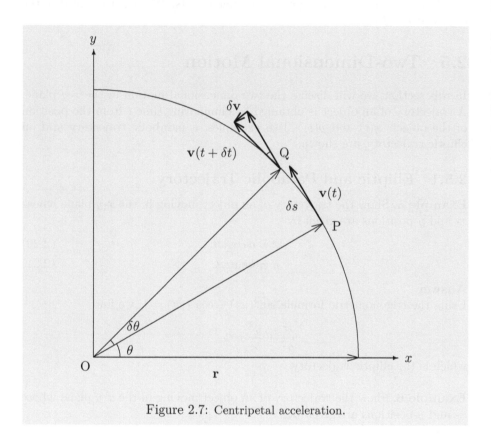

Figure 2.7: Centripetal acceleration.

Suppose that an object at point P has the velocity $\mathbf{v}(t)$ at time t, and an object at point Q has the velocity $\mathbf{v}(t + \delta t)$ at time $t + \delta t$. Denoting an arc $\overset{\frown}{PQ}$ as δs and the angle $\angle POQ$ as $\delta\theta$, we find

$$\delta s = r\delta\theta. \tag{2.25}$$

Dividing (2.25) through by δt and taking the limit $\delta t \to 0$ yields,

$$\lim_{\delta t \to 0} \frac{\delta s}{\delta t} = r \lim_{\delta t \to 0} \frac{\delta \theta}{\delta t},$$

$$v = r\omega, \tag{2.26}$$

where ω is referred to as the magnitude of an angular velocity $\boldsymbol{\omega}$. The angular velocity $\boldsymbol{\omega}$, the position vector $\mathbf{r}$, and the tangential velocity $\mathbf{v}$ obey the right-hand rule. Then the vector form of (2.26) is

$$\mathbf{v} = \boldsymbol{\omega} \times \mathbf{r}. \tag{2.27}$$

As the angle between $\mathbf{v}(t)$ and $\mathbf{v}(t + \delta t)$ is equal to $\delta\theta$, the magnitude of the velocity change δv is

$$\delta v = v\delta\theta. \tag{2.28}$$

Dividing (2.28) through by δt and taking the limit $\delta t \to 0$, the magnitude of acceleration at time t is obtained

$$\lim_{\delta t \to 0} \frac{\delta v}{\delta t} = v \lim_{\delta t \to 0} \frac{\delta\theta}{\delta t},$$

$$\alpha = v\omega = r\omega^2 = \frac{v^2}{r}. \tag{2.29}$$

The direction of the acceleration is from point P to the rotation center O in the limit $\delta t \to 0$, so that the acceleration of the uniform circular motion is referred to as the centripetal acceleration. It is written in the vector form as

$$\boldsymbol{\alpha} = -r\omega^2\mathbf{e}_r = -\frac{v^2}{r}\mathbf{e}_r, \tag{2.30}$$

where $\mathbf{e}_r$ is the basis vector in the radial direction.

Next, we will discuss the projection of the circular motion to the x and y-axes. The transformation relations between plane polar coordinates and two-dimensional Cartesian coordinates are given by (1.1) and (1.2), as

$$x = r\cos\theta = r\cos\omega t, \tag{2.31}$$
$$y = r\sin\theta = r\sin\omega t, \tag{2.32}$$

but for the constant radius in this case. The motion is just the harmonic oscillation which will be discussed in Chapter 3. The x- and y-components of the velocity are obtained by differentiating (2.31) and (2.32) with respect to t,

$$v_x = -r\omega\sin\omega t = -v\sin\omega t = -\omega y, \tag{2.33}$$
$$v_y = r\omega\cos\omega t = v\cos\omega t = \omega x. \tag{2.34}$$

Problem 7. Show that the x- and y-components of the velocity of uniform circular motion (2.33) and (2.34) are obtained geometrically, applying (2.26) at $\theta = 0$ and $\theta = \pi/2$ in Fig. 2.7.

2.6   Acceleration in Plane Polar Coordinates

In this section, we will derive the acceleration in plane polar coordinates. Suppose that an object at point P moves to point Q in a small time increment δt. Let the position of point P be (r, θ) and Q be $(r+\delta r, \theta+\delta\theta)$. The velocity at point P is written as

$$\mathbf{v} = \frac{dr}{dt}\mathbf{e}_r + r\frac{d\theta}{dt}\mathbf{e}_\theta, \tag{2.35}$$

where $\mathbf{e}_r$ and $\mathbf{e}_\theta$ are the basis vectors of the r, θ-directions in the plane polar coordinates. Differentiating (2.35) with respect to t, we obtain the acceleration,

$$\frac{d\mathbf{v}}{dt} = \frac{d^2r}{dt^2}\mathbf{e}_r + \frac{dr}{dt}\frac{d\mathbf{e}_r}{dt} + \frac{dr}{dt}\frac{d\theta}{dt}\mathbf{e}_\theta + r\frac{d^2\theta}{dt^2}\mathbf{e}_\theta + r\frac{d\theta}{dt}\frac{d\mathbf{e}_\theta}{dt}. \tag{2.36}$$

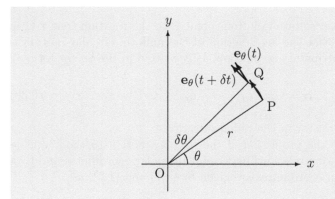

Figure 2.8: Time change of the basis vector $\mathbf{e}_\theta$.

We will consider the derivatives of the basis vectors. The basis vector $\mathbf{e}_\theta$ does not change for the motion in the r-direction, but changes for the motion in the θ-direction. We suppose that $\mathbf{e}_\theta$ is located at (r, θ) and moves to $(r, \theta + \delta\theta)$ in a small time increment δt. We can find that $\delta\mathbf{e}_\theta$ is $\delta\theta$ in magnitude and is directed in the negative r-direction, so that

$$\delta\mathbf{e}_\theta = -\delta\theta\mathbf{e}_r.$$

Dividing the above equation through by δt and taking the limit $\delta t \to 0$, we obtain

$$\frac{d\mathbf{e}_\theta}{dt} = \lim_{\delta t \to 0}\frac{\delta\mathbf{e}_\theta}{\delta t} = -\frac{d\theta}{dt}\mathbf{e}_r. \tag{2.37}$$

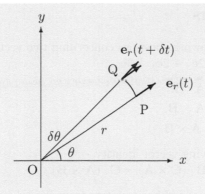

Figure 2.9: Time change of the basis vector $\mathbf{e}_r$.

The basis vector $\mathbf{e}_r$ does not change for the motion in the r-direction, but changes for the motion in the θ-direction. In Fig. 2.9 we suppose that $\mathbf{e}_r$ is located at (r, θ) and moves to $(r, \theta + \delta\theta)$ in a small-time increment δt. We can see that $\delta\mathbf{e}_r$ is $\delta\theta$ in magnitude and is directed in the positive θ-direction, so that

$$\delta\mathbf{e}_r = \delta\theta\,\mathbf{e}_\theta.$$

Dividing the above equation through by δt and taking the limit $\delta t \to 0$, we get

$$\frac{d\mathbf{e}_r}{dt} = \lim_{\delta t \to 0} \frac{\delta\mathbf{e}_r}{\delta t} = \frac{d\theta}{dt}\mathbf{e}_\theta. \tag{2.38}$$

Substituting from (2.37) and (2.38) into (2.36), we get

$$\begin{aligned}
\frac{d\mathbf{v}}{dt} &= \frac{d^2r}{dt^2}\mathbf{e}_r + 2\frac{dr}{dt}\frac{d\theta}{dt}\mathbf{e}_\theta + r\frac{d^2\theta}{dt^2}\mathbf{e}_\theta - r\left(\frac{d\theta}{dt}\right)^2\mathbf{e}_r \\
&= \left\{\frac{d^2r}{dt^2} - r\left(\frac{d\theta}{dt}\right)^2\right\}\mathbf{e}_r + \left(2\frac{dr}{dt}\frac{d\theta}{dt} + r\frac{d^2\theta}{dt^2}\right)\mathbf{e}_\theta.
\end{aligned} \tag{2.39}$$

The second term of the right-hand side of (2.39) is the general form of the centripetal acceleration. We will use this result in Chapters 3 and 11. As the deduction of the acceleration in spherical coordinates is somewhat complicated, we will derive it in Appendix A. It should be kept in mind that additional accelerations arise in orthogonal curvilinear coordinates, because the direction of basis vectors continuously changes.

2.7   Problems

1. Answer the following questions concerning two vectors $\mathbf{A} = 3\mathbf{e}_x - 4\mathbf{e}_y + 5\mathbf{e}_z$ and $\mathbf{B} = -2\mathbf{e}_x + 2\mathbf{e}_y - 3\mathbf{e}_z$.

   (1) Calculate $\mathbf{A} \cdot \mathbf{B}$.

   (2) Calculate $\mathbf{A} \times \mathbf{B}$.

   (3) Normalize $\mathbf{A} \times \mathbf{B}$.

2. Prove the following vector identity,
   $\mathbf{A} \cdot (\mathbf{B} \times \mathbf{C}) = \mathbf{B} \cdot (\mathbf{C} \times \mathbf{A}) = \mathbf{C} \cdot (\mathbf{A} \times \mathbf{B})$.

3. Make two-dimensional basis vectors from linearly independent two vectors $\mathbf{A}$ and $\mathbf{B}$. Let the first basis vector be $\mathbf{A}/|\mathbf{A}|$.
   (Hint: If you decompose $\mathbf{B}$ into two vectors, one is parallel and the other is orthogonal to $\mathbf{A}$, and subtract the parallel vector from $\mathbf{B}$, there remains a vector orthogonal to $\mathbf{A}$.)

4. A vehicle is travelling at $50.0\,\mathrm{km\,h^{-1}}$ on a straight road. A driver brakes and stops the vehicle at $5.00\,\mathrm{m}$ braking. Calculate the magnitude and the direction of the acceleration to significant digits of three, supposing that the acceleration is constant during the braking.

5. Usain Bolt, the most famous athlete in the early 21st century, won the men's final $100\,\mathrm{m}$ race in the 2009 Berlin World Championships in Athletics on 16 August 2009. At that time, he established the world record of $9.58\,\mathrm{s}$, breaking his own previous world record of $9.69\,\mathrm{s}$ in the 2008 Beijing Olympic Games. Precise records are as follows: the starting reaction time was $0.15\,\mathrm{s}$ and the lap times of every $10\,\mathrm{m}$ were $1.89\,\mathrm{s}$, $0.99\,\mathrm{s}$, $0.90\,\mathrm{s}$, $0.86\,\mathrm{s}$, $0.83\,\mathrm{s}$, $0.82\,\mathrm{s}$, $0.81\,\mathrm{s}$, $0.82\,\mathrm{s}$, $0.83\,\mathrm{s}$, and $0.83\,\mathrm{s}$. Answer the following questions.

   (1) Calculate the mean speed of every $10\,\mathrm{m}$ to significant digits of three. In the above calculation, subtract the start reaction time from the lap time and use the net lap time in the first $10\,\mathrm{m}$.

   (2) Calculate the mean acceleration between $5\,\mathrm{m}$–$15\,\mathrm{m}$, $15\,\mathrm{m}$–$25\,\mathrm{m}$, $\cdots$, $85\,\mathrm{m}$–$95\,\mathrm{m}$ to significant digits of three.

   (3) We will idealize Bolt's run as follows: he ran in the constant acceleration from start to $20\,\mathrm{m}$, achieving the top speed at $20\,\mathrm{m}$ and kept it to the goal. Calculate the constant acceleration and the top speed to significant digits of three.

6. The Earth is rotating about the Earth's axis with the period of 23 hours 56 minutes and 4.09 seconds ($86164.09\,\mathrm{s}$). Calculate the angular velocity of the Earth to significant digits of five. Next, calculate the centripetal acceleration at any point on the equator to significant digits of three, supposing that the radius of the Earth is $6.38 \times 10^6\,\mathrm{m}$.

7. The orbital period of the Moon is 27 days 7 hours and 43.2 minutes. Calculate the magnitude of the tangential velocity and the centripetal acceleration to significant digits of three, supposing that the orbit of the Moon is a circle of radius $r = 3.84 \times 10^8$ m.

Chapter 3

Force and Motion

In this chapter, we will study Newton's three laws of motion. Newton's second law, or the equation of motion, gives the relationship between the changing rate of linear momentum of a body and the force exerting on it. Thanks to this equation, we can understand the laws of orbital motion of planets, launch rockets to the Moon, and send space probes off to other planets in the solar system. The equation of motion is truly the essence of classical mechanics.

3.1   Newton's Three Laws of Motion

1. *Newton's first law of motion* (the law of inertia)
    When no external forces are exerting on a *particle*[1] (a point mass) or the total force exerting on a particle is zero, the particle either remains at rest or moves with a constant velocity.

2. *Newton's second law of motion* (the equation of motion)
    The changing rate of momentum of a particle is proportional to the force exerting on it. This law is described in the form of an ordinary differential equation[2] as

$$\frac{d\mathbf{p}}{dt} = \mathbf{F}, \tag{3.1}$$

$$\mathbf{p} = m\mathbf{v}, \tag{3.2}$$

    where $\mathbf{p}$, $\mathbf{F}$, $\mathbf{v}$, and m are momentum, force, velocity, and mass of the particle, respectively.

3. *Newton's third law of motion* (the law of action and reaction)
    When two bodies are interacting, they exert forces equal in magnitude

---

[1]The idealization of an object which has no spatial extent but has only mass.

[2]Exactly speaking, this equation is a proportional relation. The unit of force is defined for the proportional coefficients so as to become unity.

DOI: 10.1201/9781003310068-3

and opposite in direction on each other. Supposing that $\mathbf{F}_{BA}$ is the force acting on body B by body A and $\mathbf{F}_{AB}$ is vice versa, the relation

$$\mathbf{F}_{BA} = -\mathbf{F}_{AB}$$

holds (Fig. 3.1).

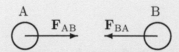

Figure 3.1: Law of action and reaction.

When the mass of a particle is constant, (3.1) becomes,

$$m\frac{d\mathbf{v}}{dt} = m\boldsymbol{\alpha} = \mathbf{F},\tag{3.3}$$

or

$$m\frac{d^2\mathbf{r}}{dt^2} = \mathbf{F}.\tag{3.4}$$

All problems of classical mechanics can be solved by the equation of motion or the transformed versions of it.

The unit of force is defined as N (Newton) after Isaac Newton (Fig. 3.1), and the force $1\,\mathrm{N}$ accelerates a particle of mass $1\,\mathrm{kg}$ by $1\,\mathrm{m\,s^{-2}}$.

$$[F] = [\mathrm{N}] = [\mathrm{kg\,m\,s^{-2}}].$$

The component equations of (3.4) in Cartesian coordinates are

$$m\frac{d^2x}{dt^2} = m\alpha_x = F_x,\tag{3.5}$$

$$m\frac{d^2y}{dt^2} = m\alpha_y = F_y,\tag{3.6}$$

$$m\frac{d^2z}{dt^2} = m\alpha_z = F_z,\tag{3.7}$$

where F_x, F_y, F_z are x, y, z-components of $\mathbf{F}$.

The Gravity of the Earth
Near the Earth's surface, the gravity of the Earth exerting on a particle of

★★★★★★★★★★★★ Sir Isaac Newton (1642–1727) ★★★★★★★★★★★★

Figure 3.2: Sir Isaac Newton was an English physicist, mathematician, astronomer, and Christian theologian. He was born on December 25, 1642, at Woolsthorpe Manor in Woolsthorpe-by-Colsterworth, Lincolnshire. He was admitted to Trinity College, Cambridge, in June 1661 and was awarded a scholarship in 1664. He learned teachings and writings of Aristotle, Descartes, Galileo, and Kepler. In 1665, he discovered the generalized binominal theorem. Soon after he obtained the B.A. degree in August 1665, Cambridge University was closed as the Great Plague was prevailing in London, so he returned to Woolsthorpe for two years, during which he developed the theories on integration and differentiation, optics and the law of universal gravitation. He returned to Cambridge in April, 1667 and became a fellow of Trinity College in October, 1667. Isaac Barrow, the first Lucasian Professor, was impressed by Newton's study and awarded the M.A. to him in 1668. One year later, Isaac Barrow abdicated Lucasian Professor in favor of Isaac Newton. Newton described the law of universal gravitation and Newton's three laws in "*Philosophiæ Naturalis Principia Mathematica*" published in 1687, and established classical mechanics.

★★

unit mass is 9.80 N, normal to the Earth's surface. The Earth's gravity depends on the altitude,[3] the latitude, and the distribution of material in the

---

[3] The difference between the Earth's gravity at the Earth's surface and at the altitude of 30 km is less than 1%.

Earth's crust, and we will discuss precisely the Earth's gravity in Chapter 11. From the equation of motion, we find that the value of the acceleration due to gravity is $9.80\,\mathrm{m\,s^{-2}}$ and is usually designated by g.

Example 1. Suppose that a bob of mass m is attached to one end of a massless and inextensible string. Pulling up the other end of the string, the bob reaches a constant acceleration α. Obtain the tension of the string S, letting the magnitude of the acceleration due to gravity be g.

Answer

Taking the z-axis vertically upward, the total force exerting on the bob is $S - mg$. Then the equation of motion is

$$m\alpha = S - mg.$$

Therefore,

$$S = m(g + \alpha).$$

Problem 1. A vehicle of mass $1.50 \times 10^3\,\mathrm{kg}$ including the mass of a crew is travelling at $50.0\,\mathrm{km\,h^{-1}}$ on a straight road. A driver finding an obstacle on the road brakes urgently and stops the vehicle after $14.0\,\mathrm{m}$ travel. Calculate the frictional force exerting on the tires, supposing that the acceleration is constant and the driving force becomes zero during the braking.

Problem 2. The maximum speed of the fast ball of Shohei Ohtani, the most valuable player (MVP) of the Major League Baseball in 2021, is $1.65 \times 10^2\,\mathrm{km\,h^{-1}}$. The ball takes $1.00 \times 10^{-2}\,\mathrm{s}$ to stop in a catcher's mitt. Calculate the average acceleration of the ball in the catcher's mitt and the average force exerted by the catcher on the ball. Suppose that the trajectory of the ball is straight and its mass is $1.45 \times 10^2\,\mathrm{g}$.

3.2 Falling Motion under Gravity

In this section, we will discuss the falling motion of a particle under the Earth gravity and the air resistance.

3.2.1 The Case of No Air Resistance

Supposing that the air resistance is negligible and taking the y-axis vertically downward, the equation of motion concerning a falling motion of a particle is

$$m\frac{d^2y}{dt^2} = mg. \tag{3.8}$$

Dividing (3.8) through by m and integrating the result with respect to t under the condition that $u = u_0$ at $t = 0$, we obtain

$$\frac{dy}{dt} = v = v_0 + gt. \tag{3.9}$$

Further, integrating (3.9) with respect to t under the condition that $y = y_0$ at $t = 0$, we find

$$y = y_0 + v_0 t + \frac{1}{2} g t^2. \tag{3.10}$$

Falling motion from rest exerted no forces but gravity is called *free-fall*. The relation between the falling distance d and the falling time of free-fall is given by

$$d = y - y_0 = \frac{1}{2} g t^2. \tag{3.11}$$

Equation (3.11) is called the *law of a falling body*, which was discovered by Galileo Galilei (Fig. 3.3) in 1592.

Problem 3. A tourist released a stone from the observation deck of the Ulm Tower in Germany, which is the highest church tower in the world (161.53 m in height). The stone rebounded on the stone pavement below 5.4 s after the release. How high is the deck of the tower? Obtain the height to significant digits of two, supposing that the air resistance is negligible and the value of the acceleration due to gravity is $9.8\,\mathrm{m\,s^{-2}}$. (Caution: This is a hypothetical experiment, never try it!)

3.2.2 The Case of Air Resistance Proportional to Falling Speed

In general, when a body falls in a viscous fluid under gravity, a resistive force exerts on the body. When the falling speed is slow, the resistive force is proportional to the falling speed (*Stokes' law*). Taking the y-axis vertically downward and the y-component of velocity as v, the equation of motion becomes

$$m \frac{dv}{dt} = mg - kv. \tag{3.12}$$

Dividing (3.12) through by m and transposing terms including v to the left-hand side, we obtain

$$\frac{dv}{dt} + \frac{k}{m} v = g. \tag{3.13}$$

Multiplying (3.13) by the integrating factor $\exp(kt/m)$ and arranging terms of the left-hand side, we obtain

$$\exp\left(\frac{k}{m} t\right) \frac{dv}{dt} + \exp\left(\frac{k}{m} t\right) \frac{k}{m} v = g \exp\left(\frac{k}{m} t\right),$$

$$\frac{d}{dt} \left\{ v \exp\left(\frac{k}{m} t\right) \right\} = g \exp\left(\frac{k}{m} t\right).$$

$\star\star\star\star\star\star\star\star\star\star\star\star\star$ Galileo Galilei (1564–1642) $\star\star\star\star\star\star\star\star\star\star\star\star\star\star\star\star$

Figure 3.3: Galileo Galilei was an Italian astronomer, physicist, mathematician, engineer, and natural philosopher. He was born on February 15, 1564, in Pisa, Grand Duchy of Toscana. In the field of astronomy, he improved the telescope (Galilean telescope) and observed the phase of Venus, found four satellites of Jupiter (later named as the Galilean satellites), and observed and analyzed sunspots. Based on his observations he supported heliocentrism, which made his later life very severe. In the field of physics, he discovered the Galilean transformation, the isochronism of a simple pendulum, the law of falling bodies, and the law of inertia of moving bodies. In the field of technology, he invented a thermometer and the geometric and military compass, which brought a lot of money to him. Beyond his great discoveries and inventions, his greatest contribution is the establishment of the method for modern science, so he is called the father of modern science or physics. His beloved elder daughter, Virginia (Sister Maria Seleste who had been taking care of him), died on April 2, 1634, and was buried with him in his tomb later. Galileo lost the light of his eyes in 1637 and died on January 8, 1642, in Arcetri, Grand Duchy of Toscana. Concerning the motion of falling bodies, the famous and symbolic story is told that Galileo dropped two balls of different weight, but made of the same material, from the Leaning Tower of Pisa simultaneously, and observed that the two balls fell to the ground at the same time. Now the story is considered fiction, and the accepted story is that Galileo used a slope with rails and rolled two balls of different weight but the same material on it. He found the law of a falling body; the falling time does not depend on the weight of bodies so long as the air resistance is negligible, and the falling distance from rest is proportional to the square of the falling time.

$\star\star\star\star\star\star\star\star\star\star\star\star\star\star\star\star\star\star\star\star\star\star\star\star\star\star\star\star\star\star\star\star\star\star\star\star\star\star\star\star\star\star\star\star\star\star\star\star\star\star$

Integrating with respect to t, we get

$$v \exp\left(\frac{k}{m}t\right) = \frac{mg}{k}\exp\left(\frac{k}{m}t\right) + C,$$

$$v = \frac{mg}{k} + C\exp\left(-\frac{k}{m}t\right). \tag{3.14}$$

Equation (3.14) is the general solution of (3.13) and C is an integration constant. C is obtained by applying the condition that $v = v_0$ at $t = 0$ to (3.14),

$$C = v_0 - \frac{mg}{k}.$$

Substituting C into (3.14), we get the particular solution

$$v = \frac{mg}{k} + \left(v_0 - \frac{mg}{k}\right)\exp\left(-\frac{k}{m}t\right). \tag{3.15}$$

In the limit $t \to \infty$ the falling speed approaches,

$$v_\infty = \frac{mg}{k}. \tag{3.16}$$

Namely, when the air resistance is proportional to the falling speed, it approaches a constant speed called *terminal velocity*. Falling speeds of precipitating particles and parachutists approach their terminal velocity.

Problem 4. If you want to know only the terminal velocity of a falling body in a viscous fluid without knowing time evolution of the falling speed, you don't need to solve the differential equation. How can you do this?

3.2.3 The Case of Air Resistance Proportional to the Square of Falling Speed

When the falling speed in a viscous fluid is fast, the resistive force is proportional to the square of the falling speed. Taking the y-axis vertically downward and the y-component of the velocity as v, the equation of motion becomes,

$$m\frac{dv}{dt} = mg - kv^2. \tag{3.17}$$

Dividing (3.17) through by m and transforming the resultant equation, we obtain a differential equation of the variable separation form

$$\frac{1}{2\sqrt{g}}\left(\frac{1}{\sqrt{g} + \sqrt{k/m}\,v} + \frac{1}{\sqrt{g}\ \ \sqrt{k/m}\,v}\right)dv = dt. \tag{3.18}$$

Integrating both sides with respect to each variable, we get

$$\sqrt{\frac{m}{k}}\log\left|\sqrt{g} + \sqrt{\frac{k}{m}}v\right| - \sqrt{\frac{m}{k}}\log\left|\sqrt{g} - \sqrt{\frac{k}{m}}v\right| = 2\sqrt{g}t + C',$$

$$\frac{\sqrt{g} + \sqrt{k/m}\,v}{\sqrt{g} - \sqrt{k/m}\,v} = \pm\exp\left(2\sqrt{\frac{kg}{m}}t + \sqrt{\frac{k}{m}}C'\right) = C\exp\left(2\sqrt{\frac{kg}{m}}t\right). \tag{3.19}$$

If the initial speed is 0, we obtain

$$C = 1. \tag{3.20}$$

Substituting (3.20) into (3.19), we get

$$v = \sqrt{\frac{mg}{k}} \frac{1 - \exp\left(-2\sqrt{kg/m}\,t\right)}{1 + \exp\left(-2\sqrt{kg/m}\,t\right)}. \tag{3.21}$$

In the limit $t \to \infty$, the falling speed approaches

$$v_\infty = \sqrt{\frac{mg}{k}}. \tag{3.22}$$

We find the terminal velocity when the air resistance is proportional to the square of the falling speed.

3.3   Parabolic Motion

We will consider the motion of a particle projected with the initial velocity $\mathbf{V}_0$ and the elevation angle θ from the level surface neglecting the air resistance (Fig. 3.4). Taking the y-axis vertically upward, the x-axis in the horizontal and projecting direction, and the origin O as the projecting point, the x- and y-component equations of Newton's second law become

$$m\frac{d^2x}{dt^2} = 0, \tag{3.23}$$

$$m\frac{d^2y}{dt^2} = -mg, \tag{3.24}$$

where g is the magnitude of the acceleration due to gravity. Integrating (3.23) with respect to t considering that $u = V_0 \cos\theta$ at $t = 0$, we find

$$\frac{dx}{dt} = u = V_0 \cos\theta. \tag{3.25}$$

Integrating (3.25) with respect to t considering that $x = 0$ at $t = 0$, we get

$$x = V_0 \cos\theta\, t. \tag{3.26}$$

Integrating (3.24) with respect to t considering that $v = V_0 \sin\theta$ at $t = 0$, we find

$$\frac{dy}{dt} = v = -gt + V_0 \sin\theta, \tag{3.27}$$

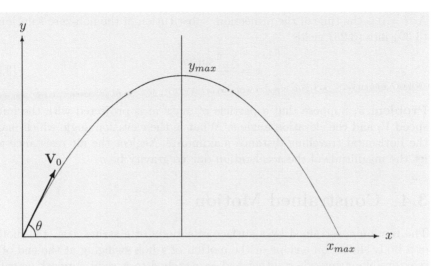

Figure 3.4: Parabolic motion of a particle, where x_{max} designates the maximum reached distance and y_{max} the maximum reached height.

Integrating (3.27) with respect to t taking into account that $y = 0$ at $t = 0$, we get

$$y = -\frac{1}{2}gt^2 + V_0 \sin \theta \, t. \tag{3.28}$$

Eliminating t by (3.26) and (3.28), we obtain the orbit of the particle

$$y = -\frac{g}{2V_0{}^2\cos^2\theta}\left(x - \frac{V_0{}^2\cos\theta\sin\theta}{g}\right)^2 + \frac{V_0{}^2\sin^2\theta}{2g}. \tag{3.29}$$

Equation (3.29) represents a parabola.

Example 2. Suppose that a particle of mass m is projected with the initial velocity V_0 and the elevation angle θ from the Earth's surface. Obtain the distance of the falling point from the projection point, neglecting the air resistance and letting the magnitude of the acceleration due to gravity be g.

Answer
Substituting $y = 0$ into (3.28) to obtain the time in which the particle falls to the Earth's surface,

$$0 = -\frac{1}{2}gt^2 + V_0 \sin \theta \, t,$$

$$t = 0, \quad \frac{2V_0 \sin \theta}{g}. \tag{3.30}$$

As $t = 0$ is the time of the projection, substitution of the non-zero solution of
(3.30) into (3.26) yields

$$x = \frac{V_0{}^2 \sin 2\theta}{g}. \tag{3.31}$$

Problem 5. Suppose that a particle of mass m is projected with the initial
speed V_0 and the elevation angle θ. What is the elevation angle which makes
the horizontal travelling distance maximum? Neglect the air resistance and
let the magnitude of the acceleration due to gravity be g.

3.4   Constrained Motion

The motion constrained by a surface of a body or a string, e.g., the motion
of a body sliding on a slope or the motion of a bob swinging at the end of an
inextensible string whose other end is attached to a rigid support, is called
the constrained motion.

3.4.1   Sliding Motion on a Frictionless Slope

When a body slides down along a frictionless slope with the angle θ to the
level surface, the exerting forces on the body are the resistive force normal to
the slope and the gravity. The gravity should be resolved into parallel and
vertical components to the slope (Fig. 3.5). The former is the same magnitude
and opposite of the vertical resistive force, and the latter drives the body to
slide down with the constant acceleration. Therefore, if we take the x-axis
parallel and downward to the slope and the y-axis upward and normal to the
slope, we find the equations of motion as

$$m\frac{d^2x}{dt^2} = mg \sin\theta, \tag{3.32}$$

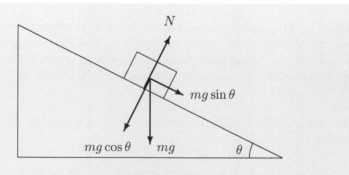

Figure 3.5: Sliding motion on a frictionless slope.

$$0 = N - mg\cos\theta. \tag{3.33}$$

The motion along the slope is equivalent to the falling motion under the gravity $g\sin\theta$, which is called the *reduced gravity*.

3.4.2 Sliding Motion on a Frictional Slope

Forces preventing the motion of bodies are called resistive forces, which act generally opposite to the direction of motion. The frictional force exerting at the interface of the bodies is one of the resistive forces. The magnitude of the frictional force is proportional to the vertical resistive force and depends on the properties of the body surface. The proportional coefficient is called the frictional coefficient, which differs largely whether the body is in motion or at rest. The frictional coefficient in motion is called the *kinematic frictional coefficient* μ' and one at rest is called the *static frictional coefficient* μ. In general, $\mu > \mu'$ and

$$R = \mu N \quad \text{at rest,} \tag{3.34}$$

$$R = \mu' N \quad \text{in motion.} \tag{3.35}$$

Taking the x-axis parallel and downward to the slope and the y-axis upward and normal to the slope as shown in Fig. 3.6, the x- and y-components of the equation of motion become

$$m\frac{d^2x}{dt^2} = mg\sin\theta - \mu'N, \tag{3.36}$$

$$0 = N - mg\cos\theta. \tag{3.37}$$

Example 3. Suppose that a body of mass m is at rest on a frictional slope with the angle θ to the level surface. Obtain the condition under which the

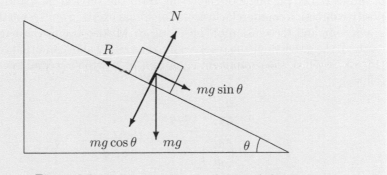

Figure 3.6: Sliding motion on a frictional slope.

body begins to slide, letting the static frictional coefficient be μ and the magnitude of the acceleration due to gravity be g.

Answer
The condition that the body begins to slide is

$$R \leq mg \sin \theta. \tag{3.38}$$

The vertical component of the equation of motion is

$$0 = N - mg \cos \theta. \tag{3.39}$$

The relation between R and N is

$$R = \mu N. \tag{3.40}$$

Using (3.38), (3.39), and (3.40), we find

$$\mu \leq \tan \theta. \tag{3.41}$$

Therefore, the static frictional coefficient may be obtained by increasing the elevation angle until the body begins to slide.

3.4.3   Simple Pendulum

Suppose that a particle of mass m is attached to a rigid support O via a massless and inextensible string of length l. Keeping the system at rest, the string becomes vertical and the particle stays at the lowest position, or the equilibrium position. Pulling the particle from its equilibrium position, keeping the string tight, and releasing the particle, the particle begins to oscillate in the vertical plane around its equilibrium position. This oscillatory system is referred to as a simple pendulum. A plane polar coordinate system is adequate to describe the motion because the particle moves along the circle of radius l. Let the angle of the string from the vertical be θ and taking the positive direction counterclockwise. The forces exerting on the particle are the gravity mg and the tension of the string S. Making use of the parallelogram method, the gravity may be resolved into the radial and azimuthal components (Fig. 3.7). Then, the component equations of Newton's second law become as follows,

$$ml \frac{d^2\theta}{dt^2} = -mg \sin \theta, \tag{3.42}$$

$$ml \left(\frac{d\theta}{dt} \right)^2 = S - mg \cos \theta, \tag{3.43}$$

where use is made of the acceleration terms in plane polar coordinates given by (2.39). Now we assume a small amplitude oscillation, namely $\theta \ll 1$.

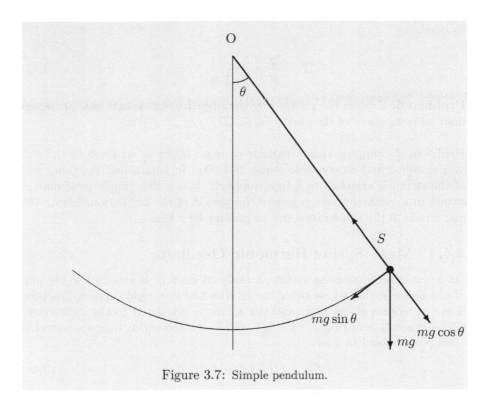

Figure 3.7: Simple pendulum.

Expanding $\sin \theta$ in a Taylor series around $\theta = 0$ and neglecting all terms of order θ^2 and higher, (3.42) becomes

$$\frac{d^2\theta}{dt^2} + \frac{g}{l}\theta = 0. \tag{3.44}$$

Assuming an exponential type solution $\theta \propto \exp(\lambda t)$ and substituting it into (3.44), we obtain

$$\lambda = \pm \tilde{\imath}\sqrt{\frac{g}{l}} = \pm \tilde{\imath}\omega, \tag{3.45}$$

where $\tilde{\imath} = \sqrt{-1}$ is the imaginary unit. Therefore, the general solution of (3.44) is

$$\theta = \theta_0 \cos(\omega t - \phi_0), \tag{3.46}$$

where θ_0, ω and ϕ_0 are the amplitude, the angular frequency,[4] and the initial phase, respectively. The period, the time required to return from some phase

---

[4] Angular frequency is often abbreviated to just frequency.

to itself, is

$$T = \frac{2\pi}{\omega} = 2\pi\sqrt{\frac{l}{g}}. \tag{3.47}$$

Problem 6. Explain the reason why the negative sign is attached to the azimuthal component of the gravity in (3.42).

Problem 7. Suppose that a particle of mass $50.0\,\mathrm{g}$ is attached to the end of a massless and inextensible string of $1.00\,\mathrm{m}$ in length and the other end of the string is attached to a rigid support. When this simple pendulum is swung in a vertical plane, calculate the period of the oscillation, letting the magnitude of the acceleration due to gravity be $9.80\,\mathrm{m\,s^{-2}}$.

3.4.4   Mass–Spring Harmonic Oscillator

On a level and frictionless surface, a body of mass m is attached to the end of a massless spring whose other end is attached to a rigid support. Suppose that the motion of the body and the spring is restricted to the x-direction. Giving a small deviation $x = l - l_0$ to the body, the spring exerts a restoring force proportional to x as

$$F = -k(l - l_0) = -kx, \tag{3.48}$$

where k is the spring constant, l_0 is the natural length, and l is the length of spring exerted by an external force. The relationship (3.48) is called *Hooke's law*. The equation of motion describing the oscillatory motion is

$$m\frac{d^2x}{dt^2} = -kx. \tag{3.49}$$

The general solution of (3.49) is given by

$$x = x_0 \cos(\omega t - \phi_0), \tag{3.50}$$

where $\omega = \sqrt{k/m}$ is called the angular frequency and ϕ_0 is the initial phase. The period of the oscillatory motion is

$$T = \frac{2\pi}{\omega} = 2\pi\sqrt{\frac{m}{k}}. \tag{3.51}$$

Problem 8. Suppose that we connect two springs of length l and different spring constants k_1 and k_2 in series. Prove the following relationship

$$\frac{1}{k} = \frac{1}{k_1} + \frac{1}{k_2}, \tag{3.52}$$

where k is the equivalent spring constant, supposing that the serial springs are one spring of length $2l$.

3.5 Centripetal Force in Uniform Circular Motion

The acceleration of a uniform circular motion is directed to the center of a circle and the magnitude of it is constant as was discussed in subsection 2.5.2. Therefore, the centripetal force of constant magnitude is required for the centripetal acceleration. Examples of this force include the gravitational force exerting on planets in their orbital motion, the pulling force by a hammer-thrower acting on a hammer and a frictional force exerting on tires of a car cornering a curve, and so on.

The equation of motion for a particle of mass m in a uniform circular motion of radius r, the tangential speed v, and the angular velocity ω is

$$m\frac{v^2}{r}(-\mathbf{e}_r) = mr\omega^2(-\mathbf{e}_r) = \mathbf{F}, \qquad (3.53)$$

where $\mathbf{e}_r$ is the basis vector in the radial direction.

Problem 9. A train is cornering a curved track with radius $r = 2.50 \times 10^2$ m. The track is banked θ from the level surface, so that trains can travel safely at $50.0 \, \mathrm{km \, h^{-1}}$. Calculate θ letting the magnitude of the acceleration due to gravity be $9.80 \, \mathrm{m \, s^{-2}}$ (Fig. 3.8).

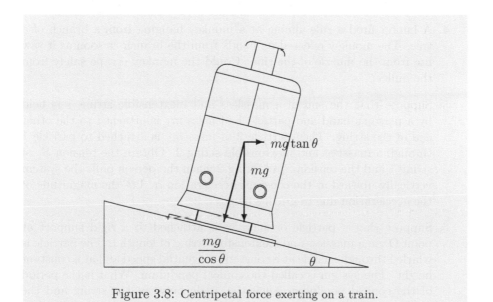

Figure 3.8: Centripetal force exerting on a train.

3.6   Problems

1. What is the maximum height reached by a particle which is projected vertically upward from the Earth's surface at the initial speed V_0? Next, obtain the time for the particle to fall to the ground, letting the magnitude of the acceleration due to gravity be g.

2. Suppose that a particle of mass m is projected from the Earth's surface at the initial speed V_0 and the elevation angle θ. What is the maximum height reached by the particle and how far does it travel until falling to the ground, letting the magnitude of the acceleration due to gravity be g?

3. The Japanese hammer-thrower, Koji Murofushi, got a bronze medal at the 2012 Summer Olympics in London with the record of 78.71 m. Answer the following questions, supposing the mass of a hammer is 7.26 kg, the elevation angle of the projection is 45°, and neglecting the air resistance. Let the value of the acceleration due to gravity be $9.80\,\mathrm{m\,s^{-2}}$.

   (1) Calculate the initial speed of the hammer at the projection, supposing that the initial position is at the Earth's surface.

   (2) Calculate the force pulling the wire of the hammer by Murofushi, supposing that the length of the wire is 1.20 m and the distance from the central axis of Murofushi's body to his palm is 1.10 m.

4. A hunter fired a rifle aiming at a monkey hanging from a branch of a tree. The monkey released its hands from the branch as soon as it saw fire from the muzzle of the rifle. Could the monkey escape safely from the bullet?

5. Suppose that the end of a massless and inextensible string 1 is held by a person's hand and particle 1 of mass m_1 is attached to the other end of the string. Then particle 2 of mass m_2 is attached to particle 1 through a massless and inextensible string 2. Obtain the tension S_1 of string 1 and the tension S_2 of string 2 when the person pulls the system vertically upward at the constant acceleration α. Let the magnitude of the acceleration due to gravity be g.

6. Suppose that a particle of mass m is attached to a rigid support at point O via a massless and inextensible string of length l. The particle is whirled through a circle at a constant tangential speed and at a constant height. The system is called the conical pendulum. What is the period of the conical pendulum when the angle between the string and the plumb line passing through O is θ?

7. The Moon is orbiting about the center of mass of the Earth–Moon system at the period of 27 days 7 hours and 43.7 minutes. Answer the

following questions, letting the distance between the Earth and the Moon be 3.84×10^8 m and the mass of the Moon be 7.35×10^{22} kg.

(1) Calculate the centripetal force exerting on the Moon, supposing that the lunar orbit is a circle.

(2) The centripetal force is due to the universal gravitation between the Earth and the Moon. Obtain the ratio of the magnitude of the acceleration due to gravity at the center of the Moon to that of the Earth's surface, $g_0 = 9.80 \, \mathrm{m \, s^{-2}}$.

8. Two springs of length l and with different spring constants k_1 and k_2 are connected in parallel. Prove the relationship $k = k_1 + k_2$, supposing that two springs are equivalent to a spring of length l and spring constant k.

Chapter 4

Inertial Force

So far, it is presumed that the observer of a moving body is on the reference frame at rest somewhere in the universe, which is called the absolute reference frame. How is the motion of a body observed on a reference frame moving relative to the absolute reference frame? Does the equation of motion remain valid in these reference frames or not? A reference frame which moves in the constant velocity relative to the absolute reference frame is called an inertial frame. The equation of motion is valid in an inertial frame so long as the force does not depend on the velocity. A reference frame in an acceleration motion to the absolute reference frame is referred to as a non-inertial reference frame, and the equation of motion is still valid adding inertial forces or apparent forces to the real forces. In the uniformly rotating reference frame, two inertial forces, the centrifugal force, and Coriolis force exist. Foucault pendulum is precisely discussed at the end of this chapter.

4.1 Relative Motion

How is the motion of a body moving at the velocity $\mathbf{V}$ seen by an observer on a reference frame moving at the velocity $\mathbf{V}_0$ relative to the absolute reference frame? From the observer's viewpoint, the velocity $-\mathbf{V}_0$ should be added to the velocity of the body, because the observer recognizes himself being at rest. Therefore, the body moves at the velocity $\mathbf{V} - \mathbf{V}_0$ on the observer's reference frame. For instance, a passenger boarding a Shin–Kansen train travelling at $300 \, \text{km} \, \text{h}^{-1}$ sees the outside view moving backward at $300 \, \text{km} \, \text{h}^{-1}$.

Problem 1. An observer is driving a car eastward on a straight road at $50.0 \, \text{km} \, \text{h}^{-1}$. A bus is travelling northward at $50.0 \, \text{km} \, \text{h}^{-1}$ on a road crossing the former at a right angle. What is the direction and how fast is the motion of the bus seen by the observer?

DOI: 10.1201/9781003310068-4

4.2 Inertial Frames and Non-Inertial Frames

A reference frame moving in the uniform velocity relative to the absolute reference frame is called the inertial reference frame, in which the equation of motion remains valid so long as forces do not depend on the velocity. The reference frame in the acceleration motion relative to the absolute reference frame is referred to as the non-inertial reference frame, in which inertial forces must be added to the real forces for the equation of motion to remain valid. Two inertial forces in the rotating system, the centrifugal force and the Coriolis force, are exactly derived and the Foucault pendulum will be discussed in detail in the following section.

4.2.1 Inertial Frames

As shown in Fig. 4.1, letting the constant velocity of the inertial frame be $\mathbf{V}_0$, the position vector of a body in the absolute reference frame at time t be $\mathbf{r}$, and the position vector of the body in the inertial frame at t be $\mathbf{r}'$, the relationship between two position vectors is

$$\mathbf{r} = \mathbf{r}_0 + \mathbf{V}_0 t + \mathbf{r}'. \tag{4.1}$$

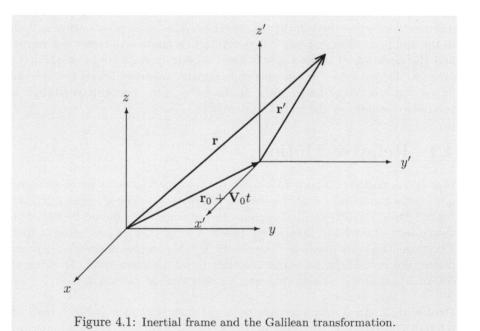

Figure 4.1: Inertial frame and the Galilean transformation.

Equation (4.1) gives the transformation between two reference frames and is called the *Galilean transformation*.[1] Differentiating (4.1) with respect to t, we obtain

$$\frac{d\mathbf{r}}{dt} = \frac{d\mathbf{r}'}{dt} + \mathbf{V}_0. \tag{4.2}$$

Differentiating (4.2) with respect to t yields

$$\frac{d^2\mathbf{r}}{dt^2} = \frac{d^2\mathbf{r}'}{dt^2}. \tag{4.3}$$

Thus, the accelerations are equal in both reference frames. Therefore, the equation of motion is generally valid in the inertial frames so long as the forces do not depend on the velocity. In this case, the equation of motion is said to be invariant for the Galilean transformation.

4.2.2 Non-Inertial Frames

A reference frame moving in the acceleration $\boldsymbol{\alpha}_0$ relative to the absolute reference frame is called the *non-inertial reference frame*. Supposing that the velocity of the non-inertial frame at time t is $\mathbf{V}_0$, the velocity of a body in the absolute reference frame at t is $\mathbf{v}$, and the velocity of the body in the non-inertial frame at t is $\mathbf{v}'$, the relationship between $\mathbf{v}$ and $\mathbf{v}'$ is

$$\mathbf{v} = \mathbf{V}_0 + \mathbf{v}', \tag{4.4}$$

$$\mathbf{V}_0 = \mathbf{V}_0(0) + \int_0^t \boldsymbol{\alpha}_0 dt, \tag{4.5}$$

where $\mathbf{V}_0(0)$ is the initial velocity of the non-inertial reference frame. Differentiating (4.4) with respect to t, we get

$$\frac{d\mathbf{v}}{dt} = \frac{d\mathbf{v}'}{dt} + \boldsymbol{\alpha}_0. \tag{4.6}$$

Then, the equation of motion in the non-inertial reference frame becomes

$$m\frac{d\mathbf{v}'}{dt} = m\frac{d\mathbf{v}}{dt} - m\boldsymbol{\alpha}_0, \tag{4.7}$$

From (3.3) and (4.7), we find

$$m\frac{d\mathbf{v}'}{dt} = \mathbf{F} - m\boldsymbol{\alpha}_0. \tag{4.8}$$

It is clear that the equation of motion in the non-inertial reference frame is valid by adding the *inertial force* (the *apparent force*) $-m\boldsymbol{\alpha}_0$ to the real forces.

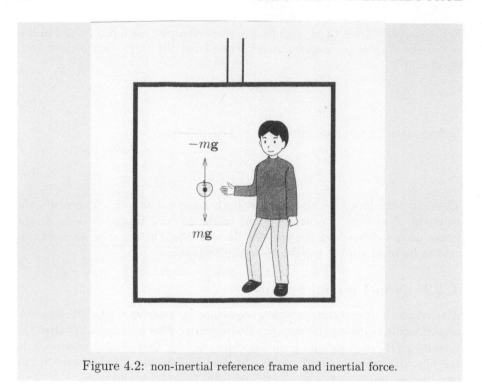

Figure 4.2: non-inertial reference frame and inertial force.

Example 1. A lift cabin is descending at an acceleration the same as that due to gravity **g**. If a person in the cabin released an apple from rest, the apple would be at rest in the air observed by the person in the cabin (Fig. 4.2). An observer on the Earth's surface observes that the person and the apple are descending with the acceleration **g**. An inertial force $-m\mathbf{g}$ must be added to the gravity force for the equation of motion to be valid in the reference frame bound to the lift cabin. A reference frame bound to the rotating Earth is not an inertial frame, but in this example the time scale is too short for the apple to perceive the Earth's rotation, so that the reference frame bound to the Earth is considered to be an approximate inertial reference frame.

---

[1]People who believed in geocentrism considered that the Earth was not in an orbital motion because a stone released from the tower fell to the ground just below the point of release. Galileo refuted this idea illustrating that a bird's egg dropped from a nest at the top of a mast of a sailing ship fell just below the nest.

4.3 Inertial Forces in a Rotating System

Let the Cartesian components of an arbitrary vector $\mathbf{A}$ on an inertial reference frame be (A_x, A_y, A_z), we obtain

$$\mathbf{A} = A_x \mathbf{e}_x + A_y \mathbf{e}_y + A_z \mathbf{e}_z, \tag{4.9}$$

where $(\mathbf{e}_x, \mathbf{e}_y, \mathbf{e}_z)$ are Cartesian basis vectors on the inertial frame. Suppose that Cartesian components of $\mathbf{A}$ on a rotating reference frame are (A'_x, A'_y, A'_z), we find

$$\mathbf{A} = A'_x \mathbf{e}_{x'} + A'_y \mathbf{e}_{y'} + A'_z \mathbf{e}_{z'}, \tag{4.10}$$

where $(\mathbf{e}_{x'}, \mathbf{e}_{y'}, \mathbf{e}_{z'})$ are the basis vectors on the rotating reference frame. Letting the total derivative operator on the inertial frame be d_a/dt, we find

$$
\begin{aligned}
\frac{d_a \mathbf{A}}{dt} &= \mathbf{e}_x \frac{d_a A_x}{dt} + \mathbf{e}_y \frac{d_a A_y}{dt} + \mathbf{e}_z \frac{d_a A_z}{dt} \\
&= \mathbf{e}_{x'} \frac{dA'_x}{dt} + \mathbf{e}_{y'} \frac{dA'_y}{dt} + \mathbf{e}_{z'} \frac{dA'_z}{dt} + \frac{d\mathbf{e}_{x'}}{dt} A'_x + \frac{d\mathbf{e}_{y'}}{dt} A'_y + \frac{d\mathbf{e}_{z'}}{dt} A'_z, \quad (4.11)
\end{aligned}
$$

where

$$\frac{d\mathbf{A}}{dt} = \mathbf{e}_{x'} \frac{dA'_x}{dt} + \mathbf{e}_{y'} \frac{dA'_y}{dt} + \mathbf{e}_{z'} \frac{dA'_z}{dt}, \tag{4.12}$$

and

$$\frac{d\mathbf{e}_{x'}}{dt} = \mathbf{\Omega} \times \mathbf{e}_{x'}, \quad \frac{d\mathbf{e}_{y'}}{dt} = \mathbf{\Omega} \times \mathbf{e}_{y'}, \quad \frac{d\mathbf{e}_{z'}}{dt} = \mathbf{\Omega} \times \mathbf{e}_{z'}. \tag{4.13}$$

Therefore, the relationship of the total derivative between two reference frames becomes

$$\frac{d_a \mathbf{A}}{dt} = \frac{d\mathbf{A}}{dt} + \mathbf{\Omega} \times \mathbf{A}. \tag{4.14}$$

Applying (4.14) to the position vector $\mathbf{r}$, we obtain

$$
\begin{aligned}
\frac{d_a \mathbf{r}}{dt} &= \frac{d\mathbf{r}}{dt} + \mathbf{\Omega} \times \mathbf{r}, \\
\mathbf{v}_a &= \mathbf{v} + \mathbf{\Omega} \times \mathbf{r}.
\end{aligned} \tag{4.15}
$$

Applying (4.14) to $\mathbf{v}_a$, we get

$$\frac{d_a \mathbf{v}_a}{dt} = \frac{d\mathbf{v}_a}{dt} + \mathbf{\Omega} \times \mathbf{v}_a. \tag{4.16}$$

Substituting (4.15) into (4.16), we find

$$
\begin{aligned}
\frac{d_a \mathbf{v}_a}{dt} &= \frac{d}{dt}(\mathbf{v} + \mathbf{\Omega} \times \mathbf{r}) + \mathbf{\Omega} \times (\mathbf{v} + \mathbf{\Omega} \times \mathbf{r}) \\
&= \frac{d\mathbf{v}}{dt} + 2\mathbf{\Omega} \times \mathbf{v} - \Omega^2 \mathbf{R}.
\end{aligned} \tag{4.17}
$$

Use is made of the following vector identity in the derivation of (4.17),

$$\mathbf{\Omega} \times (\mathbf{\Omega} \times \mathbf{r}) = \mathbf{\Omega} \times (\mathbf{\Omega} \times \mathbf{R}) = -\Omega^2 \mathbf{R},$$

where $\mathbf{R}$ is a position vector from the axis of rotation. The equation of motion in the inertial frame is

$$m\frac{d_a \mathbf{v}_a}{dt} = \sum_{i=1}^{n} \mathbf{F}_i. \tag{4.18}$$

From (4.17) and (4.18), we obtain the equation of motion in the rotating reference frame,

$$m\frac{d\mathbf{v}}{dt} = \sum_{i=1}^{n} \mathbf{F}_i - 2m\mathbf{\Omega} \times \mathbf{v} + m\Omega^2 \mathbf{R}. \tag{4.19}$$

The second term and the third term of the right-hand side of (4.19) are inertial forces in the rotating frame at the constant angular velocity $\mathbf{\Omega}$. The former is called the *Coriolis force* and the latter is referred to as the *centrifugal force*. While the centrifugal force acts on all bodies on the rotating frame, the Coriolis force acts only on moving bodies. In the Northern Hemisphere, the Coriolis force acts on a moving body to deflect vertically rightward from the moving direction.

4.3.1   The Coriolis Force

Let's decompose the Coriolis force to the x, y, z-components in local Cartesian coordinates centering at the latitude θ and an arbitrary longitude on the Earth's surface. Taking the x, y, z-directions as eastward, northward, and upward, respectively, the Coriolis force $\mathbf{F}_{\mathrm{Co}}$ becomes

$$\mathbf{F}_{\mathrm{Co}} = -2m\mathbf{\Omega} \times \mathbf{v} = -2m \begin{vmatrix} \mathbf{e}_x & \mathbf{e}_y & \mathbf{e}_z \\ 0 & \Omega\cos\theta & \Omega\sin\theta \\ u & v & w \end{vmatrix}$$

$$= \mathbf{e}_x(2mv\Omega\sin\theta - 2mw\Omega\cos\theta) + \mathbf{e}_y(-2mu\Omega\sin\theta)$$
$$+ \mathbf{e}_z(2mu\Omega\cos\theta). \tag{4.20}$$

The horizontal components of the Coriolis force at latitude θ are proportional to the vertical component of the angular velocity of the Earth's rotation at that latitude, and $f = 2\Omega\sin\theta$ is called the *Coriolis parameter*.

Example 2. Zero gravity
Recently, we see live broadcasts of a crew in a cabin of a space shuttle. They can move around in the cabin as if they were swimming in water. And a liquid droplet is floating in the air in the shape of a sphere due to the surface tension. It seems that the gravity of the Earth does not exert on any bodies

in the cabin. This situation is called the state of *zero gravity*, but exactly speaking the term zero gravity is not true. The gravity only balances with the centrifugal force, so the total force exerting any bodies in the cabin is zero in the reference frame bound to the space shuttle.

4.3.2 Foucault Pendulum

Léon Foucault (Fig. 4.3), a French physicist, performed an experiment using a very large simple pendulum, called later the Foucault pendulum, in January 1851, at the Panthéon, Paris (Fig. 4.4). A bob of mass 28.0 kg was attached to the end of a wire of length 67.0 m whose other end was attached to the dome of the Panthéon. The pendulum oscillated very slowly and the oscillatory plane is invariant to the absolute reference frame. Then the oscillatory plane was observed to rotate clockwise relative to the Earth's surface.[2] Foucault succeeded in showing the rotation of the Earth in an easy-to-see experiment. The original apparatus used in the 1851 experiment is kept now in Métiers Museum, Paris.

Let's obtain the period that the oscillatory plane of the Foucault pendulum turns around a full circle. Supposing that the pendulum is settled at latitude θ in the Northern Hemisphere, where the vertical component of the angular velocity of the Earth's rotation is $f = 2\Omega \sin \theta$. Let the length of the pendulum be l, the mass of a bob be m, and the angle of the pendulum from the vertical be ϕ. We will take the x-axis as the oscillatory direction, the z-axis vertically upward, and the y-axis so as to obey the right-hand rule. So long as the amplitude of the oscillation is small, the x- and y-components of the equation of motion are

$$m\frac{d^2x}{dt^2} = mf\frac{dy}{dt} - mg\frac{x}{l}, \tag{4.21}$$

$$m\frac{d^2y}{dt^2} = -mf\frac{dx}{dt} - mg\frac{y}{l}. \tag{4.22}$$

Dividing (4.21) and (4.22) through by m and letting $\omega^2 = g/l$, we obtain

$$\frac{d^2x}{dt^2} - f\frac{dy}{dt} + \omega^2 x = 0, \tag{4.23}$$

$$\frac{d^2y}{dt^2} + f\frac{dx}{dt} + \omega^2 y = 0. \tag{4.24}$$

Multiplying (4.24) by the imaginary unit $\tilde{\imath}$ and adding the resultant equation to (4.23), we obtain a second-order ordinary differential equation,

$$\frac{d^2z}{dt^2} + \tilde{\imath}f\frac{dz}{dt} + \omega^2 z = 0, \tag{4.25}$$

---

[2]In the southern hemisphere, the oscillatory plane rotates counterclockwise.

★★★★★★★★★★★★★★★ Léon Foucault (1819–1868) ★★★★★★★★★★★★★★★

Figure 4.3: Léon Foucault was born on September 18, 1819, in Paris. As he
was sickly, he was educated by private teachers at home. He studied medicine
and was soon interested in physics. He had a keen observational sense and
acute insight into natural phenomena in addition to an understanding of high
technology enabling him to make experimental apparatus. He met Armand
Fizeau (1819–1896), a French physicist, and began to study with him. In 1845,
they succeeded in taking precise photographs of the solar surface, which seems
to be the foundation of later investigations. The first is the measurement of
light speed. They soon began to work independently, and Fizeau obtained the
light speed of 313,000 km s^{-1} in 1849. While Foucault obtained the light speed
of 298,000 km s^{-1} in 1862, which is only 0.6% in error of the most reliable value
today. The second is the experiment of the Foucault pendulum, whose idea
came from an equipment to synchronize a camera with solar movement. A
pendulum was used in the equipment, and he found that the oscillatory plane
of the pendulum rotated relative to the Earth's surface. The third is the proof
of the wave properties of light, showing that light travels faster in air than in
water. He noticed the wave properties of light through the interference between
incoming and outgoing solar light. The fourth is the invention of the so-called
Foucault knife-edge test, which is a method to judge whether the mirror of a
reflecting telescope is perfectly parabolic. The test is still used today by amateur
and small commercial telescope makers because of its easiness and cheapness.
In 1855, he discovered electric eddy currents in a copper disk rotating in the
magnetic field. In the same year, he was awarded the Copley Medal from the
Royal Society for his remarkable researches.

★★★

where $z = x + \tilde{i}y$. Assuming an exponential type solution $z = z_0 \exp(\tilde{i}\mu t)$ and
substituting it into (4.25), we get

$$\mu^2 + f\mu - \omega^2 = 0,$$
$$\mu = -\frac{f}{2} \pm \sqrt{(f/2)^2 + \omega^2}. \tag{4.26}$$

Figure 4.4: The Foucault pendulum in the Panthéon, Paris.

The general solution of (4.25) becomes

$$z = \{\alpha \exp\left(\tilde{i}\omega't\right) + \beta \exp\left(-\tilde{i}\omega't\right)\} \exp\left(-\tilde{i}\frac{f}{2}t\right), \tag{4.27}$$

where $\omega' = \sqrt{(f/2)^2 + \omega^2}$.
Letting the initial conditions be

$$z = a, \quad \frac{dz}{dt} = 0 \quad \text{at} \quad t = 0. \tag{4.28}$$

Applying the initial condition (4.28) to (4.27), we find

$$\alpha = \frac{\omega' + f/2}{2\omega'}a, \tag{4.29}$$

$$\beta = \frac{\omega' - f/2}{2\omega'}a. \tag{4.30}$$

Substituting from (4.29) and (4.30) into (4.27) and separating the resultant equation to the real and imaginary parts,

$$x = a\cos\omega't\cos\frac{f}{2}t + \frac{f}{2\omega'}a\sin\omega't\sin\frac{f}{2}t, \tag{4.31}$$

$$y = -a\cos\omega't\sin\frac{f}{2}t + \frac{f}{2\omega'}a\sin\omega't\cos\frac{f}{2}t. \tag{4.32}$$

The oscillatory motion observed on the absolute reference frame (X, Y) is obtained by letting $f = 0$ in (4.31) and (4.32),

$$X = a \cos \omega' t, \tag{4.33}$$

$$Y = \frac{f}{2\omega'} a \sin \omega' t. \tag{4.34}$$

Substituting from (4.33) and (4.34) into (4.31) and (4.32), we get

$$x = \cos \frac{f}{2} t X + \sin \frac{f}{2} t Y, \tag{4.35}$$

$$y = -\sin \frac{f}{2} t X + \cos \frac{f}{2} t Y. \tag{4.36}$$

Expressing (4.35) and (4.36) in a matrix form, we find

$$\begin{pmatrix} x \\ y \end{pmatrix} = \begin{pmatrix} \cos ft/2 & \sin ft/2 \\ -\sin ft/2 & \cos ft/2 \end{pmatrix} \begin{pmatrix} X \\ Y \end{pmatrix}. \tag{4.37}$$

Comparing (4.37) with (1.8), we can find that the coordinate system (x, y) is transformed by rotating the coordinate system (X, Y) by $ft/2$ counterclockwise about the origin. Therefore, the oscillatory plane of the Foucault pendulum rotates at the angular velocity $-f/2$ relative to the Earth's surface.[3] The period in which the oscillatory plane of the Foucault pendulum rotates a full circle is called *one Foucault pendulum day*,[4] which is given by

$$T_{\mathrm{F}} = \frac{2\pi}{f/2} = \frac{2\pi}{\Omega \sin \theta}. \tag{4.38}$$

Thus, one Foucault pendulum day is just the *sidereal day* at both poles and infinity at the equator. Next, we will find the trajectory of the oscillatory motion of the bob. Eliminating t from (4.33) and (4.34), we get

$$\frac{X^2}{a^2} + \frac{Y^2}{f/2\omega' a^2} = 1. \tag{4.39}$$

Therefore, the bob of the Foucault pendulum traces an elliptic trajectory observed on the absolute reference frame.

4.4   Problems

1. When Foucault demonstrated the experiment of the Foucault pendulum in January 1851 at the Panthéon, Paris ($48.9°$ N), the length of the wire was $67.0\,\mathrm{m}$.

---

[3]The oscillatory plane rotates clockwise in the northern hemisphere where $f > 0$ and counterclockwise in the southern hemisphere.

[4]There is an analogous phenomenon in the oceans and great lakes called the inertial oscillation whose period is one-half pendulum day.

(1) Calculate the period of the pendulum, letting the magnitude of the acceleration due to gravity be $9.80\,\mathrm{m\,s^{-2}}$.

(2) The bob of the Foucault pendulum traces the elliptic trajectory. What is the ratio of the semiminor axis to the semimajor axis?

2. What is one Foucault pendulum day at Paris $(48°51'44"\mathrm{N})$?

3. A person releases a stone from the observation deck $(4.50 \times 10^2\,\mathrm{m}$ in height) of the Tokyo Sky Tree $(6.34 \times 10^2\,\mathrm{m}$ in total height). What is the displacement of the stone from the position just below the released point when it collides with the ground and what is the direction of the deflection? Neglect the air resistance and the effect of winds. Let the latitude of the Tokyo Sky Tree be $35.7°\,\mathrm{N}$ and the magnitude of the acceleration due to gravity be $9.80\,\mathrm{m\,s^{-2}}$.

4. An H–2 rocket was launched due eastward from Tanegashima Space Center $(30.40°\mathrm{N})$, Japan. The rocket fell into the Pacific Ocean after $5.00 \times 10^3\,\mathrm{km}$ travel at a horizontal speed $1.00 \times 10^3\,\mathrm{m\,s^{-1}}$. What was the displacement and the direction of deflection of the rocket from its eastward path?

Chapter 5

Work and Energy

In this chapter, the first transformation of the equation of motion is performed. The resultant equation is called the energy equation, which relates the changing rate of kinetic energy of a particle and the work done by external forces exerting on it. Some problems of classical mechanics are easily solved using the energy equation rather than starting from the equation of motion. Potential energy is defined for the conservative force. When all forces exerting on a particle are conservative, the sum of kinetic energy and potential energy is conserved, which is referred to as the law of mechanical energy conservation.

5.1 Transformation of the Equation of Motion

If the mass of a particle is constant, the equation of motion is written as

$$m\frac{d\mathbf{v}}{dt} = \mathbf{F}. \tag{5.1}$$

Operating the scalar product of $\mathbf{v}$ on (5.1), we get

$$m\mathbf{v} \cdot \frac{d\mathbf{v}}{dt} = \mathbf{v} \cdot \mathbf{F} = \frac{d\mathbf{r}}{dt} \cdot \mathbf{F},$$

$$\frac{d}{dt}\left(\frac{1}{2}mv^2\right) = \frac{dK}{dt} = \mathbf{F} \cdot \frac{d\mathbf{r}}{dt}, \tag{5.2}$$

where $K = 1/2\,mv^2$ is referred to as the kinetic energy. Integrating (5.2) with respect to t from t_1 to t_2, we obtain

$$K_2 - K_1 = \int_{t_1}^{t_2} \mathbf{F} \cdot \frac{d\mathbf{r}}{dt}dt = \int_{\mathbf{r}_1}^{\mathbf{r}_2} \mathbf{F} \cdot d\mathbf{r} = W_{21}, \tag{5.3}$$

$$W_{21} = \int_{\mathbf{r}_1}^{\mathbf{r}_2} \mathbf{F} \cdot d\mathbf{r}, \tag{5.4}$$

DOI: 10.1201/9781003310068-5

where K_1 and K_2 are kinetic energy at time t_1 and t_2, and $\mathbf{r}_1$ and $\mathbf{r}_2$ are position vectors at t_1 and t_2. W_{21} is called the work done by the external force moving the particle from point 1 to point 2. Equation (5.3) states that the increase of the kinetic energy is equal to the work done by the force on the particle. In physics, the work is zero even if the force is exerting but no displacement occurs in the direction of the force. When the force is not exerting on a particle or the force is perpendicular to the direction of motion, the work done by the force is zero so that the kinetic energy of the particle is conserved.

Example 1. Using (5.3), find the speed of a particle of mass m when it falls y from the origin. Suppose that the particle is at rest at the origin and the magnitude of the acceleration due to gravity is g.

Answer
Let's take the y-axis vertically downward and the falling speed of the particle be v. After falling y from the origin, the work done by the gravity is

$$W = \int_0^y mg\,dy = mgy.$$

As the work is equal to the increase of the kinetic energy of the particle,

$$\frac{1}{2}mv^2 - 0 = mgy,$$
$$v = \sqrt{2gy}.$$

Problem 1. A particle of mass m is moving at the velocity $\mathbf{v}$ on a reference frame rotating at the constant angular velocity $\mathbf{\Omega}$. What is the work done by the Coriolis force?

5.2   Conservative Forces and Potential Energy

When the work done by a force on a particle to move it from point 1 to point 2 depends not on the path but only on the position of two points, the force is said to be conservative. In the field of the Earth's gravity, the work done by gravity on a particle falling vertically y and the work done on a particle sliding down $y/\sin\theta$ (θ is the elevation angle of a slope) along a frictionless slope are equal. Thus, gravity is one of the conservative forces. In addition to gravity, the gravitational force, Coulomb force, the magnetic force, and the elastic force are conservative forces.

Suppose that a particle is moved from point 1 to point 2 by a force which is equal and opposite to the conservative force $\mathbf{F}$. Then the conservative force has the potential to move the particle from point 2 to point 1, the work of which is the same as the work done by the opposite force. This potential work stored to the particle is called the potential energy. The reference point of the potential energy is arbitrary, so that only the difference of the potential energy between two points has physical meaning.

Let the potential energy at point 1 be U_1, the potential energy at point 2 be U_2, the difference of the potential energy is equal to the work done by the opposite force,

$$U_2 - U_1 = -\int_1^2 \mathbf{F}\cdot d\mathbf{r}, \tag{5.5}$$

where $\mathbf{F}$ is the conservative force. As the gradient of U along the path is given by ∇U, we find

$$U_2 - U_1 = \int_1^2 \nabla U \cdot d\mathbf{r}. \tag{5.6}$$

From (5.5) and (5.6), we obtain

$$\mathbf{F} = -\nabla U. \tag{5.7}$$

Component equations of (5.7) in Cartesian coordinates are

$$F_x = -\frac{\partial U}{\partial x}, \tag{5.8}$$

$$F_y = -\frac{\partial U}{\partial y}, \tag{5.9}$$

$$F_z = -\frac{\partial U}{\partial z}. \tag{5.10}$$

Problem 2. A body of mass m is held at rest on a frictionless slope with the angle θ to a level surface. Suppose that the body is pulled up distance l along the slope by a force equal and opposite to the parallel component of the Earth gravity g. Obtain the work done on the body.

Example 2. Suppose a semi-sphere with a smooth inner surface of radius a. Let the center of the semi-sphere be O and the lowest point of the inner sphere be Q. A particle of mass m is held at rest at point R on the inner sphere. When the particle is released, obtain the work done by gravity to move it from R to Q, letting the angle $\angle$ROQ $= \theta_0$ $(0 < \theta_0 < \pi/2)$ and the magnitude of the acceleration due to gravity be g. What is the speed of the particle at Q?

Answer
Letting an arbitrary point on the inner sphere be P and the angle $\angle$POQ $= \theta$, the tangential component of the gravity exerting on the particle is $mg\sin\theta$. As the line element of the tangential direction of the sphere is $-ad\theta$,[1] the

---

[1] The minus sign means that the direction of the tangential component of gravity and the positive direction of θ is opposite.

work done by gravity is

$$W = -\int_{\theta_0}^{0} mg \sin \theta a d\theta$$

$$= mga \Big[\cos \theta \Big]_{\theta_0}^{0} = mga(1 - \cos \theta_0).$$

This work is equal to the work done by gravity on the particle falling vertically downward from R to the level of Q. The tangential speed of the particle at Q is obtained from (5.3).

$$\frac{1}{2}mv^2 - 0 = mga(1 - \cos \theta_0),$$

$$v = \sqrt{2ag(1 - \cos \theta_0)}.$$

Problem 3. What is the potential energy of a particle of mass m and at height h from the Earth's surface? Let the reference level be the Earth's surface and the magnitude of the acceleration due to gravity be g.

5.3   Potential Energy of a Spring

A body of mass m is at rest on a level and frictionless surface and is connected to a rigid support via a spring of spring constant k. The work to pull the body against the elastic force by a distance x from the equilibrium point is

$$W = \int_{0}^{x} kx dx = \frac{1}{2}kx^2. \tag{5.11}$$

Therefore, letting the reference point of the potential energy be the equilibrium point, the potential energy possessed by the spring is

$$U = \frac{1}{2}kx^2 . \tag{5.12}$$

The elastic force of the spring is obtained using (5.8)

$$F_x = -\frac{\partial U}{\partial x} = -kx. \tag{5.13}$$

Equation (5.11) is called the *Hooke's law*.

5.4   The Law of Mechanical Energy Conservation

When a force exerting on a particle is conservative, (5.3) becomes

$$K_2 - K_1 = \int_{1}^{2} \mathbf{F} \cdot d\mathbf{r} = -(U_2 - U_1),$$

$$U_1 + K_1 = U_2 + K_2, \tag{5.14}$$

with the aid of (5.5).

$E = U + K$ is called the mechanical energy. Equation (5.14) states that the mechanical energy is conserved as long as the exerting force is conservative. This law is called the *law of mechanical energy conservation*.

Example 3. Suppose that a particle of mass m is projected vertically downward at the initial speed v_0. Obtain the speed of the particle after falling a distance h, letting the magnitude of the acceleration due to gravity be g.

Answer
Let's take the y-axis vertically downward and the projecting position as the reference point of the potential energy. As gravity is the conservative force, the mechanical energy is conserved at the initial and final positions. Therefore,

$$\frac{1}{2}mv_0{}^2 + 0 = \frac{1}{2}mv^2 - mgh,$$
$$v = \sqrt{v_0{}^2 + 2gh}.$$

5.5   The Unit of Work and Energy

The work done by the force $1\,\mathrm{N}$ moving a body a distance $1\,\mathrm{m}$ in the direction of the force is defined as $1\,\mathrm{J}$ (Joule). The unit of the work and energy is named after James Joule (Fig. 5.1), who was an English physicist and studied the relationship between mechanical energy, heat, and electric energy. The unit of work and energy is expressed by the SI base units,

$$[W] = [Fs] = \left[m\frac{d^2x}{dt^2}s \right],$$
$$[\mathrm{J}] = [\mathrm{kg\,m^2\,s^{-2}}]$$

The changing rate of the work is called the power, and is given by

$$P = \frac{dW}{dt}. \tag{5.15}$$

Thus, the unit of the power is

$$[P] = [\mathrm{J\,s^{-1}}] = [\mathrm{W\,(Watt)}],$$

which is named after James Watt (1736–1819), who was a Scottish mechanical engineer and invented the technology of steam engines.

Problem 4. A body of mass $1.0\,\mathrm{kg}$ is moving at a constant speed $1.0\,\mathrm{m\,s^{-1}}$. What is the kinetic energy of the body?

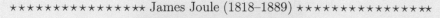

★★★★★★★★★★★★★★★ James Joule (1818–1889) ★★★★★★★★★★★★★★★

Figure 5.1: James Prescott Joule was an English physicist, born in Salford, Lancashire, on December 24, 1818. He was the son of a wealthy brewer and was a sickly child, so he was educated by private teachers, one of whom was John Dalton. When he grew up he managed the family brewery and studied physics as a hobby. In 1840, he started to investigate the efficiency of electric motors, and discovered what came to be called Joule's first law in 1841; i.e., the heat which is produced by electric current is proportional to the product of the electric resistance and the square of the electric current. He investigated the relationship between the mechanical energy and heat and found the convertibility of the mechanical energy to heat in 1845, using his most famous experimental apparatus, spinning a paddle wheel by a falling weight in a water-filled insulated vessel. In 1852, he collaborated with William Thomson (later to become Lord Kelvin) and they discovered what is now called the Joule–Thomson effect; i.e., the temperature of gas decreases when it expands freely. In 1870, he was awarded the Copley Medal for his experimental research on the dynamic theory of heat. In 1872, he became the president of the British Academy of Science. He was awarded the Albert Medal of the Royal Society of Arts in 1880, for establishing the true relationship between heat, electricity, and mechanical energy. He passed away in Sale, Cheshire, on October 11, 1889.

★★

5.6   The Mechanical Equivalent of Heat

James Joule started to investigate the efficiency of electric motors, and discovered *Joule's first law* in 1841; i.e., the heat produced by the electric current is proportional to the product of the electric resistance and the square of the electric current. He investigated the relationship between mechanical energy and heat, and found the convertibility of mechanical energy to heat in 1845, using his most famous experimental apparatus in which paddlewheels were rotated by a falling weight in a water-filled insulated vessel. The relationship between mechanical energy and heat is

$$E = JQ, \tag{5.16}$$

where E [J] is mechanical energy, Q [cal] is heat, and $J = 4.1855$ [J cal^{-1}] is referred to as the *mechanical equivalent of heat*. Joule continued to study various energies and established the *law of total energy conservation* (the *first law of thermodynamics*).

5.7 Problems

1. A particle of mass m is attached to a rigid support on a ceiling via a massless and inextensible string of length l. After pulling the particle tight and the angle between the string and the plumb line is kept θ_0, the particle is released from rest. What is the tangential speed of the particle at the lowest point? Let the magnitude of the acceleration due to gravity be g.

2. Suppose that a stone is released from the top of a tower of 90.0 m in height. Calculate the speed of the stone just before colliding with the Earth's surface, letting the value of the acceleration due to gravity be 9.80 m s^{-2} and neglecting the air resistance.

3. A particle of mass m is projected at the initial speed V_0 and the elevation angle θ from the Earth's surface. Obtain the maximum height reached by the particle, letting the magnitude of the acceleration due to gravity be g.

4. A vehicle of mass 1.50×10^3 kg, including the mass of the crew, is travelling on a straight road at 50.0 km h^{-1}. A driver finding an obstacle on the road brakes urgently and stops the vehicle after 5.00 m. Using the energy equation, obtain the resistive force exerting on the vehicle supposing that the driving force is zero during braking. Calculate the kinematic frictional coefficient letting the value of the acceleration due to gravity be 9.80 m s^{-2}.

5. Suppose that a particle of mass m is flicked at the tangential speed V_0 from the lowest point of a frictionless spherical shell with the inner radius a. Obtain the condition of V_0 for the particle to return to the lowest point without separating from the inner surface of the spherical shell. Let the magnitude of the acceleration due to gravity be g.

6. Suppose that a particle of mass m is put on the top T of a frictionless
sphere of radius a and having the center at O. Given an infinitesimal
displacement, the particle begins to slide on the spherical surface and
leaves the surface at point P. What is the angle $\angle$TOP, letting the magnitude of the acceleration due to gravity be g?

Chapter 6

Oscillatory Motion

Simple oscillatory motions were discussed in Chapter 3. In this chapter, more complicated oscillations, such as damped oscillations, forced oscillations, and coupled oscillations, are discussed. When we discuss forced oscillations, we must solve a second-order inhomogeneous ordinary differential equation, so that the general method of solving it is precisely shown. In the damped oscillation, three kinds of damping, damped oscillation, over damping, and critical damping exist. In the forced oscillation, if the frequency of the external forcing is equal to the intrinsic frequency of the system the amplitude goes to infinity, which is called resonance. When the resistive force works in the forced oscillation, the amplitude remains finite during the occurrence of resonance.

6.1 Damped Oscillations

On a flat and frictionless bottom of a vessel filled with a viscous fluid, a rectangular body of mass m is attached to a rigid support via a spring of spring constant k and natural length l_0. Suppose that the resistive force is proportional to the speed of the body and the proportional coefficient is given by $2m\gamma$. When the body is given a small displacement from its equilibrium position, the body begins to oscillate and the equation of motion becomes

$$m\frac{d^2x}{dt^2} = 2m\gamma\frac{dx}{dt} - kx, \qquad (6.1)$$

taking the x-axis as the oscillating direction of the body. Putting $\omega = \sqrt{k/m}$, (6.1) becomes

$$\frac{d^2x}{dt^2} + 2\gamma\frac{dx}{dt} + \omega^2x = 0. \qquad (6.2)$$

DOI: 10.1201/9781003310068-6

Assuming an exponential type solution $x \propto \exp \lambda t$ and substituting it into (6.2), we get

$$\lambda^2 + 2\gamma\lambda + \omega^2 = 0,$$
$$\lambda = -\gamma \pm \sqrt{\gamma^2 - \omega^2}. \tag{6.3}$$

The behavior of the solution is classified into three cases owing to the large/small relation between ω and γ.

1. The case that $\gamma > \omega$

When (6.3) is two different real roots, the particular solutions of (6.2) are

$$x \propto \exp\left(-\gamma + \sqrt{\gamma^2 - \omega^2}\right) t, \quad x \propto \exp\left(-\gamma - \sqrt{\gamma^2 - \omega^2}\right) t.$$

The general solution of (6.2) is given by the linear combination of two particular solutions,

$$x = \exp\left(-\gamma t\right) \left\{ \alpha \exp\left(\sqrt{\gamma^2 - \omega^2}\, t\right) + \beta \exp\left(-\sqrt{\gamma^2 - \omega^2}\, t\right) \right\}, \tag{6.4}$$

where α and β are real numbers. Applying the condition

$$x = x_0, \quad \frac{dx}{dt} = 0, \quad \text{at} \quad t = 0 \tag{6.5}$$

to (6.4), we obtain

$$x = \frac{1}{2} x_0 \exp\left(-\gamma t\right) \left\{ \left(1 + \frac{\gamma}{\sqrt{\gamma^2 - \omega^2}}\right) \exp\left(\sqrt{\gamma^2 - \omega^2}\, t\right) \right.$$
$$\left. + \left(1 - \frac{\gamma}{\sqrt{\gamma^2 - \omega^2}}\right) \exp\left(-\sqrt{\gamma^2 - \omega^2}\, t\right) \right\}. \tag{6.6}$$

Equation (6.6) shows that when the resistive force is large ($\gamma > \omega$), the motion is exponentially damped without oscillation. This motion is referred to as the *over damping*.

2. The case that $\gamma < \omega$

When (6.3) is two different imaginary roots, the particular solutions of (6.2) are

$$x \propto \exp\left(-\gamma t\right) \exp\left(\tilde{\imath} \sqrt{\omega^2 - \gamma^2}\, t\right), \quad x \propto \exp\left(-\gamma t\right) \exp\left(-\tilde{\imath} \sqrt{\omega^2 - \gamma^2}\, t\right).$$

The general solution of (6.2) is given by the linear combination of two particular solutions,

$$x = \exp\left(-\gamma t\right) \left\{ \alpha \exp\left(\tilde{\imath} \sqrt{\omega^2 - \gamma^2}\, t\right) + \beta \exp\left(-\tilde{\imath} \sqrt{\omega^2 - \gamma^2}\, t\right) \right\}, \tag{6.7}$$

where α and β are complex numbers as,

$$\alpha = \alpha_r + \tilde{i}\alpha_i, \quad \beta = \beta_r + \tilde{i}\beta_i.$$

Taking the real part of (6.7) and arranging terms, we obtain

$$\Re[x] = A \exp(-\gamma t) \cos\left(\sqrt{\omega^2 - \gamma^2}\, t - \phi\right), \tag{6.8}$$

where

$$A = \sqrt{(\alpha_r + \beta_r)^2 + (\beta_i - \alpha_i)^2}, \quad \phi = \tan^{-1}\left(\frac{\beta_i - \alpha_i}{\alpha_r + \beta_r}\right).$$

Applying the conditions

$$x = x_0, \quad \frac{dx}{dt} = 0, \quad \text{at} \quad t = 0, \tag{6.9}$$

to (6.8), we obtain

$$x = \frac{\omega x_0}{\sqrt{\omega^2 - \gamma^2}} \exp(-\gamma t) \cos\left(\sqrt{\omega^2 - \gamma^2}\, t - \tan^{-1}\frac{\gamma}{\sqrt{\omega^2 - \gamma^2}}\right). \tag{6.10}$$

Equation (6.10) shows the oscillatory motion whose amplitude damps exponentially with time. This motion is called the *damped oscillation*.

3. The case that $\gamma = \omega$

When (6.3) is the equal root, one of the particular solutions of (6.2) is,

$$x = \alpha \exp(-\gamma t). \tag{6.11}$$

Next, we must find another particular solution of (6.2). We will do this by making use of the *variation of constants*. Namely, assuming that the coefficient of the particular solution (6.11) is not a constant but a function of t, we substitute it into (6.2). Thus, we obtain the second order ordinary differential equation with respect to $\alpha(t)$,

$$\frac{d^2\alpha}{dt^2} - 0.$$

The general solution is

$$\alpha = A' + Bt,$$

where A' (the prime is for later convenience) and B are integral constants. Then another particular solution of (6.2) is

$$x = (A' + Bt) \exp(-\gamma t). \tag{6.12}$$

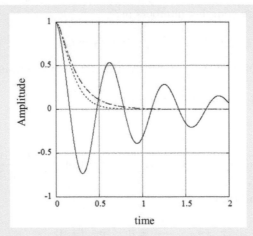

Figure 6.1: The behavior of the damped oscillation; the initial conditions
are $x = 1.0$, $dx/dt = 0$. The ordinate is the nondimensional amplitude
and the abscissa is the time. The *solid line* indicates the damped oscilla-
tion ($\omega = 10.0\,\mathrm{s}^{-1}, \gamma = 1.0\,\mathrm{s}^{-1}$), the *dash-dotted line* is the over damping
($\omega = 10.0\,\mathrm{s}^{-1}, \gamma = 12.5\,\mathrm{s}^{-1}$), and the *dotted line* is the critical damping
($\omega = \gamma = 10.0\,\mathrm{s}^{-1}$).

The general solution of (6.2) is given by the linear combination of (6.11) and
(6.12),

$$x = (A + Bt) \exp{(-\gamma t)}. \tag{6.13}$$

Applying the conditions

$$x = x_0, \qquad \frac{dx}{dt} = 0, \quad \text{at} \quad t = 0, \tag{6.14}$$

to (6.13), we obtain

$$x = x_0(1 + \gamma t) \exp{(-\gamma t)}. \tag{6.15}$$

When the resistive force satisfies the special condition $\gamma = \omega$, the oscillatory
motion does not occur and the displacement damps very quickly. This motion
is referred to as the *critical damping*, because the condition separates the
damped oscillation and the over damping. The behavior of three cases of
damped oscillation is shown in Fig. 6.1. The door damper is an example of
the application of the critical damping in daily life. The spring constant is
adjusted to satisfy the condition of the critical damping so that a door shuts
smoothly and quickly.

6.2 Forced Oscillations

In this section, the oscillatory system on which an external force is exerting is discussed. When the frequency of the external force is equal to the intrinsic frequency of the oscillatory system, a *resonance* occurs.

6.2.1 The Case of Non-Resistive Force

A body of mass m is attached to a rigid support via a spring of spring constant k. Suppose that the external force of the angular frequency ω_0 is exerted on the oscillatory body. Then the equation of motion describing the oscillatory motion is

$$m\frac{d^2 x}{dt^2} + kx = mF_0 \cos \omega_0 t. \tag{6.16}$$

Defining $\omega = \sqrt{k/m}$ and replacing the forcing term by the complex function,[1] we obtain

$$\frac{d^2 x}{dt^2} + \omega^2 x = F_0 \exp{(\tilde{\iota}\omega_0 t)}. \tag{6.17}$$

Equation (6.17) is the second-order inhomogeneous ordinary differential equation. The general solution is given by the sum of a complementary function[2] and a particular solution. The complementary function is already obtained in (3.50) and is written again,

$$x = x_0 \cos{(\omega t - \phi)}. \tag{6.18}$$

Next, we will find the particular solution. We denote the differential operator as

$$D \equiv \frac{d}{dt}.$$

Then (6.17) becomes

$$(D^2 + \omega^2)x = (D + \tilde{\iota}\omega)(D - \tilde{\iota}\omega) = F_0 \exp{(\tilde{\iota}\omega_0 t)}. \tag{6.19}$$

We define the operator of the left side of (6.19) as,

$$F(D) = (D^2 + \omega^2) = (D + \tilde{\iota}\omega)(D - \tilde{\iota}\omega). \tag{6.20}$$

---

[1] This method is much easier for obtaining the particular solution than the method of the integration by parts.

[2] A general solution of a homogeneous equation i.e., putting an inhomogeneous term of an inhomogeneous equation equal to zero.

Denoting the inverse operator of $F(D)$ by $F^{-1}(D)$, and operating it on (6.19), we find

$$F^{-1}(D)F(D)x = F^{-1}(D)F_0 \exp(\tilde{\imath}\omega_0 t),$$

$$x = \frac{1}{(D+\tilde{\imath}\omega)(D-\tilde{\imath}\omega)} F_0 \exp(\tilde{\imath}\omega_0 t) = \frac{1}{D+\tilde{\imath}\omega} u, \qquad (6.21)$$

where

$$u = \frac{1}{D-\tilde{\imath}\omega} F_0 \exp(\tilde{\imath}\omega_0 t).$$

Operating $(D - \tilde{\imath}\omega)$ on the above equation, we get

$$\frac{du}{dt} - \tilde{\imath}\omega u = F_0 \exp(\tilde{\imath}\omega_0 t).$$

Multiplying the above equation by the integrating factor $\exp(-\tilde{\imath}\omega t)$, and integrating the resultant equation with respect to t, we find

$$\frac{d}{dt}[u \exp(-\tilde{\imath}\omega t)] = F_0 \exp\tilde{\imath}(\omega_0 - \omega)t,$$

$$u \exp(-\tilde{\imath}\omega t) = \frac{-\tilde{\imath}F_0}{\omega_0 - \omega} \exp\tilde{\imath}(\omega_0 - \omega)t,$$

$$u = \frac{-\tilde{\imath}F_0}{\omega_0 - \omega} \exp(\tilde{\imath}\omega_0 t). \qquad (6.22)$$

Substituting from (6.22) into (6.21), we obtain

$$x = \frac{1}{D+\tilde{\imath}\omega} \frac{-\tilde{\imath}F_0}{\omega_0 - \omega} \exp(\tilde{\imath}\omega_0 t).$$

Operating $(D + \tilde{\imath}\omega)$ on the above equation, we get

$$\frac{dx}{dt} + \tilde{\imath}\omega x = \frac{-\tilde{\imath}F_0}{\omega_0 - \omega} \exp(\tilde{\imath}\omega_0 t). \qquad (6.23)$$

Multiplying (6.23) by the integrating factor $\exp(\tilde{\imath}\omega t)$, we find

$$\frac{d}{dt}[x \exp(\tilde{\imath}\omega t)] = \frac{-\tilde{\imath}F_0}{\omega_0 - \omega} \exp\tilde{\imath}(\omega_0 + \omega)t,$$

Integrating the above equation with respect to t, we get

$$x = \frac{-F_0}{\omega_0^2 - \omega^2} \exp(\tilde{\imath}\omega_0 t).$$

Taking the real part of the above solution, we obtain

$$\Re[x] = \frac{-F_0}{\omega_0^2 - \omega^2} \cos(\omega_0 t). \qquad (6.24)$$

The general solution of (6.17) is given by the sum of the complementary function (6.18) and the particular solution (6.24),

$$x = x_0 \cos{(\omega t - \phi)} - \frac{F_0}{\omega_0{}^2 - \omega^2} \cos{(\omega_0 t)}. \tag{6.25}$$

The result shows that when the angular frequency of the external forcing is equal to the intrinsic frequency of the oscillatory system, the amplitude goes to infinity. This behavior of the system is referred to as a resonance. It is shown in the next subsection that the amplitude remains finite in the existence of resistive force.

6.2.2 Forced Oscillation with the Resistive Force Proportional to Speed

A body of mass m is attached to a rigid support via a spring of spring constant k. Suppose that the external force of the angular frequency ω_0 is exerting on the oscillatory body. Further, we assume that a resistive force is proportional to the speed of the body. Then the equation of motion describing the oscillatory motion is

$$m\frac{d^2x}{dt^2} + 2m\gamma\frac{dx}{dt} + kx = mF_0 \cos{\omega_0 t}. \tag{6.26}$$

Dividing both sides through by m and replacing the forcing term by the complex function, we obtain

$$\frac{d^2x}{dt^2} + 2\gamma\frac{dx}{dt} + \omega^2 x = F_0 \exp{(\tilde{\imath}\omega_0 t)}, \tag{6.27}$$

where $\omega = \sqrt{k/m}$. The general solution is given by the sum of the complementary function and the particular solution of (6.27). The complementary function separates into three cases owing to the large/small relation between ω and γ.

1. The case that $\gamma > \omega$

The complementary function is given by (6.4), and we will find the particular solution. Equation (6.27) becomes

$$(D^2 + 2\gamma D + \omega^2)x = F_0 \exp{(\tilde{\imath}\omega_0 t)}. \tag{6.28}$$

We define the differential operator of (6.28) as

$$F(D) = \left(D + \gamma + \sqrt{\gamma^2 - \omega^2}\right)\left(D + \gamma - \sqrt{\gamma^2 - \omega^2}\right). \tag{6.29}$$

Denoting the inverse operator of $F(D)$ by $F^{-1}(D)$ and operating it on (6.28), we obtain

$$F^{-1}(D)F(D)x = F^{-1}(D)F_0 \exp(\tilde{\imath}w_0 t),$$

$$x = \frac{1}{D + \gamma + \sqrt{\gamma^2 - \omega^2}} \frac{1}{D + \gamma - \sqrt{\gamma^2 - \omega^2}} F_0 \exp(\tilde{\imath}w_0 t)$$

$$= \frac{1}{D + \gamma + \sqrt{\gamma^2 - \omega^2}} u, \tag{6.30}$$

where

$$u = \frac{1}{D + \gamma - \sqrt{\gamma^2 - \omega^2}} F_0 \exp(\tilde{\imath}w_0 t).$$

Operating $\left(D + \gamma - \sqrt{\gamma^2 - \omega^2}\right)$ on the above equation, we obtain

$$\frac{du}{dt} + \left(\gamma - \sqrt{\gamma^2 - \omega^2}\right) u = F_0 \exp(\tilde{\imath}w_0 t).$$

Multiplying the above equation by the integrating factor $\exp\left(\gamma - \sqrt{\gamma^2 - \omega^2}\right) t$ and integrating the resulting equation with respect to t, we get

$$u = \frac{F_0}{\gamma - \sqrt{\gamma^2 - \omega^2} + \tilde{\imath}w_0} \exp(\tilde{\imath}w_0 t). \tag{6.31}$$

Substituting from (6.31) into (6.30) yields

$$x = \frac{1}{D + \gamma + \sqrt{\gamma^2 - \omega^2}} \frac{F_0}{\gamma - \sqrt{\gamma^2 - \omega^2} + \tilde{\imath}w_0} \exp(\tilde{\imath}w_0 t).$$

Operating $\left(D + \gamma + \sqrt{\gamma^2 - \omega^2}\right)$ on the above equation, we find

$$\frac{dx}{dt} + \left(\gamma + \sqrt{\gamma^2 - \omega^2}\right) x = \frac{F_0}{\gamma - \sqrt{\gamma^2 - \omega^2} + \tilde{\imath}w_0} \exp(\tilde{\imath}w_0 t). \tag{6.32}$$

Multiplying (6.32) by the integrating factor $\exp\left(\gamma + \sqrt{\gamma^2 - \omega^2}\right) t$, we obtain

$$\frac{d}{dt}\left[x \exp\left(\gamma + \sqrt{\gamma^2 - \omega^2}\right) t\right] = \frac{F_0}{\gamma - \sqrt{\gamma^2 - \omega^2} + \tilde{\imath}w_0}$$

$$\times \exp\left(\gamma + \sqrt{\gamma^2 - \omega^2} + \tilde{\imath}w_0\right) t.$$

Integrating both sides with respect to t, we get

$$x = \frac{F_0}{\omega^2 - \omega_0{}^2 + 2\tilde{\imath}\gamma w_0} \exp(\tilde{\imath}w_0 t) = \frac{F_0(\omega^2 - \omega_0{}^2 - 2\tilde{\imath}\gamma w_0)}{(\omega^2 - \omega_0{}^2)^2 + 4\gamma^2 w_0{}^2} \exp(\tilde{\imath}w_0 t).$$

Taking the real part of the above equation yields

$$\Re[x] = \frac{F_0}{\sqrt{(\omega^2 - \omega_0^2)^2 + 4\gamma^2\omega_0^2}} \cos(\omega_0 t - \phi'), \tag{6.33}$$

where $\phi' - \tan^{-1}\{2\gamma\omega_0/(\omega^2 - \omega_0^2)\}$. The general solution is given by the sum of the complementary function (6.4) and the particular solution (6.33).

$$x = \exp(-\gamma t)\left\{\alpha\exp\left(\sqrt{\gamma^2 - \omega^2}t\right) + \beta\exp\left(-\sqrt{\gamma^2 - \omega^2}t\right)\right\}$$

$$+ \frac{F_0}{\sqrt{(\omega^2 - \omega_0^2)^2 + 4\gamma^2\omega_0^2}} \cos(\omega_0 t - \phi'). \tag{6.34}$$

In the limit $t \to \infty$, the complementary function damps and the forced oscillation remains. As the amplitude of the forcing term is a monotonically decreasing function of ω_0, a resonance does not occur.

2. The case that $\gamma < \omega$

The complementary function is given by (6.7), and we will find the particular solution. Equation (6.27) becomes

$$(D^2 + 2\gamma D + \omega^2)x = F_0\exp(\tilde{\imath}\omega_0 t). \tag{6.35}$$

We define the differential operator of (6.35) as

$$F(D) = \left(D + \gamma + \tilde{\imath}\sqrt{\omega^2 - \gamma^2}\right)\left(D + \gamma - \tilde{\imath}\sqrt{\omega^2 - \gamma^2}\right). \tag{6.36}$$

Denoting the inverse operator of $F(D)$ by $F^{-1}(D)$ and operating it on (6.35), we obtain

$$F^{-1}(D)F(D)x = F^{-1}(D)F_0\exp(\tilde{\imath}\omega_0 t),$$

$$x = \frac{1}{D + \gamma + \tilde{\imath}\sqrt{\omega^2 - \gamma^2}}\frac{1}{D + \gamma - \tilde{\imath}\sqrt{\omega^2 - \gamma^2}}F_0\exp(\tilde{\imath}\omega_0 t)$$

$$= \frac{1}{D + \gamma + \tilde{\imath}\sqrt{\omega^2 - \gamma^2}}u, \tag{6.37}$$

where

$$u = \frac{1}{D + \gamma - \tilde{\imath}\sqrt{\omega^2 - \gamma^2}}F_0\exp(\tilde{\imath}\omega_0 t).$$

Operating $\left(D + \gamma - \tilde{\imath}\sqrt{\omega^2 - \gamma^2}\right)$ on the above equation, we get

$$\frac{du}{dt} + \left(\gamma - \tilde{\imath}\sqrt{\omega^2 - \gamma^2}\right)u = F_0\exp(\tilde{\imath}\omega_0 t).$$

Multiplying the above equation by the integrating factor $\exp\left(\gamma - \tilde{\imath}\sqrt{\omega^2 - \gamma^2}\right)t$, we get

$$\frac{d}{dt}\left[u\exp\left(\gamma - \tilde{\imath}\sqrt{\omega^2 - \gamma^2}\right)t\right] = F_0\exp\left(\gamma - \tilde{\imath}\sqrt{\omega^2 - \gamma^2} + \tilde{\imath}\omega_0\right)t,$$

Integrating both sides with respect to t, we obtain

$$u = \frac{F_0}{\gamma - \tilde{\imath}\left(\sqrt{\omega^2 - \gamma^2} - \omega_0\right)}\exp\left(\tilde{\imath}\omega_0 t\right). \tag{6.38}$$

Substituting from (6.38) into (6.37) yields

$$x = \frac{1}{D + \gamma + \tilde{\imath}\sqrt{\omega^2 - \gamma^2}}\frac{F_0}{\gamma - \tilde{\imath}\left(\sqrt{\omega^2 - \gamma^2} - \omega_0\right)}\exp\left(\tilde{\imath}\omega_0 t\right).$$

Operating $\left(D + \gamma + \tilde{\imath}\sqrt{\omega^2 - \gamma^2}\right)$ on the above equation, we obtain

$$\frac{dx}{dt} + \left(\gamma + \tilde{\imath}\sqrt{\omega^2 - \gamma^2}\right)x = \frac{F_0}{\gamma - \tilde{\imath}\left(\sqrt{\omega^2 - \gamma^2} - \omega_0\right)}\exp\left(\tilde{\imath}\omega_0 t\right). \tag{6.39}$$

Multiplying (6.39) by the integrating factor $\exp\left(\gamma + \tilde{\imath}\sqrt{\omega^2 - \gamma^2}\right)t$, we get

$$\frac{d}{dt}\left[x\exp\left(\gamma + \tilde{\imath}\sqrt{\omega^2 - \gamma^2}\right)t\right] = \frac{F_0}{\gamma - \tilde{\imath}\left(\sqrt{\omega^2 - \gamma^2} - \omega_0\right)}$$

$$\times \exp\left(\gamma + \tilde{\imath}\sqrt{\omega^2 - \gamma^2} + \tilde{\imath}\omega_0\right)t,$$

Integrating both sides with respect to t, we obtain

$$x = \frac{F_0}{(\gamma + \tilde{\imath}\omega_0)^2 + (\omega^2 - \gamma^2)}\exp\left(\tilde{\imath}\omega_0 t\right)$$

$$= \frac{\omega^2 - \omega_0^2 - 2\tilde{\imath}\gamma\omega_0}{(\omega^2 - \omega_0^2)^2 + 4\gamma^2\omega_0^2}F_0[\cos\left(\omega_0 t\right) + \tilde{\imath}\sin\left(\omega_0 t\right)]. \tag{6.40}$$

Taking the real part of (6.40), we find

$$\Re[x] = \frac{F_0}{(\omega^2 - \gamma_0^2)^2 + 4\gamma^2\omega_0^2}[(\omega^2 - \omega_0^2)\cos\left(\omega_0 t\right) + 2\gamma\omega_0\sin\left(\omega_0 t\right)]$$

$$= \frac{F_0}{\sqrt{(\omega^2 - \omega_0^2)^2 + 4\gamma^2\omega_0^2}}\cos\left(\omega_0 t - \phi'\right), \tag{6.41}$$

where $\phi' = \tan^{-1}\left\{2\gamma\omega_0/(\omega^2 - \omega_0^2)\right\}$. The general solution is given by the sum of the complementary function (6.8) and the particular solution (6.41).

$$x = A\exp\left(-\gamma t\right)\cos\left(\sqrt{\omega^2 - \gamma^2}t - \phi\right)$$

$$+ \frac{F_0}{\sqrt{(\omega^2 - \omega_0^2)^2 + 4\gamma^2\omega_0^2}}\cos\left(\omega_0 t - \phi'\right). \tag{6.42}$$

In the limit $t \to \infty$, the oscillatory motion expressed by the complementary function damps and the forced oscillation remains. The amplitude has the maximum so that the resonance occurs. The dependence of the amplitude on ω_0/ω is shown in Fig. 6.2 for $\gamma/\omega = 0.1, 0.25, 0.5$.

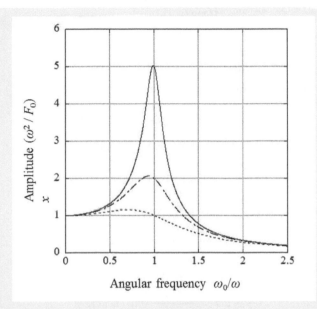

Figure 6.2: The dependence of the amplitude on the frequency of external forcing. The ordinate is the nondimensional amplitude $\omega^2 x/F_0$ and the abscissa is the nondimensional angular frequency ω_0/ω. The *solid line* shows the case of $\gamma/\omega_0 = 0.1$, the *dash-dotted line* indicates $\gamma/\omega_0 = 0.25$, and the *dotted line* is $\gamma/\omega_0 = 0.5$.

3. The case that $\gamma = \omega$

The complementary function is given by (6.13), and we will find the particular solution. Equation (6.27) becomes

$$(D^2 + 2\omega D + \omega^2)x = F_0 \exp(\tilde{\imath}\omega_0 t). \tag{6.43}$$

We define the differential operator of (6.43) as

$$F(D) = (D^2 + 2\omega D + \omega^2)x = (D + \omega)^2. \tag{6.44}$$

Denoting the inverse operator of $F(D)$ by $F^{-1}(D)$ and operating it on (6.35), we obtain

$$F^{-1}(D)F(D)x = F^{-1}(D)F_0 \exp(\tilde{\imath}\omega_0 t),$$

$$x = \frac{1}{D+\omega}\frac{1}{D+\omega}F_0 \exp\left(\tilde{\imath}\omega_0 t\right) = \frac{1}{D+\omega}u, \qquad (6.45)$$

where

$$u = \frac{1}{D+\omega}F_0 \exp\left(\tilde{\imath}\omega_0 t\right).$$

Operating $(D+\omega)$ on the above equation, we get

$$\frac{du}{dt} + \omega u = F_0 \exp\left(\tilde{\imath}\omega_0 t\right).$$

Multiplying the above equation by the integrating factor $\exp\left(\omega t\right)$ yields

$$\frac{d}{dt}\left[u\exp\left(\omega t\right)\right] = F_0 \exp\left(\omega + \tilde{\imath}\omega_0\right)t,$$

Integrating both sides with respect to t, we obtain

$$u = \frac{F_0}{\omega + \tilde{\imath}\omega_0} \exp\left(\tilde{\imath}\omega_0 t\right). \qquad (6.46)$$

Substituting from (6.46) into (6.45), we get

$$x = \frac{1}{D+\omega}\frac{F_0}{\omega + \tilde{\imath}\omega_0} \exp\left(\tilde{\imath}\omega_0 t\right).$$

Operating $(D+\omega)$ on the above equation yields

$$\frac{dx}{dt} + \omega x = \frac{F_0}{\omega + \tilde{\imath}\omega_0} \exp\left(\tilde{\imath}\omega_0 t\right). \qquad (6.47)$$

Multiplying the above equation by the integrating factor $\exp\left(\omega t\right)$, we obtain

$$\frac{d}{dt}\left[x\exp\left(\omega t\right)\right] = \frac{F_0}{\omega + \tilde{\imath}\omega_0} \exp\left(\omega + \tilde{\imath}\omega_0\right)t,$$

Integrating both sides with respect to t, we get

$$x = \frac{F_0}{\omega^2 - \omega_0{}^2 + 2\tilde{\imath}\omega\omega_0} \exp\left(\tilde{\imath}\omega_0 t\right). \qquad (6.48)$$

Taking the real part of (6.48), we find

$$\Re\{x\} = \frac{F_0}{\omega^2 + \omega_0{}^2} \cos\left(\omega_0 t - \phi'\right), \qquad (6.49)$$

where $\phi' = \tan^{-1}\{2\omega\omega_0/(\omega^2 - \omega_0{}^2)\}$. The general solution is given by the sum of the complementary function (6.13) and the particular solution (6.50).

$$x = (A + Bt)\exp\left(-\gamma t\right) + \frac{F_0}{\omega^2 + \omega_0{}^2} \cos\left(\omega_0 t - \phi'\right). \qquad (6.50)$$

In the limit $t \to \infty$, the oscillatory motion expressed by the complementary function damps and the forced oscillation remains. The amplitude of the forcing term is a monotonically decreasing function of ω_0, so that a resonance does not occur.

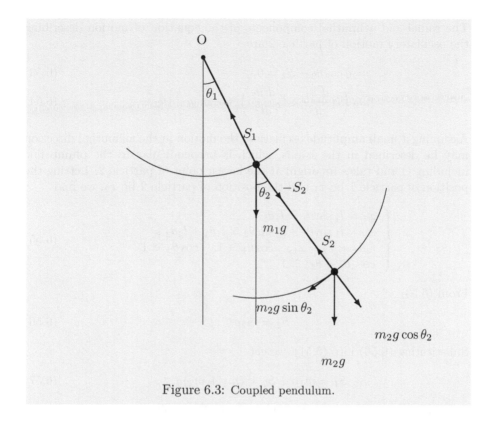

Figure 6.3: Coupled pendulum.

6.3 Coupled Pendulums

Suppose that particle 1 of mass m_1 is attached to rigid support O through massless and inextensible string 1 of length l_1, and particle 2 of mass m_2 is attached to particle 1 through massless and inextensible string 2 of length l_2 (Fig. 6.3). This oscillatory system is referred to as a coupled pendulum. When a small displacement is given to the particles in a vertical plane including point O, the system begins to oscillate. Letting the angle of string 1 from the plumb line be θ_1, the angle of string 2 from the plumb line be θ_2, the tension exerting on string 1 be S_1 and the tension exerting on string 2 be S_2, the radial and azimuthal components of the equation of motion describing the oscillatory motion of particle 1 become

$$m_1 g \cos \theta_1 - S_1 + S_2 \cos (\theta_2 - \theta_1) = 0, \tag{6.51}$$

$$m_1 l_1 \frac{d^2 \theta_1}{dt^2} = -m_1 g \sin \theta_1 + S_2 \sin (\theta_2 - \theta_1). \tag{6.52}$$

The radial and azimuthal components of the equation of motion describing the oscillatory motion of particle 2 are

$$m_2 g \cos\theta_2 - S_2 = 0, \tag{6.53}$$

$$m_2 \left(l_1 \frac{d^2\theta_1}{dt^2} + l_2 \frac{d^2\theta_2}{dt^2} \right) = -m_2 g \sin\theta_2. \tag{6.54}$$

Assuming a small amplitude oscillation, the motion in the azimuthal direction may be described in the x-axis, which is perpendicular to the plumb line including O and takes its origin at the lowest point of particle 2. Letting the position of particle 1 be x_1 and the position of particle 2 be x_2, we find

$$\begin{cases} x_1 = l_1 \sin\theta_1 \cong l_1\theta_1 , \\ x_2 = l_1 \sin\theta_1 + l_2 \sin\theta_2 \cong l_1\theta_1 + l_2\theta_2 , \\ l_2\theta_2 = x_2 - x_1, \quad \cos\theta_1 \cong 1, \quad \cos\theta_2 \cong 1 . \\ \cos(\theta_2 - \theta_1) \cong 1 . \end{cases} \tag{6.55}$$

From (6.53)

$$S_2 = m_2 g. \tag{6.56}$$

Substituting (6.56) into (6.51), we get

$$S_1 = m_1 g + S_2 = (m_1 + m_2)g. \tag{6.57}$$

Substituting from (6.55) and (6.56) into (6.52), we obtain

$$m_1 \frac{d^2 x_1}{dt^2} = -m_1 g \frac{x_1}{l_1} + m_2 g \left(\frac{x_2 - x_1}{l_2} - \frac{x_1}{l_1} \right),$$

$$\frac{d^2 x_1}{dt^2} = -\frac{g}{l_1}(1+\mu)x_1 + \mu \frac{g}{l_2}(x_2 - x_1), \tag{6.58}$$

where $\mu = m_2/m_1$. Substituting from (6.55) into (6.54) yields,

$$\frac{d^2 x_2}{dt^2} = -\frac{g}{l_2}(x_2 - x_1), \tag{6.59}$$

We assume an exponential type solution with the angular frequency ω as

$$x_1 = \alpha_1 \exp \tilde{\iota}\omega t, \quad x_2 = \alpha_2 \exp \tilde{\iota}\omega t.$$

Substituting the above solutions into (6.58) and (6.59), we find

$$\left\{ (1+\mu)\frac{g}{l_1} + \mu\frac{g}{l_2} - \omega^2 \right\} \alpha_1 - \mu\frac{g}{l_2}\alpha_2 = 0, \tag{6.60}$$

$$-\frac{g}{l_2}\alpha_1 + \left(\frac{g}{l_2} - \omega^2 \right) \alpha_2 = 0. \tag{6.61}$$

Equations (6.60) and (6.61) are linear algebraic equations of two unknowns α_1 and α_2, and the necessary condition for α_1 and α_2 to have nontrivial solutions is that the determinant of coefficients of unknowns becomes zero,

$$\begin{vmatrix} (1+\mu)\dfrac{g}{l_1} + \mu\dfrac{g}{l_2} - \omega^2 & -\mu\dfrac{g}{l_2} \\ -\dfrac{g}{l_2} & \dfrac{g}{l_2} - \omega^2 \end{vmatrix} = 0. \qquad (6.62)$$

Expanding (6.62) yields

$$\omega^2 = \frac{1}{2}\left[(1+\mu)\left(\frac{g}{l_1} + \frac{g}{l_2}\right) \pm \sqrt{(1+\mu)^2\left(\frac{g}{l_1} + \frac{g}{l_2}\right)^2 - 4(1+\mu)\frac{g^2}{l_1 l_2}} \right] \quad (6.63)$$

The oscillatory motion corresponding to two frequencies is referred to as a normal mode oscillation. The general solutions of x_1 and x_2 are given by the linear combination of two normal modes. The normal mode oscillation is illustrated for the case that $l_1 = l_2 = l$, $m_1 = m_2 = m$. We obtain from (6.63)

$$\omega^2 = \left(2 \pm \sqrt{2}\right)\frac{g}{l}. \qquad (6.64)$$

1. The case that $\omega^2 = \left(2 - \sqrt{2}\right) g/l$

Substituting ω^2 into (6.61), we get

$$\frac{\alpha_1}{\alpha_2} = \sqrt{2} - 1. \qquad (6.65)$$

Particle 1 and particle 2 oscillate in phase and the ratio of the amplitude of the former to the latter is 0.41. The period is

$$T = \frac{2\pi}{\sqrt{2 - \sqrt{2}}}\sqrt{\frac{l}{g}}. \qquad (6.66)$$

2. The case that $\omega^2 = \left(2 + \sqrt{2}\right) g/l$

Substituting ω^2 into (6.61) yields,

$$\frac{\alpha_1}{\alpha_2} = -(\sqrt{2} + 1). \qquad (6.67)$$

Particle 1 and particle 2 oscillate out of phase and the ratio of the amplitude of the former to the latter is 2.41. The period is

$$T = \frac{2\pi}{\sqrt{2 + \sqrt{2}}}\sqrt{\frac{l}{g}}. \qquad (6.68)$$

The normal mode oscillations are illustrated in Fig. 6.4.

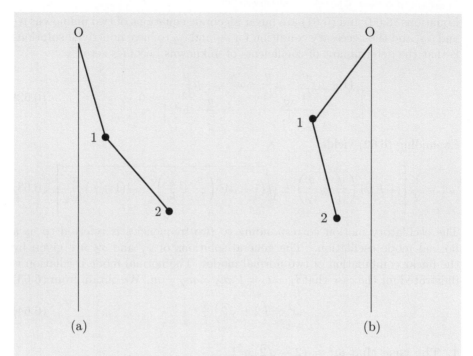

Figure 6.4: Normal mode oscillation for the case of (a) $\omega^2 = (2 - \sqrt{2})g/l$ and (b) $\omega^2 = (2 + \sqrt{2})g/l$.

6.4 Coupled Oscillations

As shown in Fig. 6.5, body 1 of mass m is connected to body 2 of mass m via spring 2 of spring constant k' on a level and frictionless surface. Body 1 is attached to rigid support 1 via spring 1 of spring constant k and body 2 is attached to rigid support 2 via spring 3 of spring constant k. The three springs have the natural length l. Taking the origin of the x-axis at the position of rigid support 1, and letting the displacement of body 1 be x_1 and the displacement of body 2 be x_2, the forces exerting on body 1 are $-kx_1$ by spring 1 and $k'(x_2 - x_1)$ by spring 2, and the forces exerting on body 2 are $-k'(x_2 - x_1)$ by spring 2 and $-kx_2$ by spring 3. Then the equations of motion describing the oscillatory motion of body 1 and body 2 are

$$m_1 \frac{d^2 x_1}{dt^2} = -kx_1 + k'(x_2 - x_1), \tag{6.69}$$

$$m_2 \frac{d^2 x_2}{dt^2} = -kx_2 - k'(x_2 - x_1). \tag{6.70}$$

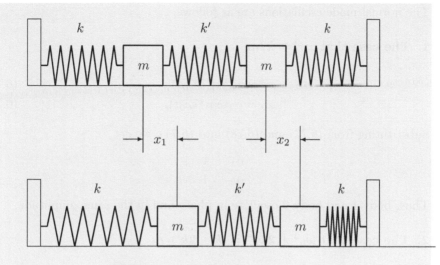

Figure 6.5: A coupled oscillation.

Defining

$$\omega_0{}^2 = \frac{k}{m}, \quad \omega_0'{}^2 = \frac{k'}{m},$$

(6.69) and (6.70) become

$$\frac{d^2 x_1}{dt^2} = -\omega_0{}^2 x_1 + \omega_0'{}^2 (x_2 - x_1), \tag{6.71}$$

$$\frac{d^2 x_2}{dt^2} = -\omega_0{}^2 x_2 - \omega_0'{}^2 (x_2 - x_1). \tag{6.72}$$

The normal mode oscillation may be obtained by the same method as that of the coupled pendulum of the previous section, but we will obtain the normal mode oscillation by another method. Adding and subtracting (6.71) and (6.72), we obtain

$$\frac{d^2 (x_1 + x_2)}{dt^2} = -\omega_0{}^2 (x_1 + x_2), \tag{6.73}$$

$$\frac{d^2 (x_2 - x_1)}{dt^2} = -(\omega_0{}^2 + 2\omega_0'{}^2)(x_2 - x_1), \tag{6.74}$$

where $x_1 + x_2$ and $x_2 - x_1$ correspond to the normal mode oscillations. The general solutions are,

$$x_1 + x_2 = A_1 \cos(\omega_0 t - \phi_1), \tag{6.75}$$

$$x_2 - x_1 = A_2 \cos\left(\sqrt{\omega_0{}^2 + 2\omega_0'{}^2} t - \phi_2\right). \tag{6.76}$$

The normal mode oscillations are as follows.

1. The case that $\omega_0{}^2 = k/m$

$$x_1 = \alpha_1 \exp(\tilde{\imath}\omega_0 t), \tag{6.77}$$
$$x_2 = \alpha_2 \exp(\tilde{\imath}\omega_0 t). \tag{6.78}$$

Substituting from (6.77) and (6.78) into (6.71), we get

$$\frac{\alpha_1}{\alpha_2} = \frac{1}{1}.$$

Thus, body 1 and body 2 oscillate in phase and in the same amplitude.

2. The case that $\omega_0{}^2 + 2\omega_0'{}^2 = (k + 2k')/m$

$$x_1 = \alpha_1 \exp\left(\tilde{\imath}\sqrt{\omega_0{}^2 + 2\omega_0'{}^2}t\right), \tag{6.79}$$

$$x_2 = \alpha_2 \exp\left(\tilde{\imath}\sqrt{\omega_0{}^2 + 2\omega_0'{}^2}t\right). \tag{6.80}$$

Substituting from (6.79) and (6.80) into (6.71), we get

$$\frac{\alpha_1}{\alpha_2} = \frac{-1}{1}.$$

Body 1 and body 2 oscillate out of phase and with the same amplitude. We will find the general solutions by subtracting (6.76) from (6.75),

$$x_1 = \frac{1}{2}\left[A_1 \cos(\omega_0 t - \phi_1) - A_2 \cos\left(\sqrt{\omega_0{}^2 + 2\omega_0'{}^2}t - \phi_2\right)\right]. \tag{6.81}$$

Adding (6.75) and (6.76), we get

$$x_2 = \frac{1}{2}\left[A_1 \cos(\omega_0 t - \phi_1) + A_2 \cos\left(\sqrt{\omega_0{}^2 + 2\omega_0'{}^2}t - \phi_2\right)\right]. \tag{6.82}$$

It is obvious that the general solutions are the linear combination of the normal modes.

6.5   Problems

1. In the case of heavy damping, confirm that the particular solution is given by (6.6) under the conditions that $x = x_0, dx/dt = 0$ at $t = 0$.

2. In the case of damped oscillation, confirm that the particular solution is given by (6.10) under the conditions that $x = x_0, dx/dt = 0$ at $t = 0$.

3. In the case of critical damping, confirm that the particular solution is given by (6.15) under the conditions that $x = x_0, dx/dt = 0$ at $t = 0$.

4. Suppose that a massless and extensible string of length $4l$ is pulled by the tension S on a level and frictionless surface, and three particles 1, 2, and 3 of mass m are attached to the string at positions l, $2l$, and $3l$ from one end of the string. After giving a small deviation to the system, obtain the normal mode frequencies of the oscillation and the amplitudes of the particles corresponding to each frequency.

5. Suppose that two simple pendulums of length l and mass m are attached to rigid supports with distance d and bobs are connected by a spring of spring constant k. After giving a small displacement to the system, it begins to oscillate in a vertical plane. Obtain the normal mode oscillations.

Chapter 7

Mechanics of Rigid Bodies

An ensemble of particles is referred to as a many-particle system. And a body which consists of a continuous medium and is never deformed by any external forces is called a rigid body. As is clear from the definition, the rigid body is an idealization or approximated concept of bodies. Suppose a rigid rod, and if a person pushed one end of the rod, the other end would move with time lag zero. The speed of the rod is infinity exceeding the speed of light, which violates the special theory of relativity.

The motion of many-particle systems and rigid bodies has two aspects of motion, i.e., the translational motion and the rotational motion. In this chapter, we will limit our discussion to the translational motion. The center of mass is defined as the point at which all mass of many-particle systems or rigid bodies is considered to concentrate. The translation of a center of mass may be described by the equation of motion as is for a single particle.

7.1 The Equation of Motion and the Center of Mass of Many-Particle Systems

Suppose that n particles are in motion exerting forces on each other and under the action of external forces. We will designate the external force acting on particle i as $\mathbf{F}_i$, the internal force exerting on particle i due to particle j as $\mathbf{F}_{ij}$, and the internal force exerting on particle j due to particle i as $\mathbf{F}_{ji}$. Two forces have a relationship that

$$\mathbf{F}_{ij} = -\mathbf{F}_{ji}$$

owing to Newton's third law. The equations of motion describing the motion of particle $1, 2, \cdots, i, \cdots, n$ become

$$m_1 \frac{d^2 \mathbf{r}_1}{dt^2} = \mathbf{F}_1 + \mathbf{F}_{12} + \mathbf{F}_{13} + \cdots + \mathbf{F}_{1n},$$

DOI: 10.1201/9781003310068-7

$$m_2 \frac{d^2\mathbf{r}_2}{dt^2} = \mathbf{F}_2 + \mathbf{F}_{21} + \mathbf{F}_{23} + \cdots + \mathbf{F}_{2n},$$

$$\vdots$$

$$m_i \frac{d^2\mathbf{r}_i}{dt^2} = \mathbf{F}_i + \mathbf{F}_{i1} + \mathbf{F}_{i2} + \cdots + \mathbf{F}_{in},$$

$$\vdots$$

$$m_n \frac{d^2\mathbf{r}_n}{dt^2} = \mathbf{F}_n + \mathbf{F}_{n1} + \mathbf{F}_{n2} + \cdots + \mathbf{F}_{nn-1}.$$

Making use of a sigma notation, the equation of motion for particle i is written as

$$m_i \frac{d^2\mathbf{r}_i}{dt^2} = \mathbf{F}_i + \sum_{j=1}^{n} (\mathbf{F}_{ij} - \delta_{ij}\mathbf{F}_{ij}), \tag{7.1}$$

where δ_{ij} is *Kronecker's delta* defined by

$$\delta_{ij} = \begin{cases} 1 & \text{if } i = j \\ 0 & \text{if } i \neq j \end{cases}. \tag{7.2}$$

Summing up (7.1) with respect to i from 1 to n,

$$\frac{d^2}{dt^2}\left(\sum_{i=1}^{n} m_i\mathbf{r}_i\right) = \sum_{i=1}^{n} \mathbf{F}_i, \tag{7.3}$$

where we have used the fact that $\mathbf{F}_{ij} = -\mathbf{F}_{ji}$. The mass weighted mean position vector is referred to as the center of mass, which is represented by

$$\mathbf{R} = \frac{\displaystyle\sum_{i=1}^{n} m_i\mathbf{r}_i}{\displaystyle\sum_{i=1}^{n} m_i} = \frac{\displaystyle\sum_{i=1}^{n} m_i\mathbf{r}_i}{M}, \tag{7.4}$$

where M is the total mass of the particle system. Component equations of (7.4) in Cartesian coordinates are

$$X = \frac{\displaystyle\sum_{i=1}^{n} m_i x_i}{M}, \tag{7.5}$$

$$Y = \frac{\displaystyle\sum_{i=1}^{n} m_i y_i}{M}, \tag{7.6}$$

$$Z = \frac{\displaystyle\sum_{i=1}^{n} m_i z_i}{M}. \tag{7.7}$$

From (7.3) and (7.4), we obtain

$$M\frac{d^2\mathbf{R}}{dt^2} = \sum_{i=1}^{n} \mathbf{F}_i. \tag{7.8}$$

Equation (7.8) states that if the external forces exerting on a many-particle system were known, the motion of the center of mass of the particle system would be described by the equation of motion equivalent to a particle of mass M and the position vector $\mathbf{R}$.

7.2 A Two-Particle System

Let's consider the simplest case of many-particle systems, a two-particle system. From (7.1), the equations of motion of particle 1 and particle 2 are

$$m_1\frac{d^2\mathbf{r}_1}{dt^2} = \mathbf{F}_1 + \mathbf{F}_{12}, \tag{7.9}$$

$$m_2\frac{d^2\mathbf{r}_2}{dt^2} = \mathbf{F}_2 + \mathbf{F}_{21}. \tag{7.10}$$

Adding (7.9) and (7.10),

$$m_1\frac{d^2\mathbf{r}_1}{dt^2} + m_2\frac{d^2\mathbf{r}_2}{dt^2} = \mathbf{F}_1 + \mathbf{F}_2, \tag{7.11}$$

because $\mathbf{F}_{12} = -\mathbf{F}_{21}$ due to Newton's third law. From (7.4), the position vector of the center of mass of the two-particle system is

$$\mathbf{R} = \frac{m_1\mathbf{r}_1 + m_2\mathbf{r}_2}{m_1 + m_2} = \frac{m_1\mathbf{r}_1 + m_2\mathbf{r}_2}{M}. \tag{7.12}$$

From (7.11) and (7.12), we obtain

$$M\frac{d^2\mathbf{R}}{dt^2} = \mathbf{F}_1 + \mathbf{F}_2. \tag{7.13}$$

We define the relative position vector from point 1 to point 2 as $\mathbf{r}$, namely

$$\mathbf{r} = \mathbf{r}_2 - \mathbf{r}_1. \tag{7.14}$$

Dividing (7.9) through by m_1 and dividing (7.10) through by m_2, and subtracting the former from the latter, we get

$$\mu\frac{d^2\mathbf{r}}{dt^2} = \mu\left(\frac{\mathbf{F}_2}{m_2} - \frac{\mathbf{F}_1}{m_1}\right) + \mathbf{F}_{21}, \tag{7.15}$$

where

$$\mu = \frac{m_1 m_2}{m_1 + m_2}, \tag{7.16}$$

is called the *reduced mass.* Equation (7.15) is Newton's second law for a particle of mass μ, and is solved if the forces $\mathbf{F}_1$, $\mathbf{F}_2$, and $\mathbf{F}_{21}$ were given. The position vectors of particle 1 and particle 2 are given by the solutions of (7.13) and (7.15), using the following equations,

$$\mathbf{r}_1 = \mathbf{R} - \frac{\mu}{m_1}\mathbf{r}, \tag{7.17}$$

$$\mathbf{r}_2 = \mathbf{R} + \frac{\mu}{m_2}\mathbf{r}. \tag{7.18}$$

7.3   The Center of Mass of Rigid Bodies

Because a rigid body is a continuous medium, its center of mass is obtained replacing sigma notations by integrals and mass of each particle m_i by the mass element ρdv. Therefore, the position vector of the center of mass of a rigid body is given by

$$\mathbf{R} = \frac{\displaystyle\int_V \rho \mathbf{r} dv}{\displaystyle\int_V \rho dv} = \frac{\displaystyle\int_V \rho \mathbf{r} dv}{M}, \tag{7.19}$$

where ρ is the density and dv is the volume element of the rigid body. Component equations of (7.19) in Cartesian coordinates are

$$X = \frac{\displaystyle\int_V \rho x dv}{M}, \tag{7.20}$$

$$Y = \frac{\displaystyle\int_V \rho y dv}{M}, \tag{7.21}$$

$$Z = \frac{\displaystyle\int_V \rho z dv}{M}. \tag{7.22}$$

In performing volume integrals, we should choose adequate coordinates (Cartesian, cylindrical, or spherical coordinates) corresponding to the shape of the rigid bodies.

7.4   Center of Gravity of Many-Particle Systems and Rigid Bodies

Suppose a many-particle system consisting of n particles. Let the mass and the position vector of i-th particle be m_i and $\mathbf{r}_i$, the position vector of the center of mass be $\mathbf{R}$, and the total mass of the system be M. When the many-particle system is placed in the Earth's gravity field, the action center of the total gravity coincides with its center of mass. Then the action center of the total gravity is referred to as the center of gravity. The above discussion is also applicable to rigid bodies.

We will show that a rigid body is at rest without rotation when it is hung along a plumb line including the center of gravity. In the following discussion the torque which will appear in Chapter 9 is used, so that readers should come back here after studying Chapter 9. Taking the z-axis vertically upward, the total gravity torque $\mathbf{N}$ about the origin becomes

$$\mathbf{N} = \int_V \mathbf{r} \times (-\rho g \mathbf{e}_z) dv, \tag{7.23}$$

where

$$\mathbf{r} \times (-\rho g \mathbf{e}_z) = (x\mathbf{e}_x + y\mathbf{e}_y + z\mathbf{e}_z) \times (-\rho g \mathbf{e}_z) = \rho g(x\mathbf{e}_y - y\mathbf{e}_x). \tag{7.24}$$

From (7.23) and (7.24), we get

$$\mathbf{N} = \mathbf{e}_y g \int_V \rho x \, dv - \mathbf{e}_x g \int_V \rho y \, dv. \tag{7.25}$$

From (7.25), (7.20), and (7.21), we obtain

$$\mathbf{N} = Mg(X\mathbf{e}_y - Y\mathbf{e}_x) = \mathbf{R} \times (-Mg\mathbf{e}_z). \tag{7.26}$$

It is proved by (7.26) that the sum of the torque of gravity exerting on any part of the rigid body is equal to the torque of gravity exerting on the center of mass at which all mass of the rigid body concentrates. Therefore, the rigid body may be kept at rest by exerting a force equal and opposite to the gravity at any point along the plumb line including the center of gravity. It is clear from the above discussion that the center of mass and the center of gravity have the same position but have different physical meaning.

7.5   How to Obtain the Center of Mass

In this section, a few methods to obtain the center of mass will be discussed.

7.5.1   Empirical Method

If you attach a string to any point of a rigid body and keep it at rest, the center of gravity exists along the plumb line. Next, if you attach a string to another point of the rigid body and keep it at rest, the center of gravity is along another plumb line. The center of gravity is obtained at the intersection of two lines.

7.5.2 The Method Using the Definition of the Center of Mass

The center of mass is obtained using (7.4) for a many-particle system and is obtained by (7.19) for a rigid body.

Example 1. In Cartesian coordinates, a particle of mass $2m$ is attached to (1,1,3), a particle of mass m is attached to (4,1,0), and a particle of mass $3m$ is attached to (0,3,4). Obtain the center of mass of the particle system.

Answer

$$X = \frac{1 \times 2m + 4 \times m + 0 \times 3m}{2m + m + 3m} = 1 \ ,$$

$$Y = \frac{1 \times 2m + 1 \times m + 3 \times 3m}{6m} = 2,$$

$$Z = \frac{3 \times 2m + 0 \times m + 4 \times 3m}{6m} = 3 \ .$$

The position of the center of mass is (1,2,3).

Example 2. Suppose a thin disk of radius a with homogeneous area density σ. Obtain the center of mass when a disk of radius $a/2$ is cut off closing its rim to the rim of the original disk.

Answer

Taking the x-axis to the line passing the centers of the two disks and the origin at the center of the original disk, the center of mass exists on the x-axis due to the symmetry of the body's shape. The mass of the body is

$$M = \sigma \left\{ \pi a^2 - \pi \left(\frac{a}{2} \right)^2 \right\} = \frac{3}{4} \sigma \pi a^2.$$

Letting the position of the center of mass be $(X, 0)$, and taking into account the fact that if the smaller disk was put back on the disk, the center of mass of the original disk would be the origin, then (7.5) becomes

$$0 = \frac{\frac{3}{4} \sigma \pi a^2 X + \frac{1}{4} \sigma \pi a^2 \frac{a}{2}}{\sigma \pi a^2}, \quad X = -\frac{a}{6}.$$

Example 3. Suppose a plate of a right-angled triangle with homogeneous area density σ whose right angle is adjacent to two sides of length a and b. Where is the position of the center of mass of the triangle?

Answer

Taking a side of length a as the x-axis, a side of b as the y-axis, and the original point at the vertex of a right angle, the equation of the hypotenuse becomes

$$y = -\frac{b}{a} x + b.$$

The area element between x and $x + \delta x$ is

$$\delta S = \left(-\frac{b}{a}x + b \right)\delta x.$$

Therefore, the x-position of the center of mass is

$$X = \frac{\int_0^a \sigma \left(-\frac{b}{a}x + b \right)x\,dx}{\sigma ab/2} = \frac{\sigma \left[-\frac{b}{3a}x^3 + \frac{b}{2}x^2 \right]_0^a}{\sigma ab/2} = \frac{1}{3}a.$$

Similarly, the y-position of the center of mass is

$$Y = \frac{\int_0^b \sigma \left(-\frac{a}{b}y + a \right)y\,dy}{\sigma ab/2} = \frac{\sigma \left[-\frac{a}{3b}y^3 + \frac{a}{2}y^2 \right]_0^b}{\sigma ab/2} = \frac{1}{3}b.$$

The position of the center of mass of a triangle agrees with the mathematical definition.

Example 4. Suppose that a person of mass m is standing at the center of a caster of mass M and length L (Fig. 7.1). The person begins to walk straight forward and stops at the edge of the caster. In what direction and how long does the caster move? Suppose that the center of mass of the caster is the geometric center of it and neglect the foot length of the person compared with the length of the caster.

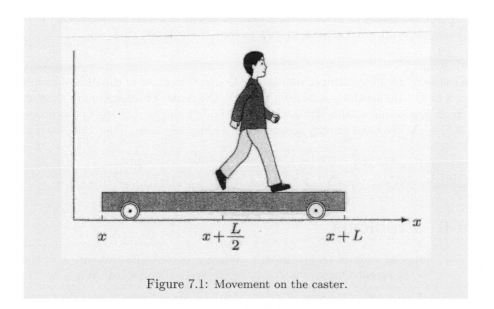

Figure 7.1: Movement on the caster.

Answer
Taking the walking direction of the person as the positive x-axis, and the position of the backward edge of the caster on the floor as the origin, and letting the position of the center of mass of the caster–person system be X, we find before the person's walk

$$X = \frac{m/2 + ML/2}{m + M} = \frac{L}{2}.$$

Letting the center of mass of the caster be $x + L/2$ when the person comes to the edge, the following equation holds because the center of mass of the caster–person system does not change before and after the person's walk.

$$X = \frac{m(x + L) + M\left(x + L/2\right)}{m + M} = x + \frac{L}{2} + \frac{mL}{2(m + M)} = \frac{L}{2},$$

$$x = -\frac{m}{2(m + M)}L.$$

Therefore, the caster moves $mL/2(m + M)$ to the opposite direction of the person's walk.

7.5.3   The Method Using the Total Torque of Gravity about the Center of Gravity

Total torque of gravity about the center of gravity becomes zero. Then the center of gravity or the center of mass is obtained using this fact.

Example 5. Suppose a disk of radius a and the homogeneous area density σ. Obtain the center of mass when a disk of radius $a/2$ and homogeneous area density σ is attached closing its rim to that of the larger disk.

Answer
Let the center of the disk of radius a be O and the center of the disk of radius $a/2$ be O'. Taking the x-axis from O to O', the center of mass G may exist on the x-axis owing to the symmetry of the body's shape. Letting the position of G be X, the total gravity torque about G becomes zero. Therefore, we find

$$\sigma \pi a^2 g X - \sigma \pi \left(\frac{a}{2}\right)^2 g \left(\frac{a}{2} - X\right) = 0, \quad X = \frac{1}{10}a,$$

where g is the magnitude of the acceleration due to gravity.

7.6   Problems

1. Suppose a log of length 6.00 m. When a person lifts the end A of the log, he exerts a force of $5.00 \times 10 \,\mathrm{kg}\,\mathrm{W}$[1] vertically upward. When he lifts

---

[1] $1 \,\mathrm{kg}\,\mathrm{W}$ is defined as the force of gravity exerting a body of mass m near the Earth's surface, which corresponds to around $9.8 \,\mathrm{N}$.

another end B, he exerts a force of $1.00 \times 10^2\,$kg W vertically upward. What is the mass and where is the position of the center of mass of the log?

2. The ratio of the mass of the Moon to the Earth's mass is 1.23×10^{-2}, and the distance between the Earth and the Moon is $3.84 \times 10^8\,$m. Obtain the center of mass of the Earth–Moon system.

3. Suppose a straight wire of length $50.0\,$cm with homogeneous linear density η. When it is bent into a right angle at $20.0\,$cm from the end, obtain the position of the center of mass of the bent wire (X, Y), taking a side of $30.0\,$cm in length as the x-axis and a side of $20.0\,$cm in length as the y-axis and neglecting the magnitude of the diameter of the wire.

4. Suppose a half circle of radius a made of wire with linear density η. Obtain the position of the center of mass, neglecting the magnitude of the diameter of the wire.

5. Suppose a half annular disk with inner radius a, outer radius b, and homogeneous area density σ. Where is the position of the center of mass? Confirm that the result coincides with that of the previous problem in the limit $b \to a$.

6. Where is the position of the center of mass of a half sphere of radius a and homogeneous density ρ?

7. Where is the position of the center of mass of a thin half spherical shell of radius a and homogeneous area density σ?

8. Where is the position of the center of mass of a half spherical shell of inner radius a and outer radius b and homogeneous density ρ? Confirm that the result coincides with that of the previous problem in the limit $b \to a$.

Chapter 8

Momentum and Impulse

In this chapter, the second transformation of the equation of motion is performed. The resultant equation is referred to as the linear momentum equation, which relates the changing rate of the linear momentum of a body and the impulse exerting on it. If no external forces exert on a body, the linear momentum is conserved. When no external forces exert on a many-body system, the total momentum is conserved because the total impulse of internal forces is zero by Newton's third law. This is called the law of linear momentum conservation. For instance, the law is useful for the problem of collision between bodies. A collision can never occur between particles having no cross section, so that we should discuss the problem of collision between bodies with finite spatial extent neglecting their freedom of rotation. The collision between small particles and a rigid body is discussed, in which the new physical concept, the scattering cross section, is introduced. The rocket motion is discussed at the end of the chapter.

8.1   Transformation of the Equation of Motion

Newton described the equation of motion in the momentum form as,

$$\frac{d}{dt}(m\mathbf{v}) = \mathbf{F}. \tag{8.1}$$

This form may be applicable to the case when the mass changes with time. So far, we have treated the case of constant mass. In this section, we will start the discussion from (8.1). Linear momentum (hereafter referred to as momentum) is defined

$$\mathbf{p} = m\mathbf{v}. \tag{8.2}$$

The unit of momentum is

$$[\mathbf{p}] = [\mathrm{kg\,m\,s^{-1}}] = [\mathrm{N\,s}].$$

DOI: 10.1201/9781003310068-8

Integrating (8.1) with respect to time t from t_1 to t_2, we get

$$\mathbf{p}(t_2) - \mathbf{p}(t_1) = \int_{t_1}^{t_2} \mathbf{F}(t)dt. \tag{8.3}$$

The right-hand side of (8.3) is called *impulse*. We define time-averaged force as

$$\langle \mathbf{F}(t) \rangle = \frac{\displaystyle\int_{t_1}^{t_2} \mathbf{F}(t)dt}{t_2 - t_1}. \tag{8.4}$$

From (8.3) and (8.4), we obtain

$$\mathbf{p}(t_2) - \mathbf{p}(t_1) = \langle \mathbf{F}(t) \rangle (t_2 - t_1). \tag{8.5}$$

When the force is independent on time, $\langle \mathbf{F}(t) \rangle = \mathbf{F}$ and (8.5) becomes

$$\mathbf{p}(t_2) - \mathbf{p}(t_1) = \mathbf{F}(t_2 - t_1). \tag{8.6}$$

The unit of impulse is

$$\left[\int_{t_1}^{t_2} \mathbf{F}(t)dt \right] = [\mathrm{N\,s}].$$

Problem 1. A vehicle of mass 1.50×10^3 kg, including the mass of a crew, is travelling at $50.0\,\mathrm{km\,h^{-1}}$ on a straight road. A driver finding an obstacle on the road brakes urgently and stops the vehicle after 2.00 s. Using (8.6), calculate the frictional force exerting on the vehicle, supposing that the frictional force is constant and the driving force is zero during braking.

8.2   Conservation of Momentum

In this section, collisions of bodies are discussed ignoring the rotational motion.

8.2.1   The Case of Many-Body System

Suppose a system of n bodies that exert forces on each other in the absence of external forces. Forces may be direct forces as in collisions or indirect long-range forces such as the gravitational force and the Coulomb force. Defining $\mathbf{F}_{ij}$ as the force that body j exerts on body i, we find

$$\mathbf{F}_{ij} = -\mathbf{F}_{ji}$$

owing to Newton's third law. The equations of motion describing the motion of body $1, 2, \cdots, i, \cdots, n$ are

$$\frac{d\mathbf{p}_1}{dt} = \mathbf{F}_{12} + \mathbf{F}_{13} + \cdots + \mathbf{F}_{1n},$$

$$\frac{d\mathbf{p}_2}{dt} = \mathbf{F}_{21} + \mathbf{F}_{23} + \cdots + \mathbf{F}_{2n},$$

$$\vdots$$

$$\frac{d\mathbf{p}_i}{dt} = \mathbf{F}_{i1} + \mathbf{F}_{i2} + \cdots + \mathbf{F}_{in},$$

$$\vdots$$

$$\frac{d\mathbf{p}_n}{dt} = \mathbf{F}_{n1} + \mathbf{F}_{n2} + \cdots + \mathbf{F}_{nn-1}.$$

Using a sigma notation, the equation of motion for body i is written

$$\frac{d\mathbf{p}_i}{dt} = \sum_{j=1}^{n} (\mathbf{F}_{ij} - \delta_{ij}\mathbf{F}_{ij}), \tag{8.7}$$

where δ_{ij} is the Kronecker's delta. Summing up (8.7) with respect to i from 1 to n,

$$\frac{d}{dt}\left(\sum_{i=1}^{n} \mathbf{p}_i\right) = 0. \tag{8.8}$$

When no external forces are exerting or the total external force is zero, the total momentum of the system is conserved (the *law of momentum conservation*). Integrating (8.8) with respect to time t from t_1 to t_2, we find

$$\sum_{i=1}^{n} m_i \mathbf{v}_i(t_1) = \sum_{i=1}^{n} m_i \mathbf{v}_i(t_2), \tag{8.9}$$

where $\mathbf{v}_i(t_1)$ and $\mathbf{v}_i(t_2)$ are the velocities at t_1 and t_2 of body i. Thus, we have obtained the integration form of the law of momentum conservation.

8.2.2 The Case of Two-Body System

Suppose that disk 1 of mass m_1 and disk 2 of mass m_2 collide along a straight line on a level and frictionless surface. Letting the velocity of disk 1 and disk 2 before a collision be $\mathbf{v}_1$ and $\mathbf{v}_2$, the velocity of disk 1 and disk 2 after the collision be $\mathbf{v}_1'$ and $\mathbf{v}_2'$, a force exerted by disk 1 on disk 2 be $\mathbf{F}_{21}$ and a force exerted by disk 2 on disk 1 be $\mathbf{F}_{12}$, we find

$$\mathbf{F}_{12} = -\mathbf{F}_{21} \tag{8.10}$$

holds by Newton's third law. As the total momentum is conserved before and
after the collision, we obtain

$$m_1 \mathbf{v}_1 + m_2 \mathbf{v}_2 = m_1 \mathbf{v}_1' + m_2 \mathbf{v}_2'. \tag{8.11}$$

Example 1. Suppose that a wooden block of mass M is suspended from a
ceiling by two massless and inextensible strings of length L (Fig. 8.1). An
arrow of mass m is flying horizontally and sticks into the block. The block
and the arrow begin to swing together, and the maximum angle between the
string and the vertical is θ. Obtain the speed of the arrow just before hitting
the block. Let the magnitude of the acceleration due to gravity be g.

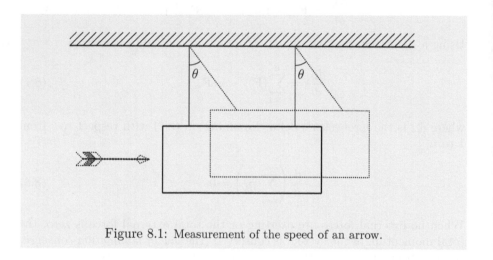

Figure 8.1: Measurement of the speed of an arrow.

Answer
Letting the speed of the wooden block just after being hit by the arrow be
V, the following equation holds owing to the law of momentum conservation,

$$mv = (m + M)V. \tag{8.12}$$

From the law of mechanical energy conservation, we obtain

$$\frac{1}{2}(m + M)V^2 = (m + M)gL(1 - \cos\theta),$$
$$V = \sqrt{2gL(1 - \cos\theta)}. \tag{8.13}$$

Substituting (8.13) into (8.12), we get

$$v = \frac{m + M}{m}\sqrt{2gL(1 - \cos\theta)}.$$

Example 2. Suppose a cannon with a buffer whose barrel has mass $M = 1.0 \times 10^3$ kg. The cannon projected a shell of mass $m = 1.0 \times 10$ kg horizontally at speed $v = 8.0 \times 10^2$ m s^{-1}. Answer the following questions.

1. What is the speed of the barrel just after the projection of the shell?

2. Obtain the average force exerted on the barrel by the shell, supposing that the time for the shell to go through the barrel is $\Delta t = 1.0 \times 10^{-2}$ s.

Answer
Taking the direction of projection as the x-axis;

1. We find due to the law of momentum conservation,

$$MV + mv = 0,$$
$$V = -\frac{mv}{M} = -\frac{(1.0 \times 10) \times (8.0 \times 10^2)}{1.0 \times 10^3} = -8.0 \,[\text{m s}^{-1}].$$

2. As the momentum change is equal to the impulse, we get

$$MV - 0 = F\Delta t \,,$$
$$F = \frac{MV - 0}{\Delta t} = \frac{1.0 \times 10 \times (-8.0 \times 10^2) - 0}{1.0 \times 10^{-2}} = -8.0 \times 10^5 \,[\text{N}] \,.$$

The negative sign means that the force is directed opposite to the projected direction of the shell.

8.3 Collision of Disks

Now the motion of disks is restricted to the translational motion, or we may consider that the rotational kinetic energy is negligibly small compared with that of the translational motion. In this section, we will discuss one-dimensional collisions.

8.3.1 Inelastic Collisions

We will consider a collision of two disks along the x-axis on a level and frictionless surface. Suppose that disk 1 of mass m_1 travelling at speed v_1 collides with disk 2 of mass m_2 travelling at speed v_2, and after the collision disk 1 and disk 2 travel at speed v_1' and v_2', respectively (Fig. 8.2). The conditions that disk 1 catches up with disk 2 and disk 1 never passes disk 2 are written as

$$v_1 > v_2, \quad v_1' \leq v_2'.$$

The law of momentum conservation is

$$m_1 v_1 + m_2 v_2 = m_1 v_1' + m_2 v_2'. \tag{8.14}$$

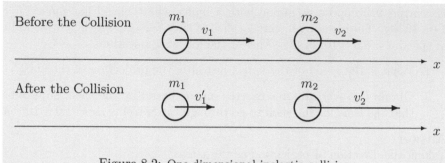

Figure 8.2: One-dimensional inelastic collision.

We have two unknown variables v_1' and v_2', while we have only one equation (8.14). Therefore, we must find another equation relating v_1' and v_2'. The equation is given by the kinematic relation known as the coefficient of restitution e. It is defined as the positive ratio of the relative speed of separation to the relative speed of approach

$$e = -\frac{v_1' - v_2'}{v_1 - v_2}. \tag{8.15}$$

It depends on the material of colliding bodies, e.g., in the case of ivory balls e is nearly equal to 1 and in the case of wet clay balls e is 0. When $0 < e < 1$, the collision is called the inelastic collision.
From (8.15)

$$v_2' = v_1' + e(v_1 - v_2). \tag{8.16}$$

Substituting (8.16) into (8.14) yields,

$$v_1' = v_1 + (1 + e)\frac{m_2}{m_1 + m_2}(v_2 - v_1). \tag{8.17}$$

Substituting (8.17) into (8.16), we obtain

$$v_2' = \frac{m_1}{m_1 + m_2}(1 + e)v_1 + \frac{m_2 - em_1}{m_1 + m_2}v_2. \tag{8.18}$$

Thus, we find the speed of two disks after the collision.

8.3.2   Elastic Collisions

Collisions of $e = 1$ are referred to as elastic collisions. Suppose that disk 1 of mass m_1 travelling at speed v_1 collides with disk 2 of mass m_2 travelling at speed v_2, and after the collision disk 1 and disk 2 travel at speed v_1' and

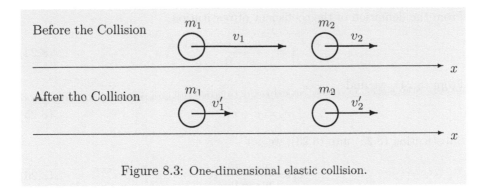

Figure 8.3: One-dimensional elastic collision.

v_2', respectively (Fig. 8.3). The conditions that disk 1 catches up with disk 2, and disk 1 never passes disk 2 are written as

$$v_1 > v_2, \quad v_1' \le v_2'.$$

Owing to the law of momentum conservation, we get

$$m_1 v_1 + m_2 v_2 = m_1 v_1' + m_2 v_2'. \tag{8.19}$$

From the definition of the coefficient of restitution,

$$1 = -\frac{v_1' - v_2'}{v_1 - v_2}. \tag{8.20}$$

Then v_1' and v_2' are obtained from (8.19) and (8.20),

$$v_1' = \frac{m_1 - m_2}{m_1 + m_2} v_1 + \frac{2m_2}{m_1 + m_2} v_2, \tag{8.21}$$

$$v_2' = \frac{2m_1}{m_1 + m_2} v_1 + \frac{m_2 - m_1}{m_1 + m_2} v_2. \tag{8.22}$$

Problem 2. Using (8.21) and (8.22), prove that the kinetic energy of two disks in the elastic collision is conserved before and after the collision.

Example 3. Suppose that disk 1 of mass m_1 travelling at speed v collides with disk 2 of mass m_2 at rest on a level and frictionless surface. When the collision is elastic, obtain a speed v_1' of disk 1 and a speed v_2' of disk 2 after the collision. Next, calculate the kinetic energy of the system before and after the collision. Suppose that the collision occurs in one-dimension.

Answer
Owing to the law of momentum conservation,

$$m_1 v_1 = m_1 v_1' + m_2 v_2'. \tag{8.23}$$

From the definition of the coefficient of restitution,

$$1 = -\frac{v_1' - v_2'}{v - 0}.$$ (8.24)

From (8.24), we find

$$v_2' = v + v_1'.$$ (8.25)

Substituting (8.25) into (8.23), we get

$$v_1' = \frac{m_1 - m_2}{m_1 + m_2}v.$$ (8.26)

Substituting (8.26) into (8.25), we obtain

$$v_2' = \frac{2m_1}{m_1 + m_2}v.$$ (8.27)

The kinetic energy of the system before the collision is

$$K = \frac{1}{2}m_1 v^2.$$ (8.28)

The kinetic energy of the system after the collision is

$$\begin{aligned} K' &= \frac{1}{2}m_1 v_1'^2 + \frac{1}{2}m_2 v_2'^2 \\ &= \frac{1}{2}m_1\left(\frac{m_1 - m_2}{m_1 + m_2}\right)^2 v^2 + \frac{1}{2}m_2\left(\frac{2m_1}{m_1 + m_2}\right)^2 v^2 \\ &= \frac{1}{2}m_1 v^2 = K. \end{aligned}$$ (8.29)

Therefore, the kinetic energy of the system is conserved before and after the collision.

8.3.3 Totally Inelastic Collisions

Suppose that two disks collide in one-dimension on a level and frictionless surface and travel together after the collision. This type of collision is referred to as totally inelastic collisions. The coefficient of restitution becomes zero because the relative speed of separation is zero (Fig. 8.4).

As the momentum of the system is conserved before and after the collision,

$$m_1 v_1 + m_2 v_2 = (m_1 + m_2)v',$$
$$v' = \frac{m_1 v_1 + m_2 v_2}{m_1 + m_2}.$$ (8.30)

Example 4. Suppose that a chain of linear density η is set in a lump on a frictionless floor. A person pulls the chain vertically upward at a constant

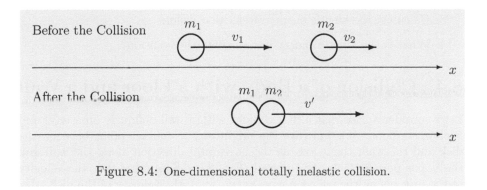

Figure 8.4: One-dimensional totally inelastic collision.

speed v. Obtain the force for pulling up the chain, when the end of the chain is at height of z from the floor. Let the magnitude of the acceleration due to gravity be g and the chain be long enough so that the other end of the chain remains on the floor.

Answer
Let's take the z-axis vertically upward and the floor surface as the origin. Suppose that the end of the chain reaches z at time t. The end goes up to $z + \Delta z$ in a small-time increment Δt, so that the following relation holds

$$v = \frac{\Delta z}{\Delta t}.$$

Letting the force at $t + \Delta t$ be $F + \Delta F$, we find

$$\eta v(z + \Delta z) - \eta v z = \left\{ F + \frac{\Delta F}{2} - \left(\eta g z + \frac{\eta g \Delta z}{2} \right) \right\} \Delta t.$$

Neglecting a term of order Δ^2, we get

$$\eta v \Delta z = (F - \eta g z) \Delta t,$$
$$F = \eta g z + \eta v \frac{\Delta z}{\Delta t} = \eta g z + \eta v^2.$$

The motion of a chain is a typical example of a totally inelastic collision, because chain rings move together after the collision.

Problem 3. Suppose that disk 2 of mass m travelling at speed v collides with disk 1 of mass m at rest on a level and frictionless surface, and they travel together at speed v' after the collision. Answer the following questions, supposing that the collision occurs in one-dimension.

1. Obtain the speed v'.

2. Calculate the kinetic energy before the collision.

3. Calculate the kinetic energy after the collision.

4. What is the kinetic energy lost through the collision?

8.4   Collision of a Body with a Floor and a Wall

Let's consider a case that a disk collides with a wall obliquely on a level and frictionless floor. We will take the origin at the contact point between the disk and the wall, the x-axis as the horizontal direction along the wall and the y-axis perpendicular to the wall (Fig. 8.5). Supposing that the velocity of the disk before the collision is $\mathbf{v}_i = (v_{ix}, v_{iy})$, the velocity of the disk after the collision is $\mathbf{v}_o = (v_{ox}, v_{oy})$, the angle between $\mathbf{v}_i$ and the y-axis is θ_i (the angle of incidence), and the angle between $\mathbf{v}_o$ and the y-axis is θ_o (the angle of reflection), we obtain

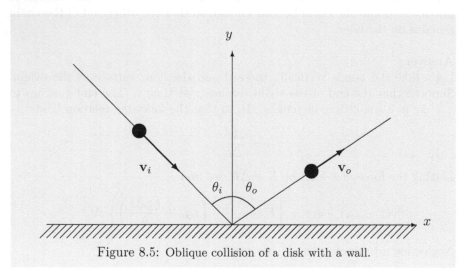

Figure 8.5: Oblique collision of a disk with a wall.

$$\theta_i = \tan^{-1}\left(\frac{v_{ix}}{v_{iy}}\right),\tag{8.31}$$

$$\theta_o = \tan^{-1}\left(\frac{v_{ox}}{v_{oy}}\right).\tag{8.32}$$

The Case of the Frictionless Rigid Wall

Suppose that a wall is frictionless and rigid. The terminology rigid means that the collision is elastic, and frictionless means that the momentum of the x-component is conserved before and after the collision. These conditions are written as follows:

$$e = -\frac{v_{oy}}{-v_{iy}} = 1,$$

$$v_{oy} = v_{iy}, \tag{8.33}$$

$$v_{ox} = v_{ix}. \tag{8.34}$$

Using (8.33) and (8.34), we get

$$\theta_i = \tan^{-1}\left(\frac{v_{ix}}{v_{iy}}\right) = \tan^{-1}\left(\frac{v_{ox}}{v_{oy}}\right) = \theta_o. \tag{8.35}$$

If the wall is frictionless and rigid, the *law of reflection* holds as the case of an optical reflection.

Example 5. A ball at rest is released from $h_1 = 3.0\,\text{m}$ in height, it rebounds from the floor, and it reaches the maximum height $h_2 = 2.4\,\text{m}$. Answer the following questions, letting the value of the acceleration due to gravity be $g = 9.8\,\text{m\,s}^{-2}$.

1. Calculate the speed of the ball v just before colliding with the floor.

2. Calculate the speed of the ball v' just after rebounding from the floor.

3. Calculate the coefficient of restitution.

Answer

1. Owing to the law of mechanical energy conservation, we find

$$\frac{1}{2}mv^2 = mgh_1,$$
$$v = \sqrt{2gh_1} = 7.7\,[\text{m\,s}^{-1}].$$

2. Owing to the law of mechanical energy conservation, we get

$$v' = \sqrt{2gh_2} = 6.9\,[\text{m\,s}^{-1}].$$

3. The coefficient of restitution is

$$e = -\frac{v' - 0}{-v - 0} = \frac{v'}{v} = 0.90.$$

8.5   Two-Dimensional Collisions

Problems of two-dimensional collisions may be solved by separating the law of momentum conservation into the x- and y-components. Here, we will consider the following example as the special case of two-dimensional collision.

Example 6. Suppose an elastic collision of two-dimensions. On a level and frictionless surface, disk 1 of mass m travelling at velocity $\mathbf{v}_1$ collides with

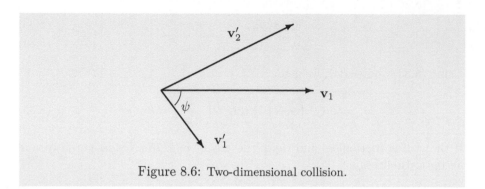

Figure 8.6: Two-dimensional collision.

disk 2 of mass m at rest. Obtain the velocity of disk 1 $\mathbf{v}_1'$ and the velocity of disk 2 $\mathbf{v}_2'$ after the collision, ignoring the rotational motion of disks.

Answer
We take the x-axis as the direction of $\mathbf{v}_1$ and the y-axis vertical to the x-axis as shown in Fig. 8.6. Let the x- and y-components of $\mathbf{v}_1$, $\mathbf{v}_1'$, and $\mathbf{v}_2'$ be $(u_1, 0)$, (u_1', v_1'), and (u_2', v_2'). We find

$$\mathbf{v}_1 = \mathbf{v}_1' + \mathbf{v}_2' \tag{8.36}$$

due to the law of momentum conservation. Resolving (8.36) to the x- and y-components, we obtain

$$u_1 = u_1' + u_2' , \tag{8.37}$$
$$0 = v_1' + v_2' . \tag{8.38}$$

We find from (8.36) that three velocity vectors make a triangle. The law of kinetic energy conservation is written as,

$$\frac{1}{2}mv_1{}^2 = \frac{1}{2}mv_1'{}^2 + \frac{1}{2}mv_2'{}^2,$$
$$v_1{}^2 = v_1'{}^2 + v_2'{}^2. \tag{8.39}$$

The three vectors $\mathbf{v}_1$, $\mathbf{v}_1'$, and $\mathbf{v}_2'$ form a triangle satisfying the *Pythagorean theorem*. Therefore, the angle between $\mathbf{v}_1'$ and $\mathbf{v}_2'$ is a right angle. Supposing that the angle between $\mathbf{v}_1$ and $\mathbf{v}_1'$ is ψ, we obtain

$$v_1' = v_1 \cos \psi, \tag{8.40}$$
$$v_2' = v_1 \cos \left(\frac{\pi}{2} - \psi \right) = v_1 \sin \psi. \tag{8.41}$$

8.6   Scattering Cross Sections

Suppose that a number of small and rigid spherical particles are projected at uniform velocity to a fixed rigid object with smooth surface. Although we can trace the trajectory of each individual particle, the statistical theory which gives the probable numbers of scattered particles to some scattered angle is useful for analyzing the results of scattering experiments. In this context, the new concepts of the *scattering cross section* and the *differential scattering cross section* are introduced. The discussion will offer a good preparation for the *Rutherford scattering* in Chapter 11, which is the experiment of the scattering of α-particles by the atomic nuclei. In this section the Earth's gravity and the air resistance are ignored.

8.6.1 Scattering by a Rigid Cylinder

Let's consider the problem of scattering by a rigid cylinder. Suppose that a rigid cylinder with radius a and height h is fixed to a level surface (x-y plane) as shown in Fig. 8.7. The x-axis is taken parallel to the level surface and to pass through the center of the cylinder O, the z-axis is taken vertically upward and passing through O, and the y-axis is taken to obey the right-hand rule. Many particles, having the uniform area density σ in the y-z plane, are projected parallel to the x-axis so that $N = 2\sigma ah$ particles collide with it. Let a particle collide with the cylinder at point P, the angle between the x-axis and OP (the incident angle) be ψ, the angle between the normal at P and the reflected direction (the reflected angle) be ψ', the angle between OP and the reflected direction be λ and the distance from P to the x-axis be b (Fig. 8.8). Customarily b is referred to as the *impact parameter* and λ is called the *scattering angle*. As the cylinder and particles are rigid bodies and the surface is smooth, the momentum before and after the collision is conserved and the incident angle is equal to the reflected angle (the *law of reflection*).

$$\psi = \psi', \tag{8.42}$$

$$|\mathbf{v}_i| = |\mathbf{v}_o|, \tag{8.43}$$

where $\mathbf{v}_i$ is the incident velocity and $\mathbf{v}_o$ is the reflected velocity. From Fig. 8.8 and (8.42), we obtain

$$\lambda = \pi - \psi - \psi' = \pi - 2\psi, \tag{8.44}$$

We can find the relationship between the impact parameter b and the scattering angle λ as

$$b = a \sin \psi = a \sin \left(\frac{\pi - \lambda}{2} \right) = a \cos \frac{\lambda}{2}. \tag{8.45}$$

Equation (8.45) gives the trajectory of a particle, so that the number of particles incident in an infinitesimal area $dS = h|db|$ is

$$dN = \sigma dS = \sigma h |db|. \tag{8.46}$$

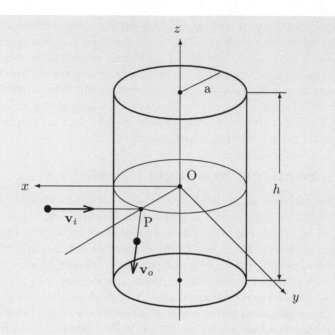

Figure 8.7: Scattering by a rigid cylinder (a bird's-eye view).

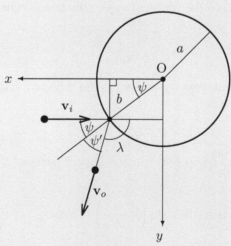

Figure 8.8: Scattering by a rigid cylinder (a plan view).

The number of scattered particles per unit angle is

$$\frac{dN}{d\lambda} = \sigma \frac{dS}{d\lambda}. \tag{8.47}$$

The physical quantity $dS/d\lambda$ is called the differential scattering cross section, and is given by

$$\frac{dS}{d\lambda} = h \frac{|db|}{d\lambda} = \frac{ha}{2} \sin \frac{\lambda}{2}. \tag{8.48}$$

Integrating (8.48) with respect to λ from 0 to 2π, we obtain the scattering cross section S,

$$S = \int_0^{2\pi} \frac{dS}{d\lambda} d\lambda = \frac{ha}{2} \int_0^{2\pi} \sin \frac{\lambda}{2} d\lambda = 2ha, \tag{8.49}$$

which is, of course, equal to the projected area of the cylinder to the y-z plane. From (8.48) and (8.49), we obtain the number of particles dN scattered into the angular range λ to $\lambda + d\lambda$,

$$dN = \frac{N}{4} \sin \frac{\lambda}{2} d\lambda. \tag{8.50}$$

Example 7. Suppose that a rigid cylinder with radius a and height h is fixed to a level surface as shown in Fig. 8.7. Many particles having the uniform area density in the y-z plane are projected parallel to the x-axis to the cylinder and 1000 particles collide with it. How many particles are scattered backward?

Answer
We can obtain the answer integrating (8.50) with respect to λ from $\pi/2$ to $3\pi/2$,

$$N_{\text{back}} = \frac{1000}{4} \int_{\pi/2}^{3\pi/2} \sin \frac{\lambda}{2} d\lambda = \frac{1000}{2} \left[-\cos \frac{\lambda}{2} \right]_{\pi/2}^{3\pi/2} = 707.$$

Thus, 707 particles are scattered backward.

8.6.2 Scattering by a Rigid Sphere

Next we will consider the problem of scattering by a rigid sphere. Suppose that a rigid sphere with radius a is fixed to a level surface and the x-axis is taken parallel to the level surface and to pass through the center of the sphere O as shown in Fig. 8.9. Many particles having the uniform area density σ in the y-z plane are projected parallel to the x-axis so that $N = \sigma \pi a^2$ particles collide with it. Let a particle collide with the sphere at point P in the x-z plane, the angle between the x-axis and OP be ψ, the angle between the normal at P and the reflected direction be ψ, the angle between OP and the reflected direction be λ and the distance from P to the x-axis be b.

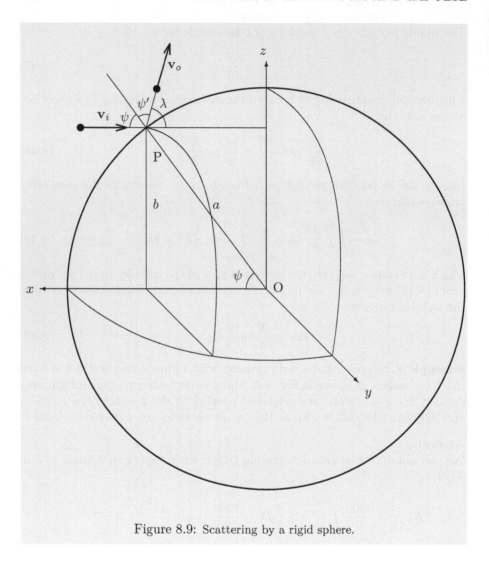

Figure 8.9: Scattering by a rigid sphere.

As the sphere is rigid and smooth, momentum is conserved before and after the collision and the incident angle is equal to the reflected angle. Namely,

$$\psi = \psi', \tag{8.51}$$

$$|\mathbf{v}_i| = |\mathbf{v}_o|. \tag{8.52}$$

From Fig. 8.9 and (8.51), we obtain

$$\lambda = \pi - \psi - \psi' = \pi - 2\psi. \tag{8.53}$$

We find the relationship between the impact parameter b and the scattering angle λ, as

$$b = a \sin \psi = a \sin \frac{\pi - \lambda}{2} = a \cos \frac{\lambda}{2}, \tag{8.54}$$

The infinitesimal area of a circular ring between radii b and $b + db$ is

$$dS = 2\pi b |db| = \pi a^2 \cos \frac{\lambda}{2} \sin \frac{\lambda}{2} d\lambda = \frac{\pi a^2}{2} \sin \lambda d\lambda. \tag{8.55}$$

The scattering cross section S is obtained by integrating (8.55) with respect to λ from 0 to π,

$$S = \int_0^\pi \frac{\pi a^2}{2} \sin \lambda d\lambda = \pi a^2, \tag{8.56}$$

which is the area of the projection of the rigid sphere to the y-z plane. The number of the scattered particles per unit angle is obtained as

$$\frac{dS}{d\lambda} = \frac{\pi a^2}{2} \sin \lambda, \tag{8.57}$$

which is called the differential scattering cross section. The number of particles scattered into the angular range λ to $\lambda + d\lambda$ is

$$dN = \sigma dS = \frac{N}{\pi a^2} \frac{\pi a^2}{2} \sin \lambda d\lambda = \frac{N}{2} \sin \lambda d\lambda. \tag{8.58}$$

In Rutherford scattering experiments, the characteristics of target objects, i.e., their positions and spatial extent, is unknown, so that the physical concept of the scattering cross section is crucial for analyzing experimental results.

Example 8. Suppose that a rigid sphere with radius a is fixed to a level surface as shown in Fig. 8.9. A number of particles having the uniform area density in the y-z plane are projected parallel to the x-axis to the sphere and 1000 particles collide with it. How many particles are scattered backward?

Answer
We can obtain the answer by integrating (8.58) with respect to λ from $\pi/2$ to π,

$$N_{\text{back}} = \frac{1000}{2} \int_{\pi/2}^\pi \sin \lambda d\lambda = \frac{1000}{2} [-\cos \lambda]_{\pi/2}^\pi = 500.$$

Thus, 500 particles are scattered backward.

8.7 Rocket Motion

We will discuss the motion of a rocket which changes its mass very rapidly by consuming fuel for propulsion. Transforming (8.1) considering time change of mass,

$$\frac{d\mathbf{p}}{dt} = m\frac{d\mathbf{v}}{dt} + \mathbf{v}\frac{dm}{dt} = \mathbf{F}, \tag{8.59}$$

At first, we will consider a rocket motion which is in one-dimensional in space free from the Earth's gravity. Suppose that the mass of the rocket is m at time t, the speed of the rocket relative to an inertial frame is u and the x-axis is taken to the travelling direction of the rocket. At time $t + \delta t$, the mass of the rocket becomes $m + \delta m$, the speed of the rocket is $u + \delta u$ and the speed of the ejected fuel gas relative to the rocket is V. Considering that the speed of the gas observed from the inertial frame is $u - V$, the law of momentum conservation is written as,

$$mu = (m + \delta m)(u + \delta u) + (-\delta m)(u - V),$$
$$m\delta u = -V\delta m, \tag{8.60}$$

neglecting a term of order δ^2. Dividing (8.60) through by δt and in the limit $\delta t \to 0$, we get

$$m\frac{du}{dt} = -V\frac{dm}{dt}. \tag{8.61}$$

Letting the mass of the rocket at $t = 0$ be m_s, the time for consuming all the fuel be τ, the mass of the rocket at $t = \tau$ be m_f and the fuel mass consumed in unit time be αm_s, we find

$$\frac{dm}{dt} = -m_s\alpha. \tag{8.62}$$

Solving (8.62) under the initial condition that $m = m_s$ at $t = 0$, we obtain

$$m = m_s(1 - \alpha t). \tag{8.63}$$

Substituting from (8.62) and (8.63) into (8.61) yields

$$\frac{du}{dt} = \frac{V\alpha}{1 - \alpha t}. \tag{8.64}$$

Solving (8.64) under the initial condition that $u = u_s$ at $t = 0$, we get

$$u = u_s - V\log_e|1 - \alpha t|. \tag{8.65}$$

Substituting $t = \tau$ into (8.63), we find

$$m_f = m_s(1 - \alpha\tau),$$
$$\tau = \frac{1}{\alpha}\left(1 - \frac{m_f}{m_s}\right). \tag{8.66}$$

Substituting $t = \tau$ into (8.65) and using (8.66), we obtain

$$u_f = u_s + V\log_e \frac{m_s}{m_f}. \tag{8.67}$$

Next, we will consider a rocket lifting vertically against the Earth's gravity. Let the magnitude of the acceleration due to gravity be g and the air resistance be negligible. The rocket motion is described adding the gravity term to the right-hand side of (8.61) and taking the x-axis vertically upward. Then

$$m\frac{du}{dt} = -V\frac{dm}{dt} - mg,$$

$$\frac{du}{dt} = -\frac{V}{m}\frac{dm}{dt} - g. \tag{8.68}$$

Substituting from (8.62) and (8.63) into (8.68), we get

$$\frac{du}{dt} = \frac{V\alpha}{1 - \alpha t} - g. \tag{8.69}$$

Solving (8.69) under the initial condition that $u = 0$ at $t = 0$, we obtain

$$u = -V\log_e |1 - \alpha t| - \int_0^t g dt. \tag{8.70}$$

At $t = \tau$, we find

$$u_f = V\log_e \frac{m_s}{m_f} - \int_0^\tau g dt, \tag{8.71}$$

using (8.66).

8.8 Problems

1. A vehicle of mass 1.50×10^3 kg, including the mass of a crew, is travelling on a straight road at 50.0 km h^{-1}. A driver finding an obstacle on the road brakes urgently and stops the vehicle after 14.0 m. Answer the following questions, assuming the driving force is zero and the frictional force is constant during braking.

    (1) Obtain the time for the vehicle to stop, and the acceleration during braking.

    (2) Obtain the frictional force exerting on the vehicle using Newton's second law.

    (3) Obtain the frictional force exerting on the vehicle using the momentum equation.

(4) Obtain the frictional force exerting on the vehicle using the energy equation.

2. Along the x-axis on a level and frictionless surface, disk 1 of mass $2m$ is travelling to the positive x-direction at speed $2v$ and disk 2 of mass $3m$ is travelling to the negative x-direction at speed v. They collide at some moment. Answer the following questions supposing that they travel together at speed v' after the collision.

   (1) Obtain the speed v' and the direction of motion of the two disks.

   (2) What is the kinetic energy lost through the collision?

3. Along the x-axis on a level and frictionless surface, disk 1 of mass m is travelling to the positive x-direction at speed $2v$ and disk 2 of mass $2m$ is travelling to the negative x-direction at speed v. They collide at some moment. Answer the following questions supposing that the kinetic energy of the system is conserved before and after the collision.

   (1) Obtain the speed v_1 and the travelling direction of disk 1 and the speed v_2 and the travelling direction of disk 2 after the collision.

   (2) Obtain the coefficient of restitution.

4. Suppose that a small ball is projected at the initial speed V_0 and the elevation angle θ from point O on a floor to a frictionless wall at distance d. What is the condition for the rebounded ball to collide directly with the floor at point O, letting the coefficient of restitution between the ball and the wall be e and the magnitude of the acceleration due to gravity be g?

5. Suppose that a rectangular box of mass M is placed on a frictionless plane at rest. A disk of mass m is placed in the box whose inside surface is also frictionless. The disk is projected at initial speed v to one wall of the box vertically.

   (1) The disk is bounced back by another wall at speed v_2 and the box moves at speed V_2. Calculate v_2 and V_2.

   (2) What is the final speed of the disk and box after an infinite number of collisions?

6. Suppose that a rigid cylinder with radius a and height h is fixed to the x-y plane as shown in Fig. 8.7. Many particles having the uniform area density in the y-z plane are projected parallel to the x-axis to the cylinder, and 1000 particles collide with it. How many particles are scattered forward? Ignore the Earth's gravity and the air resistance.

7. Suppose that a rigid sphere with radius a is fixed to the x-y plane as shown in Fig. 8.9. Many particles having the uniform area density in the y-z plane are projected parallel to the x-axis to the sphere, and 1000 particles collide with it. How many particles are scattered forward? Ignore the Earth's gravity and the air resistance.

Chapter 9

Angular Momentum Equation

In this chapter, the third transformation of the equation of motion is performed. The resultant equation is referred to as the angular momentum equation, which relates the changing rate of angular momentum of a body to torque exerting on it. Angular momentum is the rotational analogue of the linear momentum of a rotating system, while torque is the rotational analogue of force exerting on the system. When no torque is exerting on the system or the total torque is zero, its angular momentum is conserved, which is called the law of angular momentum conservation. At the end of this chapter, statics of a rigid body is discussed for several examples.

9.1 Equation of Motion for Rotational Motion

The equation of motion in the momentum form is

$$\frac{d\mathbf{p}}{dt} = \mathbf{F}. \tag{9.1}$$

Operating $\mathbf{r}\times$ on (9.1) yields

$$\mathbf{r} \times \frac{d\mathbf{p}}{dt} = \mathbf{r} \times \mathbf{F}. \tag{9.2}$$

Taking into account the following relation

$$\frac{d}{dt}(\mathbf{r} \times \mathbf{p}) = \frac{d\mathbf{r}}{dt} \times \mathbf{p} + \mathbf{r} \times \frac{d\mathbf{p}}{dt} = \mathbf{v} \times (m\mathbf{v}) + \mathbf{r} \times \frac{d\mathbf{p}}{dt} = \mathbf{r} \times \frac{d\mathbf{p}}{dt},$$

equation (9.2) becomes

$$\frac{d}{dt}(\mathbf{r} \times \mathbf{p}) = \mathbf{r} \times \mathbf{F}. \tag{9.3}$$

DOI: 10.1201/9781003310068-9

Equation (9.3) is referred to as the angular momentum equation, which is appropriate for discussing the rotational motion of a particle. Here we define the angular momentum $\mathbf{L}$, as

$$\mathbf{L} = \mathbf{r} \times \mathbf{p} = \begin{vmatrix} \mathbf{e}_x & \mathbf{e}_y & \mathbf{e}_z \\ x & y & z \\ p_x & p_y & p_z \end{vmatrix}$$
$$= \mathbf{e}_x(yp_z - zp_y) + \mathbf{e}_y(zp_x - xp_z) + \mathbf{e}_z(xp_y - yp_x). \tag{9.4}$$

Angular momentum is the rotational analogue of momentum of a particle. The direction of $\mathbf{L}$ and the rotation of the particle obey the right-hand screw rule. The unit of angular momentum is

$$[L] = [\mathrm{kg\,m\,s^{-1}\,m}] = [\mathrm{N\,m\,s}] = [\mathrm{J\,s}] \,.$$

Next we define torque $\mathbf{N}$ as,

$$\mathbf{N} = \mathbf{r} \times \mathbf{F} = \begin{vmatrix} \mathbf{e}_x & \mathbf{e}_y & \mathbf{e}_z \\ x & y & z \\ F_x & F_y & F_z \end{vmatrix}$$
$$= \mathbf{e}_x(yF_z - zF_y) + \mathbf{e}_y(zF_x - xF_z) + \mathbf{e}_z(xF_y - yF_x). \tag{9.5}$$

Torque is the rotational analogue of force. Position vector, force and torque obey the right-hand rule. Using the definitions of (9.4) and (9.5), (9.3) is written,

$$\frac{d\mathbf{L}}{dt} = \mathbf{N}. \tag{9.6}$$

The component equations of (9.6) in Cartesian coordinates are

$$\frac{d}{dt}(yp_z - zp_y) = yF_z - zF_y, \tag{9.7}$$

$$\frac{d}{dt}(zp_x - xp_z) = zF_x - xF_z, \tag{9.8}$$

$$\frac{d}{dt}(xp_y - yp_x) = xF_y - yF_x. \tag{9.9}$$

The unit of torque is

$$[N] = [\mathrm{N\,m}] = [\mathrm{J}] \,.$$

9.2 Torque and Angular Momentum

Torque and angular momentum are discussed precisely in this section.

9.2.1 Torque

Supposing that the position vector $\mathbf{r}$ and the force $\mathbf{F}$ are two-dimensional vectors on the x-y plane, the torque about the origin O becomes

$$\mathbf{N} = \mathbf{r} \times \mathbf{F}$$

$$= \begin{vmatrix} \mathbf{e}_x & \mathbf{e}_y & \mathbf{e}_z \\ x & y & 0 \\ F_x & F_y & 0 \end{vmatrix} = \mathbf{e}_z(xF_y - yF_x) = \mathbf{e}_z d |\mathbf{F}|.$$

The torque is resolved into two components $xF_y\mathbf{e}_z$ and $-yF_x\mathbf{e}_z$, the former is due to the y-component force F_y and the latter is due to the x-component force F_x (Fig. 9.1).

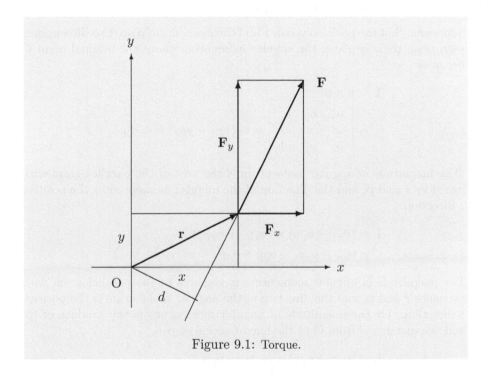

Figure 9.1: Torque.

The magnitude of the torque is the area of the parallelogram with two sides $\mathbf{r}$ and $\mathbf{F}$, and the direction of the torque is the positive z-direction. From another point of view, the magnitude of the torque is the product of the distance from O to the line of action d and the magnitude of the force $|\mathbf{F}|$.

9.2.2   A Force Couple

When two forces are equal and opposite and their lines of action do not coincide, two forces $\mathbf{F}$, $-\mathbf{F}$ are called a force couple. Taking x-y plane for two forces to be on the plane, we will obtain the torque due to a force couple about the original point O. Letting the point of action of the force $\mathbf{F}$ be $\mathbf{r}_1$, the point of action of the force $-\mathbf{F}$ be $\mathbf{r}_2$ and the distance between two lines of action be h, we find

$$\mathbf{N} = \mathbf{F} \times \mathbf{r}_1 + (-\mathbf{F}) \times \mathbf{r}_1 = \mathbf{F} \times (\mathbf{r}_1 - \mathbf{r}_2),$$
$$|\mathbf{N}| = Fh. \tag{9.10}$$

Problem 1. Show the examples of a force couple.

9.2.3   Angular Momentum

Supposing that the position vector $\mathbf{r}$ and the momentum $\mathbf{p}$ are two-dimensional vectors on the x-y plane, the angular momentum about the original point O becomes

$$\mathbf{L} = \mathbf{r} \times \mathbf{p}$$
$$= \begin{vmatrix} \mathbf{e}_x & \mathbf{e}_y & \mathbf{e}_z \\ x & y & 0 \\ p_x & p_y & 0 \end{vmatrix} = \mathbf{e}_z(xp_y - yp_x) = \mathbf{e}_z d|\mathbf{p}|.$$

The magnitude of angular momentum is the area of the parallelogram with two sides $\mathbf{r}$ and $\mathbf{p}$, and the direction of the angular momentum is the positive z-direction.

$$\mathbf{L} = (\mathbf{e}_x x + \mathbf{e}_y y) \times (\mathbf{e}_x p_x + \mathbf{e}_y p_y)$$
$$= \mathbf{e}_x x \times \mathbf{e}_y p_y + \mathbf{e}_y y \times \mathbf{e}_x p_x = \mathbf{e}_z(xp_y - yp_x).$$

The magnitude of angular momentum is the area of the parallelogram with two sides $\mathbf{r}$ and $\mathbf{p}$, and the direction of the angular momentum is the positive z-direction. Or the magnitude of angular momentum is the product of $|\mathbf{p}|$ and the distance d from O to the line of action of $\mathbf{p}$.

$$\mathbf{L} = (\mathbf{e}_x x + \mathbf{e}_y y) \times (\mathbf{e}_x p_x + \mathbf{e}_y p_y)$$
$$= \mathbf{e}_x x \times \mathbf{e}_y p_y + \mathbf{e}_y y \times \mathbf{e}_x p_x = \mathbf{e}_z(xp_y - yp_x).$$

The angular momentum is resolved into two components $xp_y\mathbf{e}_z$ and $-yp_x\mathbf{e}_z$, the former is due to the y-directional momentum p_y and the latter is due to the x-directional momentum p_x (Fig. 9.2).

Example 1. Suppose that a particle of mass m is whirled through a circle of radius r about point O at constant angular velocity ω. Obtain the magnitude of the angular momentum of the uniform circular motion.

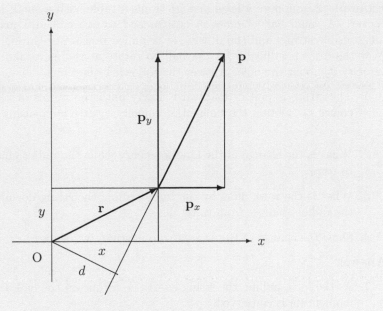

Figure 9.2: Decomposition of angular momentum.

Answer

The tangential velocity of the rotational motion is

$$v = r\omega.$$

Then the magnitude of the angular momentum becomes

$$L = mrv = mr^2\omega. \tag{9.11}$$

9.3 The Law of Angular Momentum Conservation

It is obvious from (9.6) that the angular momentum of a system rotating about point O is conserved if the torque due to external forces is zero. This law is referred to as the *law of angular momentum conservation*. The central force is such a force that the line of action passes the center of rotation and the magnitude of it depends only on radius r. Therefore, the central force does not change the angular momentum of the system.

Example 2. Suppose a level and frictionless table with a small hole at the center. A small ball of mass m is attached to one end of a massless and inextensible string, and the other end is pulled downward through the hole. Now the force is adjusted for the ball to rotate at radius r_0 and tangential velocity v_0 about the hole. Answer the following questions.

1. The string is pulled downward slowly until the radius of rotation becomes r_1. Obtain the tangential velocity v_1 at $r = r_1$, using r_0, r_1 and v_0.

2. What is the change of the kinetic energy while the radius changes from r_0 to r_1?

3. What is the work done by the force pulling the string downward while the radius changes from r_0 to r_1?

4. Show the change of the angular momentum $\Delta\omega$ using r_0, r_1, and v_0.

Answer

1. As the force pulling the string exerts no torque on the ball, the angular momentum is conserved,

$$mv_0 r_0 = mv_1 r_1,$$
$$v_1 = \frac{r_0}{r_1} v_0.$$

2. Letting the kinetic energy at radius r_0 be K_0 and the kinetic energy at radius r_1 be K_1, we obtain

$$K_1 - K_0 = \frac{1}{2}mv_1^2 - \frac{1}{2}mv_0^2 = \frac{1}{2}mv_0^2 \left\{ \left(\frac{r_0}{r_1}\right)^2 - 1 \right\}.$$

3. Letting the tangential velocity at r be v, we find

$$v = \frac{r_0}{r} v_0.$$

The centripetal force F_{cp} necessary for the uniform circular motion of radius r and the tangential velocity v is

$$F_{\text{cp}} = m\frac{v^2}{r} = m\frac{r_0^2 v_0^2}{r^3}.$$

The work done by the force F_{cp} pulling the ball from r_0 to r_1 is

$$W = \int_{r_0}^{r_1} (-F_{\text{cp}})dr = \int_{r_0}^{r_1} \left(-m\frac{r_0^2 v_0^2}{r^3} \right) dr$$

$$= mr_0^2 v_0^2 \left[\frac{1}{2r^2} \right]_{r_0}^{r_1} = \frac{1}{2}mv_0^2 \left\{ \left(\frac{r_0}{r_1}\right)^2 - 1 \right\} = K_1 - K_0.$$

4. Letting the angular velocity at r_0 be ω_0 and the angular velocity at r_1 be ω_1, we obtain

$$\omega_0 = \frac{v_0}{r_0}, \quad \omega_1 = \frac{v_1}{r_1} = \frac{1}{r_1}\frac{r_0 v_0}{r_1} = \left(\frac{r_0}{r_1}\right)^2 \omega_0,$$

$$\Delta\omega = \left(\frac{1}{r_1^2} - \frac{1}{r_0^2}\right) r_0 v_0.$$

9.4   Equation for Many-Particle Systems

In this section, the angular momentum equation for an n-particle system is derived. Operating $\mathbf{r}_i \times$ on (7.1), and summing up the resultant equation with respect to i from 1 to n, we obtain

$$\sum_{i=1}^{n} \mathbf{r}_i \times \frac{d\mathbf{p}_i}{dt} = \sum_{i=1}^{n} \mathbf{r}_i \times \mathbf{F}_i + \sum_{i=1}^{n}\sum_{j=1}^{n} \mathbf{r}_i \times (\mathbf{F}_{ij} - \delta_{ij}\mathbf{F}_{ij}). \tag{9.12}$$

Let's consider the total torque due to forces between particle i and particle j. Owing to Newton's third law,

$$\mathbf{F}_{ij} = -\mathbf{F}_{ji}.$$

Therefore,

$$\mathbf{r}_i \times \mathbf{F}_{ij} + \mathbf{r}_j \times \mathbf{F}_{ji} = (\mathbf{r}_i - \mathbf{r}_j) \times \mathbf{F}_{ij} = 0, \tag{9.13}$$

because $\mathbf{r}_i - \mathbf{r}_j$ is the position vector from particle j to particle i and is parallel to $\mathbf{F}_{ij}$. The total torque due to forces between arbitrary two particles becomes zero, so that the second term of the right-hand side of (9.12) vanishes, then

$$\frac{d}{dt}\sum_{i=1}^{n} \mathbf{r}_i \times \mathbf{p}_i = \sum_{i=1}^{n} \mathbf{r}_i \times \mathbf{F}_i,$$

$$\frac{d}{dt}\sum_{i=1}^{n} \mathbf{L}_i = \sum_{i=1}^{n} \mathbf{N}_i. \tag{9.14}$$

The changing rate of the total angular momentum of the many-particle system is equal to the total torque due to external forces. When the total torque is zero, the total angular momentum of a many-particle system is conserved.

9.5   Static Equilibrium of Rigid Bodies

When the motion of rigid bodies is restricted to the two-dimensional plane, there are three degrees of freedom of motion, i.e., two are due to the translational motion and one is due to the rotational motion about the axis vertical to the plane. Then, when a rigid body is in static equilibrium in the two-dimensional plane, the conditions, that the total force is zero and the total torque about an arbitrary axis vertical to the plane is zero, must be satisfied.

9.5.1   Conditions for Translational Motion

The condition for a rigid body not to translate is that the total force exerting on the rigid body is zero. Namely,

$$\sum_{i=1}^{n} \mathbf{F}_i = 0. \tag{9.15}$$

The component equations of (9.15) in two-dimensional Cartesian coordinates are,

$$\sum_{i=1}^{n} F_{xi} = 0, \tag{9.16}$$

$$\sum_{i=1}^{n} F_{yi} = 0. \tag{9.17}$$

9.5.2   Condition for Rotational Motion

The condition for a rigid body not to rotate is that the total torque about an arbitrary point P is zero. Namely,

$$\sum_{i=1}^{n} N_{zi} = 0. \tag{9.18}$$

Taking point P as the origin of the Cartesian coordinates, the z-component of torque is

$$\mathbf{e}_z \cdot \mathbf{N} = \mathbf{e}_z \cdot (\mathbf{r} \times \mathbf{F}) = \mathbf{e}_z \cdot \begin{vmatrix} \mathbf{e}_x & \mathbf{e}_y & \mathbf{e}_z \\ x & y & 0 \\ F_x & F_y & F_z \end{vmatrix}$$

$$= xF_y - yF_x.$$

Using the above relationship, (9.18) becomes,

$$\sum_{i=1}^{n} (x_i F_{yi} - y_i F_{xi}) = 0. \tag{9.19}$$

Equations (9.16), (9.17), and (9.19) are conditions for a rigid body to be at rest.

9.5.3   Some Examples

Example 3. A uniform ladder of mass M and length L is set against a wall (Fig. 9.3). The wall is frictionless and the floor is slipless with the static frictional coefficient μ. What is the condition of the angle θ between the floor

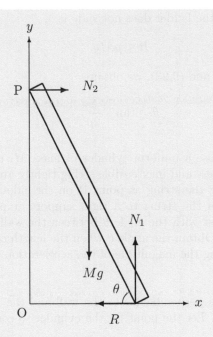

Figure 9.3: Static equilibrium of the ladder set against the wall.

and the ladder for the ladder not to slide on the floor? Let the magnitude of the acceleration due to gravity be g.

Answer
Let's take the y-axis vertically upward along the wall and the x-axis perpendicular to the wall and on the floor. Supposing that the vertical resistive force of the floor is N_1, the vertical resistive force of the wall is N_2 and the frictional force of the floor is R, the x- and y-components of the total force should be zero,

$$N_2 - R = 0, \quad N_1 - Mg = 0. \tag{9.20}$$

The total torque about the lower edge of the ladder should be zero,

$$\frac{1}{2}L\cos\theta Mg - L\sin\theta N_2 = 0. \tag{9.21}$$

From (9.21),

$$\tan\theta = \frac{Mg}{2N_2}. \tag{9.22}$$

The condition that the ladder does not slide is

$$R \leq \mu M g. \tag{9.23}$$

From (9.20), (9.22), and (9.23), we obtain

$$\theta \geq \tan^{-1} \frac{1}{2\mu}. \tag{9.24}$$

Example 4. Suppose a uniform cylinder of mass M, radius a, and height $6a$. Turning a massless and inextensible string tightly around the cylinder at height $3a$ and fixing the string at point R on the side of the cylinder, and attaching the end of the string to a rigid support at point P on the wall, the cylinder is at rest with the angle θ between the wall and the side of the cylinder (Fig. 9.4). Obtain the angle θ when the length of the string between R and P is $3a$, letting the magnitude of the acceleration due to gravity be g.

Answer
Let's take the x-axis perpendicular to the wall and the y-axis vertically upward along the wall. Let the point for the cylinder to contact the wall be Q,

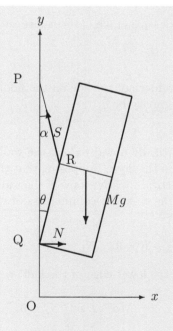

Figure 9.4: Static equilibrium of the cylinder hung from the wall.

the tension of the string be S, the vertical resistive force of the wall be N and the angle $\angle \mathrm{RPQ} = \alpha$. The x-component of the total force is zero,

$$N - S \sin \alpha = 0. \tag{9.25}$$

The y-direction of the total force is zero,

$$S \cos \alpha - Mg = 0. \tag{9.26}$$

The total torque about point Q is zero,

$$S(3a \cos \alpha + 3a \cos \theta) \sin \alpha - Mg(3a \sin \alpha + a \cos \theta) = 0. \tag{9.27}$$

As the triangle RPQ is an isosceles triangle, $\alpha = \theta$. Substituting from (9.26) into (9.27), we get

$$3a \sin \theta + 3a \sin \theta - (3a \sin \theta + a \cos \theta) = 0,$$

$$\theta = \tan^{-1} \frac{1}{3} = 18.4°.$$

Example 5. A half spherical shell of inner radius a is placed on a level surface keeping its rim level as shown in Fig. 9.5. When one end of a uniform rigid rod of length $2l(a < l < 2a)$ and mass M is put into the half spherical shell, the rod is at rest with the angle θ from a level plane. Obtain θ using a, l, and g, supposing that frictional forces do not exert between the spherical shell and the rod. Where g is the magnitude of the acceleration due to gravity.

Answer
Let the center of the sphere be O, the contact point between the end of the rod and the spherical surface be P, the contact point between the rim of the half spherical shell and the rod be Q, and the point from O vertically to the rod be H. We will take the x-axis as horizontal direction and the y-axis vertically upward. Suppose that the resistive forces at point P and point Q are N_1 and N_2. As the spherical surface is frictionless, the direction of N_1 is normal to the spherical surface and N_2 is perpendicular to the rod.
The total force of the x-direction should be,

$$N_1 \cos 2\theta - N_2 \sin \theta = 0. \tag{9.28}$$

The total force of the y-direction is zero,

$$N_1 \sin 2\theta + N_2 \cos \theta - Mg = 0. \tag{9.29}$$

The total torque about point P is zero,

$$N_2 2a \cos \theta - Mgl \cos \theta = 0. \tag{9.30}$$

Subtracting (9.30)/$\cos \theta$ from (9.29)$\times l$, we get

$$l \sin 2\theta N_1 + (l \cos \theta - 2a) N_2 = 0. \tag{9.31}$$

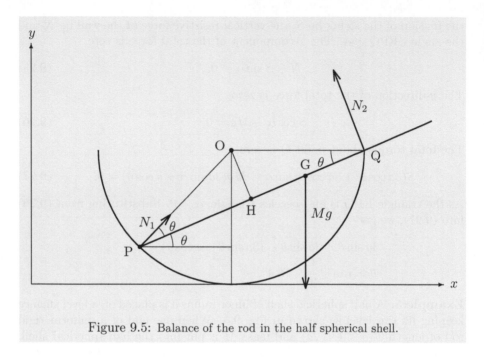

Figure 9.5: Balance of the rod in the half spherical shell.

The necessary condition for N_1, N_2 to have the nontrivial solutions is that the determinant of the coefficients of N_1 and N_2 in (9.28) and (9.31) is zero, so that

$$\begin{vmatrix} \cos 2\theta & -\sin\theta \\ l\sin 2\theta & l\cos\theta - 2a \end{vmatrix} = 0,$$

$$\cos^2\theta - \frac{l}{4a}\cos\theta - \frac{1}{2} = 0,$$

$$\theta = \cos^{-1}\frac{l}{8a}\left(1 + \sqrt{1 + \frac{32a^2}{l^2}}\right). \tag{9.32}$$

Problem 2. What is the angle θ, when $2l = 3a$ and $2l = 3.5a$ in Example 5?

9.6   Problems

1. Particle 1 of mass m and particle 2 of mass $2m$ are attached to a massless rigid rod of length l on a level and frictionless surface. The other end of the rod is fixed to a rigid support at point O, and particles are whirled through a circle about point O at constant tangential speed v. Answer the following questions supposing that particle 2 is separated at some moment.

(1) Obtain the tangential speed of particle 1 v' after separating particle 2.

(2) What is the kinetic energy before and after separating particle 2?

2. Answer the following questions supposing that a disk of mass M and radius a is rotating about the center of the disk at an angular velocity ω on a level surface. Let the kinematic frictional coefficient be μ.

(1) Obtain the torque N exerting on the disk, letting the magnitude of the acceleration due to gravity be g.

(2) What is the angular momentum of the disk?

(3) Describe the angular momentum equation.

(4) Obtain the angular velocity ω at time t, letting the angular velocity at $t = 0$ be ω_0. Using the result, obtain the time τ for the disk to be at rest.

3. Suppose that a disk of mass M and radius a is rotating about the center of the disk at an angular velocity ω on a level surface. Answer the following questions, letting the resistive force be proportional to the speed and contact area, and the proportional coefficient be k.

(1) Obtain the infinitesimal torque dN exerting on the infinitesimal circular ring between radii r and $r + \delta r$.

(2) Calculate the torque N exerting on the disk.

(3) Obtain the angular momentum L of the disk.

(4) Describe the angular momentum equation of the disk.

(5) Obtain the angular velocity ω at time t, letting the angular velocity at $t = 0$ be ω_0.

(6) Obtain the e-folding time τ.

4. A cylindrical vessel of inner radius a and height L is placed on a level surface. A uniform rod of mass M and a length $2l(< L)$ is put such that its end is in contact with the inner side of the vessel and its side is put on the rim of the vessel. The rod is at rest with the angle θ between the rod and the level surface. Answer the following questions supposing that the inner surface and the edge of the cylindrical vessel are frictionless.

(1) Obtain θ using a and l.

(2) What is the angle θ when $l = 3a$?

Chapter 10

Motion of Rigid Bodies

When external forces exert on rigid bodies, they begin translational and rotational motion in general. As for the translational motion, it was shown in Chapter 8 that the translational motion of the center of mass of a rigid body can be treated as the motion of a particle. Concerning the rotational motion, we will transform the angular momentum equation into a more suitable form for rigid bodies. In the process of the transformation of the angular momentum equation, a new physical quantity appears called the moment of inertia. The moment of inertia is the rotational analogue of the inertial mass. The larger the inertial mass of bodies is, the more their linear momentum tends to be kept constant. The larger the moment of inertia of rigid bodies is, the more their angular momentum tends to be kept constant. We derive Euler's equation for discussing the general three-dimensional rotational motion. Applying it to the Earth's rotation, we will discuss free nutation and precession of the Earth.

10.1 Rotational Motion about a Fixed Axis

Rigid bodies are not deformed by external forces so that the angular momentum equation can be transformed into a simpler form.

10.1.1 Tangential Velocity and Angular Velocity

Suppose that a particle of mass m is rotating with radius r about point O. At time t the particle is at point P, and moves to point Q in a small-time increment δt. Letting the length of the arc $\overset{\frown}{PQ}$ be δs and the angle $\angle POQ = \delta\theta$, the tangential speed v is

$$v = \lim_{\delta t \to 0} \frac{\delta s}{\delta t} = \frac{ds}{dt}. \tag{10.1}$$

DOI: 10.1201/9781003310068-10

The magnitude of the angular velocity ω is the changing rate of the angle θ, so that

$$\omega = \lim_{\delta t \to 0} \frac{\delta \theta}{\delta t} = \frac{d\theta}{dt}. \tag{10.2}$$

From the geometrical relationship,

$$\delta s = r \delta \theta. \tag{10.3}$$

Dividing (10.3) through by δt and in the limit $\delta t \to 0$, we get

$$v = r\omega. \tag{10.4}$$

As shown in Fig. 10.1, angular velocity $\boldsymbol{\omega}$, position vector $\mathbf{r}$, and velocity $\mathbf{v}$ obey the right-hand rule. Therefore, (10.4) is expressed in the vector form,

$$\mathbf{v} = \boldsymbol{\omega} \times \mathbf{r}. \tag{10.5}$$

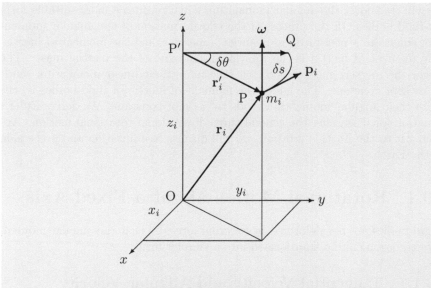

Figure 10.1: Angular momentum of a rigid body.

10.1.2   Rotational Motion of Rigid Bodies

Suppose an n-particle system, in which particles are connected each other by massless rigid rods, is rotating at an angular velocity $\boldsymbol{\omega}$ about the fixed axis.

Taking the fixed axis as the z-axis, the rotational motion of i-th particle is described by the z-component of the angular momentum equation (9.3)

$$\mathbf{e}_z \cdot \frac{d\mathbf{L}_i}{dt} = \mathbf{e}_z \cdot \mathbf{N}_i. \tag{10.6}$$

Let the mass of particle i at point P be m_i, the position vector be $\mathbf{r}_i$, the point setting down from point P vertically to the z-axis be P', and the position vector from P' to P be $\mathbf{r}'_i$. As $\mathbf{r}_i = \mathbf{r}'_i + z_i\mathbf{e}_z$, the momentum of particle i becomes

$$\mathbf{p}_i = m_i\mathbf{v}_i = m_i\boldsymbol{\omega} \times \mathbf{r}_i = m_i\boldsymbol{\omega} \times (\mathbf{r}'_i + z_i\mathbf{e}_z) = m_i\boldsymbol{\omega} \times \mathbf{r}'_i.$$

The angular momentum of particle i is

$$\mathbf{L}_i = \mathbf{r}_i \times \mathbf{p}_i = (\mathbf{r}'_i + z_i\mathbf{e}_z) \times m_i(\boldsymbol{\omega} \times \mathbf{r}'_i) = m_i r'^2_i \boldsymbol{\omega} - m_i\omega z_i\mathbf{r}'_i.$$

In the transformation of the above equation, use is made of the following vector identity and the fact that $\mathbf{r}'_i$ and $\boldsymbol{\omega}$ are orthogonal,

$$\mathbf{A} \times (\mathbf{B} \times \mathbf{C}) = \mathbf{B}(\mathbf{C} \cdot \mathbf{A}) - \mathbf{C}(\mathbf{A} \cdot \mathbf{B}).$$

The z-component of angular momentum $\mathbf{L}_i$ is

$$\mathbf{e}_z \cdot \mathbf{L}_i = m_i r'^2_i \omega. \tag{10.7}$$

Substituting from (10.7) into (10.6), we obtain

$$m_i r'^2_i \frac{d\omega}{dt} = N_{zi}. \tag{10.8}$$

Summing up (10.8) with respect to i from 1 to n,

$$I\frac{d\omega}{dt} = \sum_{i=1}^{n} N_{zi}, \tag{10.9}$$

where

$$I = \sum_{i=1}^{n} m_i r'^2_i \tag{10.10}$$

is referred to as the *moment of inertia*, which is the rotational analogue of the inertial mass. The more the moment of inertia of an n-particle system is, the more its angular momentum tends to be kept constant. When we want to calculate the moment of inertia of a rigid body, a sigma notation should be replaced by an integral in (10.10),

$$I = \int_V \rho r'^2 dv, \tag{10.11}$$

where ρ is the density of the rigid body and a function of space in general. When we calculate the moment of inertia of rigid bodies of various shapes, we should choose appropriate coordinates (Cartesian coordinates, cylindrical coordinates, and spherical coordinates) corresponding to their shapes.

Example 1. Suppose a massless rigid rod of length l. Attaching three particles of mass m at distance $l/3$, $l/2l$, and l from one end O, obtain the moment of inertia of the system about the axis perpendicular to the rod and passing O.

Answers
The moment of inertia is obtained using (10.10).

$$I = m\left(\frac{l}{3}\right)^2 + m\left(\frac{l}{2}\right)^2 + ml^2 = \frac{49}{36}ml^2.$$

10.1.3   The Moment of Inertia of Rigid Bodies of Various Shapes

In this subsection, we will calculate the moment of inertia of rigid bodies of various shapes.

(1) A Disk with Fixed Axis Perpendicular to the Plate and Passing through the Center

Suppose a uniform disk of radius a and mass M. Let's calculate the moment of inertia of the disk about the axis vertical to the circular plate and passing through its center. We use cylindrical coordinates taking the fixed axis as the z-axis. At the position (r, θ), the area element is $dS = rd\theta dr$ and the distance from the axis is r, so that the moment of inertia is obtained as follows.

$$I = \int_0^{2\pi} \int_0^a \sigma r^2 r dr d\theta$$
$$= 2\sigma\pi \int_0^a r^3 dr = 2\sigma\pi \left[\frac{1}{4}r^4\right]_0^a = \frac{1}{2}\sigma\pi a^4 = \frac{1}{2}Ma^2, \qquad (10.12)$$

where $\sigma = M/\pi a^2$ is the area density.

Problem 1. Suppose a uniform rigid rod of length l and mass M. Obtain the moment of inertia about the axis vertical to and passing through the center of the rod.

Problem 2. Suppose a uniform rigid rod of length l and mass M. Obtain the moment of inertia about the axis vertical to and passing through its one end.

(2) A Cylinder with the Fixed Axis Vertical to and Passing through the Center of the Circular Plate

Suppose a uniform cylinder of radius a, height d, and mass M. Let's calculate the moment of inertia of the cylinder about the axis vertical to and passing through the center of the circular plate. We will use cylindrical coordinates taking the fixed axis as the z-axis. At the position (r, θ, z), the volume element is $dv = rd\theta drdz$ and the distance from the axis is r, so that the moment of inertia is obtained as follows.

$$I = \int_0^d \int_0^{2\pi} \int_0^a r^2 \rho dr d\theta dz$$

$$= 2\rho d\pi \int_0^a r^3 dr = 2\rho d\pi \left[\frac{1}{4}r^4\right]_0^a = \frac{1}{2}\rho d\pi a^4 = \frac{1}{2}Ma^2, \qquad (10.13)$$

where $\rho = M/\pi a^2 d$ is the density.

(3) A Cylinder about the Fixed Axis Vertical to the Side and Passing through its Center

Suppose a uniform cylinder of radius a, height d, and mass M. Let's calculate the moment of inertia of the cylinder about the axis vertical to the side and

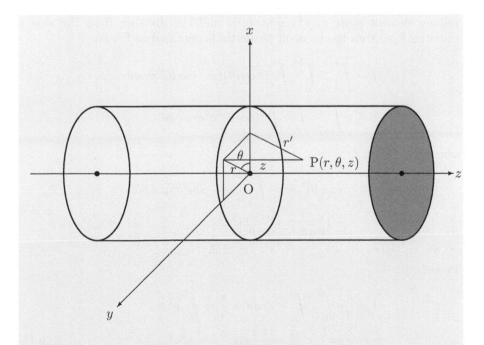

Figure 10.2: The moment of inertia of a cylinder about the fixed axis vertical to the side and passing through the center of the cylinder.

passing through its center. We will use cylindrical coordinates taking the fixed axis as the x-axis. At the position (r, θ, z), the volume element is $dv = rd\theta drdz$ and the distance from the axis is $r' = \sqrt{(r\sin\theta)^2 + z^2}$ (Fig. 10.2), so that the moment of inertia is obtained as follows.

$$I = \int_{-d/2}^{d/2} \int_0^{2\pi} \int_0^a \rho\{(r\sin\theta)^2 + z^2\}rdrd\theta dz$$

$$= \rho d \int_0^{2\pi} \int_0^a \left(\frac{1 - \cos 2\theta}{2}\right) r^3 drd\theta + \rho \left[\frac{1}{3}z^3\right]_{-d/2}^{d/2} \int_0^{2\pi} \int_0^a rdrd\theta$$

$$= \rho d\frac{1}{2}\left[\theta - \frac{1}{2}\sin 2\theta\right]_0^{2\pi} \int_0^a r^3 dr + \frac{1}{12}\rho d^3 \left[\theta\right]_0^{2\pi} \int_0^a rdr$$

$$= \frac{1}{4}\rho\pi da^4 + \frac{1}{12}\rho\pi d^3 a^2 = \frac{1}{4}Ma^2 + \frac{1}{12}Md^2, \tag{10.14}$$

where $\rho = M/\pi a^2 d$ is the density.

(4) A Sphere with the Fixed Axis Passing through its Center

Suppose a uniform sphere of radius a and mass M. Let's calculate the moment of inertia of the sphere about the axis passing its center. We will use spherical coordinates taking the fixed axis as the z-axis. At the position (r, θ, ϕ), the volume element is $dv = r^2 \cos\theta drd\phi d\theta$ and the distance from the axis is $r' = r\cos\theta$, so that the moment of inertia is obtained as follows.

$$I = \int_{-\pi/2}^{\pi/2} \int_0^{2\pi} \int_0^a (r\cos\theta)^2 \rho r^2 \cos\theta drd\phi d\theta$$

$$= \rho \int_{-\pi/2}^{\pi/2} \int_0^{2\pi} \int_0^a (\cos\theta)^3 r^4 drd\phi d\theta,$$

where

$$\int_{-\pi/2}^{\pi/2} (\cos\theta)^3 d\theta = \int_{-\pi/2}^{\pi/2} (1 - \sin^2\theta)\cos\theta d\theta$$

$$= \left[\sin\theta - \frac{1}{3}\sin\theta^3\right]_{-\pi/2}^{\pi/2} = 2 - \frac{2}{3} = \frac{4}{3}.$$

Therefore,

$$I = \frac{4}{3}\rho \int_0^{2\pi} \int_0^a r^4 drd\theta = \frac{8}{3}\rho\pi \int_0^a r^4 dr$$

$$= \frac{8}{3}\rho\pi\frac{1}{5}\left[r^5\right]_0^a = \frac{4}{3}\frac{2}{5}\pi\rho a^5 = \frac{2}{5}Ma^2, \tag{10.15}$$

where $\rho = 3M/4\pi a^3$ is the density.

10.1.4 The Parallel Axes Theorem

Let's obtain the moment of inertia of a rigid body about the axis with a distance l from the center of mass G (Fig. 10.3). Let the point setting down from G vertically to the rotational axis be O and the moment of inertia about the axis parallel to the rotating axis and passing through G be I_0. Using Cartesian coordinates in which the origin is O, the y-axis is taken as the direction from O to G, the z-axis is taken as the rotational axis and the x-axis is taken so as to obey the right-hand rule (Fig. 10.3), the moment of inertia I_0 is calculated by (10.11),

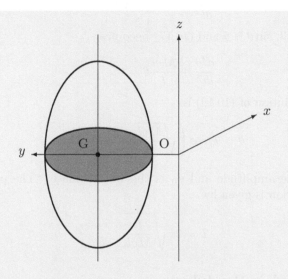

Figure 10.3: The moment of inertia of rigid bodies about the axis off the center of mass.

$$I_0 = \int\int\int \rho\{x^2 + (y-l)^2\}dxdydz = \int\int\int \rho\{x^2 + y^2 - 2ly + l^2\}dxdydz$$

$$= \int\int\int \rho(x^2+y^2)dxdydz - 2l\int\int\int \rho y dxdydz + l^2\int\int\int \rho dxdydz.$$

Remembering the definition of the center of mass in (7.7), and using the fact that $Y = l$, we obtain

$$I_0 = I - 2lYM + l^2M = I - Ml^2,$$
$$I = I_0 + Ml^2. \tag{10.16}$$

Equation(10.16) is known as the *parallel axes theorem*.

10.1.5 Physical Pendulum

A simple pendulum discussed in Chapter 3 is an idealized pendulum. However, real pendulums have the moment of inertia so that their motion should be discussed using (10.9). An oscillating rigid body about a level and fixed axis is called a physical pendulum. Letting the moment of inertia about the fixed axis be I, the mass be M, the position of the fixed axis be O, the center of mass be G, the length of $\overline{\text{GO}}$ be l, and the angle between the vertical and $\overline{\text{GO}}$ be θ, the angular momentum equation becomes

$$I\frac{d^2\theta}{dt^2} = -Mgl\sin\theta. \tag{10.17}$$

When θ is small, $\sin\theta \cong \theta$ and (10.17) becomes

$$\frac{d^2\theta}{dt^2} + \frac{Mgl}{I}\theta = 0. \tag{10.18}$$

The general solution of (10.18) is

$$\theta = \theta_0 \cos\left(\sqrt{\frac{Mgl}{I}}t - \phi_0\right), \tag{10.19}$$

where θ_0 is the amplitude and ϕ_0 is the initial phase. The period of the oscillatory motion is given by

$$T = 2\pi\sqrt{\frac{I}{Mgl}}. \tag{10.20}$$

10.1.6 Borda's Pendulum

The magnitude of the acceleration due to gravity at the Earth's surface g_0 depends on the latitude, the shape of the Earth, and the distribution of the underground materia, so that the measurement of g_0 is important from a viewpoint of geophysics, geology and probing underground resources. Borda's pendulum is a classical method of measuring g_0.[1] As the accuracy of the observed value of g_0 by Borda's pendulum is limited, it is generally used in teaching physics experiments in universities. The construction of Borda's pendulum is as follows; one end of a fine piano wire of length l is attached to a brass ball of mass M and radius a, and the other end to a rigid support called the knife-edge to minimize the friction. Letting the linear density of the piano wire be η, the moment of inertia of the pendulum is obtained using (10.16),

$$I = \frac{2}{5}Ma^2 + \frac{1}{3}\eta l^3 + Ml^2. \tag{10.21}$$

---

[1]The most accurate method in recent days is to measure falling distance and its time optically under near vacuum conditions.

The second term is negligible because it is small compared with the first and third terms.

Problem 3. In the experiment of Borda's pendulum, it is usual to ignore the moment of inertia of a piano wire. Is the treatment adequate when we want to obtain the value of the acceleration due to gravity to a significant figure of four digits, supposing a brass ball of mass $M = 2.155 \times 10^{-1}$ kg, radius $a = 2.020 \times 10^{-2}$ m, and a piano wire of length $l = 1.120$ m and the linear density of $\eta = 6.147 \times 10^{-4}$ kg m^{-1}?

10.2 Two-Dimensional Motion of Rigid Bodies

Suppose that all external forces exerting on a rigid body are in the x-y plane, and the motion is confined to the translation in the x-y plane and the rotational motion is restricted about the z-axis.

10.2.1 Governing Equations

(1) Translation of the Center of Mass

In Section 7.1, we have discussed that the motion of the center of mass of a many-particle system and a rigid body is treated analogous to that of a particle. The component equations of (7.6) are,

$$M\frac{d^2X}{dt^2} = \sum_{i=1}^{n} F_{xi}, \qquad (10.22)$$

$$M\frac{d^2Y}{dt^2} = \sum_{i=1}^{n} F_{yi}, \qquad (10.23)$$

where (X, Y) is the position of the center of mass.

(2) Angular Momentum Equation

Equation (10.9) is listed again,

$$I\frac{d\omega}{dt} = \sum_{i=1}^{n} N_{zi} = N_z, \qquad (10.24)$$

where N_z is the total torque exerting on the rigid body.

(3) Energy Equations of Rotational Motion

The kinetic energy of the volume element δv of a rigid body rotating at the angular velocity ω about the fixed axis is

$$\delta K = \frac{1}{2}\rho\delta v v^2 = \frac{1}{2}\rho\delta v(r'\omega)^2, \qquad (10.25)$$

where ρ is the density of the rigid body. Integrating (10.25) through the whole volume, the total kinetic energy is obtained using (10.11),

$$K = \int_V \delta K = \frac{1}{2}\omega^2 \int_V \rho r'^2 dv = \frac{1}{2}I\omega^2. \tag{10.26}$$

Multiplying (10.24) by ω and modifying the resultant equation yields,

$$\frac{d}{dt}\left(\frac{1}{2}I\omega^2\right) = \omega N_z = N_z \frac{d\theta}{dt}. \tag{10.27}$$

Integrating (10.27) with respect to t from t_1 to t_2, we obtain

$$K_2 - K_1 = \int_{t_1}^{t_2} N_z \frac{d\theta}{dt} dt = \int_{\theta_1}^{\theta_2} N_z d\theta = W, \tag{10.28}$$

where K_1, K_2 are the kinetic energy and θ_1, θ_2 are the azimuthal angles at t_1 and t_2. W is the work and (10.28) is the energy equation in the rotational motion.

10.2.2   Rolling Motion of Rigid Bodies on a Plane without Sliding

In this subsection, we will consider the rolling motion of a cylinder (an annular cylinder, a sphere, and a spherical shell) of mass M, moment of inertia I, and radius a without sliding on a plane. Letting the center of mass be G, the contact point with the level plane be P and the translation speed be V, we will calculate the kinetic energy of the cylinder. The kinematic condition in order that the cylinder does not slide at point P is

$$a\omega = V,$$

where ω is the magnitude of the angular velocity of the cylinder. As the total kinetic energy is the sum of the translational kinetic energy and the rotational kinetic energy, we get

$$K = \frac{1}{2}MV^2 + \frac{1}{2}I\omega^2 = \frac{1}{2}\left(M + \frac{I}{a^2}\right)V^2. \tag{10.29}$$

10.2.3   Rolling Down Motion of Rigid Bodies on a Slope without Sliding

We will consider the rolling down motion of a cylinder (a cylindrical shell, a sphere and a spherical shell) of mass M, moment of inertia I and radius a without sliding on a slope whose angle to the level surface is β. Forces exerting on the cylinder are the gravity Mg, the vertical resistive force N and the frictional force R. Taking the x-axis downward along the slope and the y-axis perpendicular to the slope, the equation of motion of the x- and y-components are

$$M\frac{d^2X}{dt^2} = Mg\sin\beta - R, \tag{10.30}$$

$$0 = N - Mg\cos\beta. \tag{10.31}$$

The angular momentum equation is

$$I\frac{d^2\theta}{dt^2} = aR. \tag{10.32}$$

The condition that the cylinder does not slide on the slope is

$$V = a\omega,$$
$$\frac{dX}{dt} = a\frac{d\theta}{dt}. \tag{10.33}$$

Differentiating (10.33) with respect to t, we get

$$\frac{d^2X}{dt^2} = a\frac{d^2\theta}{dt^2}. \tag{10.34}$$

Eliminating R from (10.30) and (10.32), and using (10.34), we obtain

$$\frac{d^2X}{dt^2} = \frac{1}{1 + I/Ma^2}g\sin\beta. \tag{10.35}$$

The magnitude of the acceleration due to gravity when a body slides down a frictionless slope is $g\sin\beta$, which is called reduced gravity. When a rigid body rolls down a slope without sliding, the translational acceleration along the slope is $1/(1 + I/Ma^2)$ times of the acceleration due to reduced gravity. Therefore, the smaller the moment of inertia of rigid bodies is, the larger the acceleration for rolling down motion becomes. In Table 10.1, we will show the translational accelerations along the slope for rigid bodies of various shapes. Frictional force is obtained from (10.30) and (10.35) as

$$R = Mg\sin\beta - M\frac{d^2X}{dt^2} = \frac{I}{Ma^2 + I}Mg\sin\beta. \tag{10.36}$$

Letting the static frictional coefficient between a rigid body and a slope be μ, the condition that the rigid body does not slide on the slope is given by

$$R \leq \mu N = \mu Mg\cos\beta,$$
$$Mg\sin\beta\frac{I}{Ma^2 + I} \leq \mu Mg\cos\beta,$$
$$\tan\beta \leq \mu\frac{Ma^2 + I}{I}. \tag{10.37}$$

Table 10.1: The translational acceleration of rigid bodies of various shapes rolling down a slope.

Shape	Moment of Inertia	Acceleration
Annular Cylinder	Ma^2	$g \sin \beta / 2$
Spherical Shell	$2Ma^2/3$	$3g \sin \beta / 5$
Cylinder	$Ma^2/2$	$2g \sin \beta / 3$
Sphere	$2Ma^2/5$	$5g \sin \beta / 7$

10.2.4 Examples of Two-Dimensional Motion of Rigid Bodies

Example 2. Suppose that one end of a massless and inextensible string is fixed to a rigid support and another end is turned around a uniform disk of mass M and radius a (Fig. 10.4). The disk is kept at rest and is released at some moment. Discuss the motion of the disk, supposing no slide between the disk and the string. Let the magnitude of the acceleration due to gravity be g.

Answer
The forces exerting on a disk are the vertical tension of the string S and the gravity Mg, so that the disk falls vertically downward without moving laterally. Taking the x-axis vertically downward, the equation of motion describing the motion of the center of mass of the disk is

$$M\frac{d^2 X}{dt^2} = Mg - S. \tag{10.38}$$

The moment of inertia about the axis passing through the center of mass and vertical to the circular plate is $I = Ma^2/2$. Letting the magnitude of the angular velocity be ω, the angular momentum equation is

$$I\frac{d\omega}{dt} = aS. \tag{10.39}$$

There is no slide between the disk and the string, so that the following relation holds,

$$a\omega = \frac{dX}{dt}. \tag{10.40}$$

Dividing (10.40) through by a and differentiating the resultant equation with respect to t, we find

$$\frac{d\omega}{dt} = \frac{1}{a}\frac{d^2 X}{dt^2}. \tag{10.41}$$

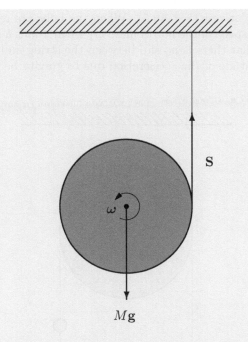

Figure 10.4: Rolling down motion of a disk turned around by a string.

Substituting (10.41) into (10.39), we get

$$S = \frac{I}{a^2} \frac{d^2 X}{dt^2}.$$

(10.42)

Substituting (10.42) into (10.38) yields,

$$M\frac{d^2 X}{dt^2} = Mg - \frac{I}{a^2} \frac{d^2 X}{dt^2},$$
$$\left(1 + \frac{I}{Ma^2}\right) \frac{d^2 X}{dt^2} = g,$$
$$\frac{d^2 X}{dt^2} = \frac{2}{3}g.$$

(10.43)

Therefore, the disk falls at the magnitude of the acceleration $2g/3$.

Example 3. Suppose that a massless and inextensible string is rolled around a fixed pulley of radius a, mass M, and the moment of inertia $I = Ma^2/2$. Attaching bobs of mass m_1 and m_2 ($m_1 > m_2$) to both ends of the string, the bobs are kept at rest (Fig. 10.5). When the bobs are released from rest, obtain the falling speed of the bob of mass m_1 when it falls h from the released point. Suppose that there is no slip between the string and the fixed pulley, and let the magnitude of the acceleration due to gravity be g.

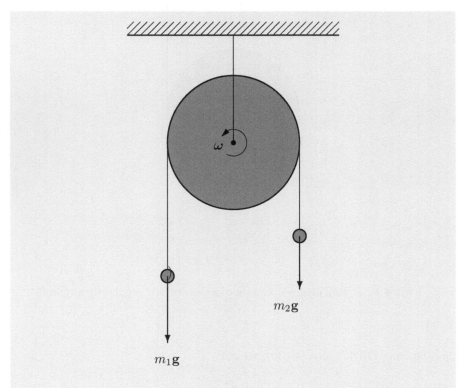

Figure 10.5: Vertical motion of bobs hung from the fixed pulley.

Answer
Taking the x-axis vertically downward, the angular velocity of the pulley corresponding to the falling speed of the bob is ω and the reference position of the potential energy of the bob is the released point,

$$\frac{1}{2}m_1v^2 + \frac{1}{2}m_2v^2 + \frac{1}{2}I\omega^2 + m_2gh - m_1gh = 0 \tag{10.44}$$

is obtained due to the law of the mechanical energy conservation. Substituting the moment of inertia of the pulley $I = Ma^2/2$ and the relation $a\omega = v$ into (10.44) yields,

$$\frac{1}{2}m_1v^2 + \frac{1}{2}m_2v^2 + \frac{1}{4}Ma^2\left(\frac{v}{a}\right)^2 + m_2gh - m_1gh = 0,$$

$$\{2(m_1 + m_2) + M\}v^2 = 4(m_1 - m_2)gh,$$

$$v = 2\sqrt{\frac{m_1 - m_2}{2(m_1 + m_2) + M}gh}. \qquad (10.45)$$

Example 4. Suppose that a small and uniform ball of radius a and mass M is at rest at the lowest point inside of a spherical shell of inner radius b. Discuss the motion of the ball when it is given a small displacement. Suppose that there is no slide between the ball and the inner surface of the spherical shell. Let the magnitude of the acceleration due to gravity be g.

Answer
Let the center of the spherical shell be O, the lowest point be A, the center of the ball be O′, angle $\angle AOO' = \theta$, the contact point of the ball, and the spherical surface be A′, the contact point of the spherical surface with point A of the ball be B, the intersection of the line OA and the extended line of O′B be C and $\angle BO'A' = \phi$ (Fig. 10.6). Owing to the condition that there is

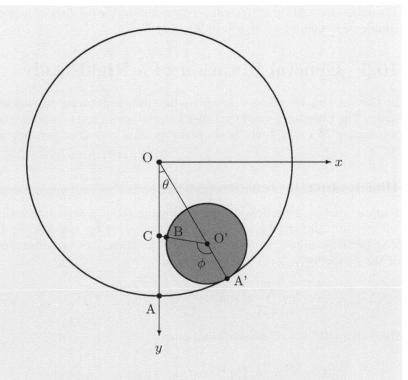

Figure 10.6: The oscillatory motion of a ball in spherical shell.

no slide between the ball and the spherical surface, we find

$$b\theta = a\phi. \tag{10.46}$$

$\angle \text{ACO}'$ is

$$\angle \text{ACO}' = (\pi - \phi) + \theta$$
$$= \pi - \left(\frac{b}{a} - 1\right)\theta. \tag{10.47}$$

Letting the moment of inertia of the ball be I, the angular momentum equation is

$$I\frac{d^2}{dt^2}\left\{\pi - \left(\frac{b}{a} - 1\right)\theta\right\} = aMg\sin\theta. \tag{10.48}$$

Taking into account that $\sin\theta \cong \theta$ and $I = 2Ma^2/5$, (10.48) becomes

$$\frac{2}{5}(b - a)\frac{d^2\theta}{dt^2} + g\theta = 0. \tag{10.49}$$

Therefore, the motion of the ball is equivalent to the oscillatory motion of the simple pendulum of string length $2(b - a)/5$.

10.3 General Rotation of a Rigid Body

In this section, we will discuss three-dimensional rotating motion of a rigid body. The Euler angles and the Euler's equations are introduced for discussing the motion of a rigid body by an observer on a coordinate system bound to it.

10.3.1 Inertia Tensor

Suppose that an n-particle system is rotating about a fixed axis with angular velocity $\boldsymbol{\omega}$. All particles are connected each other by massless rigid rods so that particles do not change their relative positions. The angular momentum of the system is

$$\mathbf{L} = \sum_{i=1}^{n} \mathbf{r}_i \times \mathbf{p}_i = \sum_{i=1}^{n} \mathbf{r}_i \times m_i(\boldsymbol{\omega} \times \mathbf{r}_i). \tag{10.50}$$

Resolving (10.50) in Cartesian components,

$$\mathbf{L} = \sum_{i=1}^{n} m_i[\mathbf{e}_x\{(y_i{}^2 + z_i{}^2)\omega_x - x_i y_i \omega_y - z_i x_i \omega_z\}$$
$$+ \mathbf{e}_y\{-x_i y_i \omega_x + (z_i{}^2 + x_i{}^2)\omega_y - y_i z_i \omega_z\}$$
$$+ \mathbf{e}_z\{-z_i x_i \omega_x - y_i z_i \omega_y + (x_i{}^2 + y_i{}^2)\omega_z\}] . \tag{10.51}$$

We can express angular momentum **L** in a matrix form as,

$$\mathbf{L} = \mathcal{I}\boldsymbol{\omega}, \tag{10.52}$$

where

$$\mathcal{I} = \begin{pmatrix} I_{xx} & I_{xy} & I_{xz} \\ I_{yx} & I_{yy} & I_{yz} \\ I_{zx} & I_{zy} & I_{zz} \end{pmatrix}, \tag{10.53}$$

$$\begin{cases} I_{xx} = \sum_{i=1}^{n} m_i(y_i^2 + z_i^2), & I_{xy} = -\sum_{i=1}^{n} m_i x_i y_i, \\ I_{xz} = -\sum_{i=1}^{n} m_i z_i x_i, & I_{yx} = -\sum_{i=1}^{n} m_i x_i y_i, \\ I_{yy} = \sum_{i=1}^{n} m_i(z_i^2 + x_i^2), & I_{yz} = -\sum_{i=1}^{n} m_i y_i z_i, \\ I_{zx} = -\sum_{i=1}^{n} m_i z_i x_i, & I_{zy} = -\sum_{i=1}^{n} m_i y_i z_i, \\ I_{zz} = \sum_{i=1}^{n} m_i(x_i^2 + y_i^2), \end{cases} \tag{10.54}$$

is a symmetric tensor and is referred to as the *inertia tensor*. In the case of a rigid body, the inertia tensor is obtained replacing sigma notations by integrals and the mass of each particle m_i by the mass element ρdv in (10.54), so that we obtain,

$$\begin{cases} I_{xx} = \int_V \rho(y^2 + z^2)dv, & I_{xy} = -\int_V \rho xy\, dv, \\ I_{xz} = -\int_V \rho zx\, dv, & I_{yx} = -\int_V \rho xy\, dv, \\ I_{yy} = \int_V \rho(z^2 + x^2)dv, & I_{yz} = -\int_V \rho yz\, dv, \\ I_{zx} = -\int_V \rho zx\, dv, & I_{zy} = -\int_V \rho yz\, dv, \\ I_{zz} = \int_V \rho(x^2 + y^2)dv. \end{cases} \tag{10.55}$$

Diagonal elements I_{xx}, I_{yy}, I_{zz} of the inertial tensor are the moment of inertia about the x, y, z-axes and the non-diagonal elements are called the *products of inertia*.

Example 5. In Cartesian coordinates, three particles of mass m are set at $(1,1,0)$, $(0,2,1)$, and $(1,0,3)$. They are connected each other by massless rigid rods so that their relative positions do not change. Calculate the inertia tensor of this system.

Answer
The inertia tensor is obtained using (10.54)

$$
\left\{
\begin{array}{l}
I_{xx} = 1^2 m + (2^2 + 1^2)m + 3^2 m = 15m, \quad I_{xy} = -1 \cdot 1m = -m\,, \\[2mm]
I_{xz} = -11 \cdot 3m = -3m, \quad I_{yx} = -1 \cdot 1m = -m\,, \\[2mm]
I_{yy} = 11^2 m + 2^2 m + (3^2 + 1^2)m = 15m, \quad I_{yz} = -1 \cdot 2m = -2m\,, \\[2mm]
I_{zx} = -11 \cdot 3m = -3m, \quad I_{zy} = -1 \cdot 2m = -2m\,, \\[2mm]
I_{zz} = 1(1^2 + 1^2)m + 2^2 m + 1^2 m = 7m\,.
\end{array}
\right.
$$

Thus, the inertia tensor is written as

$$
\mathcal{I} = m
\begin{pmatrix}
15 & -1 & -3 \\
-1 & 15 & -2 \\
-3 & -2 & 7
\end{pmatrix}.
$$

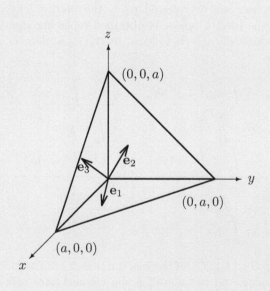

Figure 10.7: Moment of inertia of a delta cone.

Example 6. Calculate the inertia tensor of a rigid delta cone of mass M with its vertexes at the origin, (a,0,0), (0,a,0), and (0,0,a) (Fig. 10.7).

Answer
The density of the delta cone is

$$\rho = \frac{6M}{a^3}.$$

Diagonal components are obtained using (10.55)

$$I_{xx} = \rho \int_0^a dx \int_0^{a-x} dy \int_0^{a-x-y} (y^2 + z^2)dz = \frac{1}{30}\rho a^5 = \frac{1}{5}Ma^2.$$

$I_{xx} = I_{yy} = I_{zz}$ due to the symmetric shape of the delta cone. Non-diagonal components are obtained as follows

$$I_{xy} = -\rho \int_0^a dx \int_0^{a-x} dy \int_0^{a-x-y} xydz = -\frac{1}{120}\rho a^5 = -\frac{1}{20}Ma^2.$$

$I_{xy} = I_{yz} = I_{zx}$ due to the symmetric shape of the delta cone. Then the inertia tensor is

$$\mathcal{I} = \frac{Ma^2}{20} \begin{pmatrix} 4 & -1 & -1 \\ -1 & 4 & -1 \\ -1 & -1 & 4 \end{pmatrix}.$$

10.3.2 Kinetic Energy of Three-Dimensional Rotating Motion of a Rigid Body

We will calculate the kinetic energy of an n-particle system rotating at the angular velocity $\boldsymbol{\omega}$ about a fixed axis. All particles are connected each other by massless rigid rods.

$$K = \frac{1}{2}\sum_{i=1}^n m_i \mathbf{v}_i \cdot \mathbf{v}_i = \frac{1}{2}\sum_{i=1}^n m_i(\boldsymbol{\omega} \times \mathbf{r}_i) \cdot (\boldsymbol{\omega} \times \mathbf{r}_i)$$

$$= \frac{1}{2}\sum_{i=1}^n m_i \boldsymbol{\omega} \cdot \{\mathbf{r}_i \times (\boldsymbol{\omega} \times \mathbf{r}_i)\} = \frac{1}{2}\boldsymbol{\omega} \cdot \mathbf{L} = \frac{1}{2}\boldsymbol{\omega} \cdot \mathbf{I}\boldsymbol{\omega}, \qquad (10.56)$$

where use is made of the following vector identity.

$$(\boldsymbol{\omega} \times \mathbf{r}) \cdot \mathbf{A} = (\omega_y z - \omega_z y)A_x + (\omega_z x - \omega_x z)A_y + (\omega_x y - \omega_y x)A_z$$
$$= \omega_x(yA_z - zA_y) + \omega_y(zA_x - xA_z) + \omega_z(xA_y - yA_x)$$
$$= \boldsymbol{\omega} \cdot (\mathbf{r} \times \mathbf{A}).$$

10.3.3   Principal Axes and Principal Moments of Inertia

We can choose a coordinate system to diagonalize an inertial tensor. These axes are called the *principal axes*, which are bound to a rotating rigid body. Let the axes be (x_1, x_2, x_3) and the corresponding basis vectors be $(\mathbf{e}_1, \mathbf{e}_2, \mathbf{e}_3)$. Then the inertia tensor becomes a simple form as

$$\mathcal{I} = \begin{pmatrix} I_{11} & 0 & 0 \\ 0 & I_{22} & 0 \\ 0 & 0 & I_{33} \end{pmatrix} \equiv \begin{pmatrix} I_1 & 0 & 0 \\ 0 & I_2 & 0 \\ 0 & 0 & I_3 \end{pmatrix}. \tag{10.57}$$

Using (10.52) the angular momentum is expressed as

$$\mathbf{L} = \mathcal{I}\boldsymbol{\omega} = \begin{pmatrix} I_1 & 0 & 0 \\ 0 & I_2 & 0 \\ 0 & 0 & I_3 \end{pmatrix} \boldsymbol{\omega} = \begin{pmatrix} I_1\omega_1 \\ I_2\omega_2 \\ I_3\omega_3 \end{pmatrix}. \tag{10.58}$$

I_1, I_2, I_3 are called the *principal moments of inertia* and are obtained as eigenvalues λ of the following equation,

$$|\mathcal{I} - \lambda\mathcal{E}| = 0, \tag{10.59}$$

where

$$\mathcal{E} = \begin{pmatrix} 1 & 0 & 0 \\ 0 & 1 & 0 \\ 0 & 0 & 1 \end{pmatrix}$$

is an identity matrix. We will show illustratively the method to obtain the principal axes and the principal moments of inertia.

Example 7. Obtain the principal axes and the principal moment of inertia for the inertia tensor in Example 6.

Answer
Letting the *eigenvalue* be λ, the characteristic equation is

$$\begin{vmatrix} 4-\lambda & -1 & -1 \\ -1 & 4-\lambda & -1 \\ -1 & -1 & 4-\lambda \end{vmatrix} = 0,$$

$$(\lambda - 5)^2(\lambda - 2) = 0,$$

$$\lambda = 5, 5, 2.$$

The *eigenvector* $\mathbf{v}_1$ corresponding to $\lambda = 5$ is obtained by solving the equation

$$\begin{pmatrix} 4 & -1 & -1 \\ -1 & 4 & -1 \\ -1 & -1 & 4 \end{pmatrix} \mathbf{v}_1 = 5\mathbf{v}_1,$$

$$v_{13} = -v_{11}, \qquad v_{12} = 0.$$

Then we find $\mathbf{v}_1 = (1, 0, -1)$. The first basis vector is obtained by normalizing $\mathbf{v}_1$,

$$\mathbf{e}_1 = \frac{1}{\sqrt{2}}(1, 0, -1).$$

The eigenvector $\mathbf{v}_2$ corresponding to $\lambda = 2$ is obtained similarly, and the second basis vector is obtained by normalizing $\mathbf{v}_2$

$$\mathbf{e}_2 = \frac{1}{\sqrt{3}}(1, 1, 1).$$

In this case, the eigenvalues are degenerated, so the third eigenvector $\mathbf{v}_3$ is obtained by the condition that it is normal both to $\mathbf{v}_1$ and $\mathbf{v}_2$. The third basis vector is found by normalizing $\mathbf{v}_3$ as

$$\mathbf{e}_3 = \frac{1}{\sqrt{6}}(1, -2, 1).$$

Therefore, the diagonalized inertia tensor is

$$\mathcal{I} = \frac{Ma^2}{20} \begin{pmatrix} 5 & 0 & 0 \\ 0 & 5 & 0 \\ 0 & 0 & 2 \end{pmatrix}.$$

10.4 Euler Angles and Euler's Equation

In 1760, Leonhard Euler introduced three angles (called later *Euler angles*) to relate a Cartesian coordinate system (x_1, x_2, x_3) bound to a rotating rigid body to a Cartesian coordinate system (x, y, z) on an inertial reference frame. Further, he formulated Euler's equation of motion (Euler's second law of motion) for discussing the dynamics of the rotational motion of a rigid body on a reference frame bound to it.

10.4.1 Euler Angles

Let Euler angles be (α, β, γ), which correspond to three coordinates transformations (see Fig. 10.8). The first transformation is to rotate the coordinate system (x, y, z) by the angle α about the z-axis transforming to a coordinate system (x', y', z'). This coordinates transformation is expressed as

$$\begin{pmatrix} \mathbf{e}_{x'} \\ \mathbf{e}_{y'} \\ \mathbf{e}_{z'} \end{pmatrix} = \begin{pmatrix} \cos\alpha & \sin\alpha & 0 \\ -\sin\alpha & \cos\alpha & 0 \\ 0 & 0 & 1 \end{pmatrix} \begin{pmatrix} \mathbf{e}_x \\ \mathbf{e}_y \\ \mathbf{e}_z \end{pmatrix} = \mathcal{A} \begin{pmatrix} \mathbf{e}_x \\ \mathbf{e}_y \\ \mathbf{e}_z \end{pmatrix}. \qquad (10.60)$$

The second transformation is to rotate the coordinate system (x', y', z') about the x'-axis by the angle β transforming to a coordinate system (x'', y'', z'').

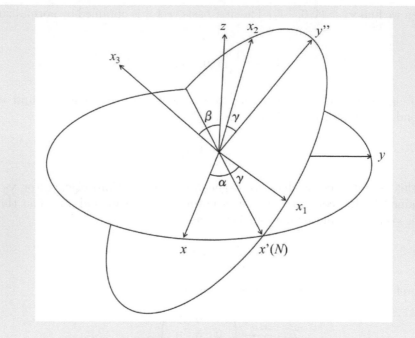

Figure 10.8: Euler Angles.

This coordinates transformation is expressed as

$$
\begin{pmatrix} \mathbf{e}_{x''} \\ \mathbf{e}_{y''} \\ \mathbf{e}_{z''} \end{pmatrix} = \begin{pmatrix} 1 & 0 & 0 \\ 0 & \cos\beta & \sin\beta \\ 0 & -\sin\beta & \cos\beta \end{pmatrix} \begin{pmatrix} \mathbf{e}_{x'} \\ \mathbf{e}_{y'} \\ \mathbf{e}_{z'} \end{pmatrix} = \mathcal{B} \begin{pmatrix} \mathbf{e}_{x'} \\ \mathbf{e}_{y'} \\ \mathbf{e}_{z'} \end{pmatrix}.
\tag{10.61}
$$

The third transformation is to rotate (x'', y'', z'') by the angle γ about the z-axis transforming to a coordinate system (x_1, x_2, x_3). This coordinates transformation is expressed as

$$
\begin{pmatrix} \mathbf{e}_1 \\ \mathbf{e}_2 \\ \mathbf{e}_3 \end{pmatrix} = \begin{pmatrix} \cos\gamma & \sin\gamma & 0 \\ -\sin\gamma & \cos\gamma & 0 \\ 0 & 0 & 1 \end{pmatrix} \begin{pmatrix} \mathbf{e}_{x''} \\ \mathbf{e}_{y''} \\ \mathbf{e}_{z''} \end{pmatrix} = \mathcal{C} \begin{pmatrix} \mathbf{e}_{x''} \\ \mathbf{e}_{y''} \\ \mathbf{e}_{z''} \end{pmatrix}.
\tag{10.62}
$$

The x'-axis is an intersection of the x-y plane and the $x_1 - x_2$ plane, so that it is referred to as the *line of node* N. Three successive coordinate transformations are unified to

$$
\begin{pmatrix} \mathbf{e}_1 \\ \mathbf{e}_2 \\ \mathbf{e}_3 \end{pmatrix} = \mathcal{C}\,\mathcal{B}\,\mathcal{A} \begin{pmatrix} \mathbf{e}_x \\ \mathbf{e}_y \\ \mathbf{e}_z \end{pmatrix} = \mathcal{D} \begin{pmatrix} \mathbf{e}_x \\ \mathbf{e}_y \\ \mathbf{e}_z \end{pmatrix}.
\tag{10.63}
$$

Nine elements of the matrix $\mathcal{D}$ are

$$\left\{ \begin{array}{l} D_{11} = \cos\gamma\cos\alpha - \sin\gamma\cos\beta\sin\alpha \ , \\ D_{12} = \cos\gamma\sin\alpha + \sin\gamma\cos\beta\cos\alpha \ , \\ D_{13} = \sin\gamma\sin\beta \ , \\ D_{21} = -\sin\gamma\cos\alpha - \cos\gamma\cos\beta\sin\alpha \ , \\ D_{22} = -\sin\gamma\sin\alpha + \cos\gamma\cos\beta\cos\alpha \ , \\ D_{23} = \cos\gamma\sin\beta \ , \\ D_{31} = \sin\beta\sin\alpha \ , \\ D_{32} = -\sin\beta\cos\alpha \ , \\ D_{33} = \cos\beta \ . \end{array} \right. \tag{10.64}$$

Operating $\mathcal{D}^{-1}$ on (10.63) from left side, we get

$$\begin{pmatrix} \mathbf{e}_x \\ \mathbf{e}_y \\ \mathbf{e}_z \end{pmatrix} = \mathcal{D}^{-1} \begin{pmatrix} \mathbf{e}_1 \\ \mathbf{e}_2 \\ \mathbf{e}_3 \end{pmatrix} . \tag{10.65}$$

Nine elements of the matrix $\mathcal{D}^{-1}$ are

$$\left\{ \begin{array}{l} D^{-1}{}_{11} = \cos\gamma\cos\alpha - \sin\gamma\cos\beta\sin\alpha \ , \\ D^{-1}{}_{12} = -\sin\gamma\cos\alpha - \cos\gamma\cos\beta\sin\alpha \ , \\ D^{-1}{}_{13} = \sin\alpha\sin\beta \ , \\ D^{-1}{}_{21} = \cos\gamma\sin\alpha + \sin\gamma\cos\beta\cos\alpha \ , \\ D^{-1}{}_{22} = -\sin\gamma\sin\alpha + \cos\gamma\cos\beta\cos\alpha \ , \\ D^{-1}{}_{23} = -\cos\alpha\sin\beta \ , \\ D^{-1}{}_{31} = \sin\beta\sin\gamma \ , \\ D^{-1}{}_{32} = \sin\beta\cos\gamma \ , \\ D^{-1}{}_{33} = \cos\beta \ . \end{array} \right. \tag{10.66}$$

From the definition of Euler angles, the meaning of them is obvious as follows:

1. α: precession or rotation of the x_3-axis about the z-axis.

2. β: nutation or bobbing up and down of the x_3-axis relative to the z-axis.

3. γ: rotation of the rigid body about the x_3-axis.

Using Euler angles, angular velocity of a rigid body is expressed as

$$\boldsymbol{\omega} = \dot{\alpha}\mathbf{e}_z + \dot{\beta}\mathbf{e}_{x'} + \dot{\gamma}\mathbf{e}_3, \tag{10.67}$$

where $\dot{\alpha} \equiv d\alpha/dt$ etc.
At first, we will express $\boldsymbol{\omega}$ by basis vectors $\mathbf{e}_x, \mathbf{e}_y, \mathbf{e}_z$. Substituting (10.60) and (10.63) into (10.67), we obtain

$$\boldsymbol{\omega} = \left(\dot{\beta}\cos\alpha + \dot{\gamma}\sin\beta\sin\alpha \right) \mathbf{e}_x + \left(\dot{\beta}\sin\alpha - \dot{\gamma}\sin\beta\cos\alpha \right) \mathbf{e}_y$$
$$+ (\dot{\alpha} + \dot{\gamma}\cos\beta)\mathbf{e}_z. \tag{10.68}$$

Next, we will express $\boldsymbol{\omega}$ by basis vectors $\mathbf{e}_1, \mathbf{e}_2, \mathbf{e}_3$. Operating $\mathcal{B}^{-1}$ on (10.61) from left side, we find

$$
\begin{pmatrix} \mathbf{e}_{x'} \\ \mathbf{e}_{y'} \\ \mathbf{e}_{z'} \end{pmatrix} = \mathcal{B}^{-1} \begin{pmatrix} \mathbf{e}_{x''} \\ \mathbf{e}_{y''} \\ \mathbf{e}_{z''} \end{pmatrix} = \mathcal{B}^{-1}\mathcal{C}^{-1} \begin{pmatrix} \mathbf{e}_1 \\ \mathbf{e}_2 \\ \mathbf{e}_3 \end{pmatrix}
$$

$$
= \begin{pmatrix} \cos\gamma & -\sin\gamma & 0 \\ \cos\beta\sin\gamma & \cos\beta\cos\gamma & -\sin\beta \\ \sin\beta\sin\gamma & \sin\beta\cos\gamma & \cos\beta \end{pmatrix} \begin{pmatrix} \mathbf{e}_1 \\ \mathbf{e}_2 \\ \mathbf{e}_3 \end{pmatrix}. \tag{10.69}
$$

Substituting (10.65) and (10.69) into (10.67), we find

$$
\boldsymbol{\omega} = \left(\dot\alpha \sin\gamma \sin\beta + \dot\beta \cos\gamma \right) \mathbf{e}_1 + \left(\dot\alpha \cos\gamma \sin\beta - \dot\beta \sin\gamma \right) \mathbf{e}_2
$$
$$
+ \left(\dot\alpha \cos\beta + \dot\gamma \right) \mathbf{e}_3. \tag{10.70}
$$

10.4.2 Euler's Equations

The angular momentum equation on an inertial reference frame is

$$
\frac{d_a \mathbf{L}}{dt} = \mathbf{N}. \tag{10.71}
$$

Using (4.14), the total derivative of $\mathbf{L}$ on the Cartesian coordinate system (x_1, x_2, x_3) rotating at angular velocity $\boldsymbol{\omega}$ is written as

$$
\frac{d\mathbf{L}}{dt} + \boldsymbol{\omega} \times \mathbf{L} = \mathbf{N}. \tag{10.72}
$$

The second term of the left-hand side is expanded as

$$
\boldsymbol{\omega} \times \mathbf{L} = \begin{vmatrix} \mathbf{e}_1 & \mathbf{e}_2 & \mathbf{e}_3 \\ \omega_1 & \omega_2 & \omega_3 \\ I_1\omega_1 & I_2\omega_2 & I_3\omega_3 \end{vmatrix}
$$
$$
= \mathbf{e}_1 \omega_2 \omega_3 (I_3 - I_2) + \mathbf{e}_2 \omega_3 \omega_1 (I_1 - I_3) + \mathbf{e}_3 \omega_1 \omega_2 (I_2 - I_1).
$$

The component equations of (10.72) are

$$
I_1 \frac{d\omega_1}{dt} + \omega_2 \omega_3 (I_3 - I_2) = N_1, \tag{10.73}
$$

$$
I_2 \frac{d\omega_2}{dt} + \omega_3 \omega_1 (I_1 - I_3) = N_2, \tag{10.74}
$$

$$
I_3 \frac{d\omega_3}{dt} + \omega_1 \omega_2 (I_2 - I_1) = N_3. \tag{10.75}
$$

Equations (10.73), (10.74), and (10.75) are referred to as *Euler's equations*.

10.4.3 Free Rotation of a Rigid Body

We will discuss a free rotation of a rigid body for special two cases.

(1) A Rigid Sphere

In this case,

$I_1 = I_2 = I_3,$

so that Euler's equations (10.73)–(10.75) become,

$$\frac{d\boldsymbol{\omega}}{dt} = 0. \tag{10.76}$$

Therefore, the angular velocity $\boldsymbol{\omega}$ is a constant in time.

(2) An Axially Symmetric Rigid Body

Choosing x_3 as the symmetric axis, we find

$I_1 = I_2 \neq I_3.$

Euler's equations are

$$I_1 \frac{d\omega_1}{dt} + (I_3 - I_1)\omega_3\omega_2 = 0, \tag{10.77}$$

$$I_1 \frac{d\omega_2}{dt} - (I_3 - I_1)\omega_3\omega_1 = 0, \tag{10.78}$$

$$I_3 \frac{d\omega_3}{dt} = 0. \tag{10.79}$$

Equation (10.79) states that ω_3 is a constant in time, so we put

$$\omega_3 = \Omega \text{ (const.)}. \tag{10.80}$$

Using (10.80), we obtain from (10.77) and (10.78),

$$\frac{d\omega_1}{dt} + \Omega'\omega_2 = 0, \tag{10.81}$$

$$\frac{d\omega_2}{dt} - \Omega'\omega_1 = 0, \tag{10.82}$$

where

$$\Omega' = \frac{I_3 - I_1}{I_1}\Omega. \tag{10.83}$$

We will define the complex angular velocity $\tilde{\omega} = \omega_1 + \tilde{\imath}\omega_2$. Multiplying (10.82) by the imaginary unit $\tilde{\imath}$ and adding the resultant equation to (10.81), we obtain the first-order ordinary differential equation in the complex dependent variable $\tilde{\omega}$.

$$\frac{d\tilde{\omega}}{dt} - \tilde{\imath}\Omega'\tilde{\omega} = 0. \tag{10.84}$$

The general solution of (10.84) is

$$\tilde{\omega} = \alpha \exp\left(\tilde{\imath}\Omega't\right).$$

Therefore,

$$\omega_1 = |\alpha| \cos(\Omega' t - \phi), \tag{10.85}$$

$$\omega_2 = |\alpha| \sin(\Omega' t - \phi). \tag{10.86}$$

From the definition of Ω', a rigid body rotates about the rotating axis counterclockwise when $I_3 > I_1$ and clockwise when $I_3 < I_1$. This motion is called the *free nutation* or the *Euler nutation*.

10.4.4 Lagrange Top

In this subsection, we will consider the motion of a top which is rotating about a symmetric principal axis x_3 under the Earth's gravity. Suppose that principal moments of inertia are I_1, I_2, and I_3 ($I_1 = I_2 \neq I_3$), mass of the top is M, the position of the center of mass is G, a supporting point of the top is O, and the distance between O and G is l.
The moments of inertia about point O are

$$\begin{cases} I_1' = I_1 + Ml^2, \\ I_2' = I_1 + Ml^2 = I_1'. \end{cases} \tag{10.87}$$

We will use Euler angles given by (10.70)

$$\boldsymbol{\omega} = \left(\dot{\alpha}\sin\gamma\sin\beta + \dot{\beta}\cos\gamma\right)\mathbf{e}_1 + \left(\dot{\alpha}\cos\gamma\sin\beta - \dot{\beta}\sin\gamma\right)\mathbf{e}_2$$
$$+ (\dot{\alpha}\cos\beta + \dot{\gamma})\mathbf{e}_3.$$

Using (10.65), the torque exerting on the top by the Earth's gravity $\mathbf{N}$ is given by

$$\mathbf{N} = l\mathbf{e}_3 \times (-Mg\mathbf{e}_z)$$
$$= -Mgl\mathbf{e}_3 \times (\sin\gamma\sin\beta\mathbf{e}_1 + \cos\gamma\sin\beta\mathbf{e}_2 + \cos\beta\mathbf{e}_3)$$
$$= Mgl\sin\beta(\cos\gamma\mathbf{e}_1 - \sin\gamma\mathbf{e}_2), \tag{10.88}$$

Substituting (10.87) and (10.88) into Euler's equations (10.73)–(10.75), we get

$$I_1'\frac{d\omega_1}{dt} + \omega_2\omega_3(I_3 - I_1') = Mgl\sin\beta\cos\gamma, \tag{10.89}$$

$$I_1'\frac{d\omega_2}{dt} + \omega_3\omega_1(I_1' - I_3) = -Mgl\sin\beta\sin\gamma, \tag{10.90}$$

$$I_3\frac{d\omega_3}{dt} = 0. \tag{10.91}$$

From (10.91), we find that $L_3 = I_3\omega_3$ is a constant of motion. Substituting (10.70) into (10.89), we find

$$I_1'\left(\ddot{\alpha}\sin\beta\sin\gamma + 2\dot{\alpha}\dot{\beta}\sin\gamma\cos\beta + \ddot{\beta}\cos\gamma - \dot{\alpha}^2\cos\gamma\cos\beta\sin\beta\right)$$

$$+ I_3 \left(\dot\alpha^2 \cos\gamma \cos\beta \sin\beta - \dot\beta \omega_3 \sin\gamma + \dot\alpha\dot\gamma \cos\gamma \sin\beta \right)$$
$$= Mgl \sin\beta \cos\gamma, \tag{10.92}$$

where $\ddot\alpha \equiv d^2\alpha/dt^2$ etc.
Substituting from (10.70) into (10.90)

$$I_1' \left(\ddot\alpha \cos\gamma \sin\beta + 2\dot\alpha\dot\beta \cos\gamma \cos\beta - \ddot\beta \sin\gamma + \dot\alpha^2 \sin\gamma \cos\beta \sin\beta \right)$$
$$- I_3 \left(\dot\alpha^2 \sin\gamma \cos\beta \sin\beta + \dot\beta \omega_3 \cos\gamma + \dot\alpha\dot\gamma \sin\gamma \sin\beta \right)$$
$$= -Mgl \sin\beta \sin\gamma . \tag{10.93}$$

Adding (10.92)$\times \sin\gamma$ to (10.93)$\times \cos\gamma$ yields

$$I_1' \left(\ddot\alpha \sin\beta + 2\dot\alpha\dot\beta \cos\beta \right) - I_3 \omega_3 \dot\beta = 0. \tag{10.94}$$

Subtracting (10.93)$\times \sin\gamma$ from (10.92)$\times \cos\gamma$, we obtain

$$I_1' \left(\ddot\beta - \dot\alpha^2 \cos\beta \sin\beta \right) + I_3 \omega_3 \dot\alpha \sin\beta = Mgl \sin\beta. \tag{10.95}$$

Letting point O be the reference point of potential energy, mechanical energy of the top is

$$E = \frac{1}{2} I_1 \left(\dot\alpha^2 \sin^2\beta + \dot\beta^2 \right) + \frac{1}{2} I_3 \left(\dot\alpha \cos\beta + \dot\gamma \right)^2 + Mgl \cos\beta. \tag{10.96}$$

Angular momentum equation about the z-axis is

$$I_z \frac{d\omega_z}{dt} = \frac{dL_z}{dt} = 0. \tag{10.97}$$

Using (10.65) and (10.70), we find

$$L_z = \mathbf{L} \cdot \mathbf{e}_z = \left(I_1'\omega_1 \mathbf{e}_1 + I_1'\omega_2 \mathbf{e}_2 + I_3\omega_3 \mathbf{e}_3 \right) \cdot \mathbf{e}_z$$
$$= \left\{ I_1' \left(\dot\alpha \sin\gamma \sin\beta + \dot\beta \cos\gamma \right) \mathbf{e}_1 + I_1' \left(\dot\alpha \cos\gamma \sin\beta - \dot\beta \sin\gamma \right) \mathbf{e}_2 \right.$$
$$\left. + I_3 \left(\dot\alpha \cos\beta + \dot\gamma \right) \mathbf{e}_3 \right\} \cdot (\sin\gamma \sin\beta \mathbf{e}_1 + \cos\gamma \sin\beta \mathbf{e}_2 + \cos\beta \mathbf{e}_3)$$
$$= \dot\alpha \left(I_1' \sin^2\beta + I_3 \cos^2\beta \right) + \dot\gamma I_3 \cos\beta. \tag{10.98}$$

L_z, E are constants of motion. From (10.91) and (10.70)

$$L_3 = I_3 \omega_3 = I_3 \left(\dot\alpha \cos\beta + \dot\gamma \right) \tag{10.99}$$

is also a constant of motion. From (10.99), we obtain

$$\dot\gamma = \frac{L_3}{I_3} - \dot\alpha \cos\beta. \tag{10.100}$$

Substituting (10.100) into (10.98), we get

$$\dot{\alpha} = \frac{L_z - L_3 \cos \beta}{I_1' \sin^2 \beta} \tag{10.101}$$

Substituting from (10.101) into (10.100), we get

$$\dot{\gamma} = \frac{L_3}{I_3} - \frac{L_z - L_3 \cos \beta}{I_1' \sin^2 \beta} \cos \beta. \tag{10.102}$$

Substituting (10.101) and (10.102) into (10.96), we find

$$E = \frac{1}{2} I_1' \dot{\beta}^2 + \frac{L_3^2}{2I_3} + Mgl \cos \beta + \frac{(L_z - L_3 \cos \beta)^2}{2I_1' \sin^2 \beta}. \tag{10.103}$$

We will change variables from β to u by letting

$$u = \cos \beta, \tag{10.104}$$

then we have

$$\dot{\beta} = -\frac{1}{\sqrt{1 - u^2}} \dot{u}. \tag{10.105}$$

Substituting (10.104) and (10.105) into (10.103), we get

$$\dot{u}^2 = \frac{2}{I_1'} \left\{ \left(E - \frac{L_3^2}{2I_3} - Mglu \right) (1 - u^2) - \frac{(L_z - L_3 u)^2}{2I_1'} \right\}$$
$$= (a - bu)(1 - u^2) - (c - du)^2 \equiv f(u), \tag{10.106}$$

where

$$\begin{cases} a = \dfrac{2E - I_3 \omega_3^2}{I_1'} \, , \\[2mm] b = \dfrac{2Mgl}{I_1'} \, , \\[2mm] c = \dfrac{L_z}{I_1'} \, , \\[2mm] d = \dfrac{L_3}{I_1'} \, . \end{cases} \tag{10.107}$$

Equations (10.101) and (10.102) are rewritten, using (10.104) and (10.107),

$$\dot{\alpha} = \frac{L_z - L_3 u}{I_1'(1 - u^2)} = \frac{c - du}{1 - u^2}, \tag{10.108}$$

$$\dot{\gamma} = \omega_3 - \frac{L_z - L_3 u}{I_1'(1 - u^2)} u = \omega_3 - \frac{(c - du)u}{1 - u^2}. \tag{10.109}$$

Performing the separation of variables, (10.106) becomes

$$dt = \frac{du}{\sqrt{(a - bu)(1 - u^2) - (c - du)^2}}. \qquad (10.110)$$

Integrating the left side with respect to t and the right side with respect to u, we obtain

$$t = \int \frac{1}{\sqrt{(a - bu)(1 - u^2) - (c - du)^2}} du = \int \frac{1}{\sqrt{f(u)}} du . \qquad (10.111)$$

Without solving the elliptic integral (10.111), we can discuss the rotational motion of a top by considering the behavior of the function $f(u)$. In addition to Lagrange tops, a symmetric top free from the Earth's gravity, a gyroscope, is considered. In the following discussion, we will assume that $\omega_3 > 0$, namely a top rotates anticlockwise about the x_3-axis. The range of an independent variable and the sign of four constants in $f(u)$ for three kinds of tops are shown in Table 10.2.

Table 10.2: The range of $u, a, b, c,$ and d for three kinds of tops when $\omega_3 > 0$.

	u	a	b	c	d
a gyroscope	$-1 \leq u \leq 1$	> 0	$= 0$	> 0	> 0
a top on a level surface	$0 < u \leq 1$	> 0	> 0	> 0	> 0
a top on a point support	$-1 < u \leq 1$	> 0	> 0	> 0	> 0

1. A gyroscope

For the case of a gyroscope, b in (10.106) is zero so that $f(u) = 0$ becomes the quadratic equation. The roots of the equation are classified to three cases: (1) two imaginary roots, (2) an equal root, and (3) two real roots. Take care that the motion is permitted when $f(u) \geq 0$ and $-1 \leq u \leq 1$. The maximum value of $f(u)$ is $f(u) = a - ac^2/(a + d^2)$ at $u - cd/(a + d^2)$.

(1) *The case of two imaginary roots*
If $a < c^2 - d^2$, $f(u) = 0$ has two imaginary roots, and any precessions and nutations do not occur.

(2) *The case of an equal root*
If $a = c^2 - d^2$, $f(u) = 0$ has an equal root, $u = cd/(a + d^2) = d/c$. The

nodding stops at colatitude

$$\beta = \cos^{-1}\left(\frac{d}{c}\right). \tag{10.112}$$

Substituting $u = d/c$ into (10.108), we find

$$\dot{\alpha} = \frac{c - d^2/c}{1 - d^2/c^2} = c = \frac{I_z}{I_1}\omega_z. \tag{10.113}$$

Substituting $u = d/c$ into (10.109), we get

$$\dot{\gamma} = \omega_3 - \frac{(c - d^2/c)d/c}{1 - d^2/c^2} = \omega_3 - d = \frac{I_1 - I_3}{I_1}\omega_3. \tag{10.114}$$

The rotating axis x_3 precesses at angular velocity $I_z\omega_z/I_1$ keeping the angle $\beta = \cos^{-1}(d/c)$ to the x_z-axis. This precession is referred to as the *regular precession*

(3) *The case of two real roots*

If $a > c^2 - d^2$, (10.106) has two real roots u_1 and u_2 ($u_1 < u_2$).

$$u_1 = \frac{1}{a + d^2}\left\{cd - \sqrt{a^2 + a(d^2 - c^2)}\right\}, \tag{10.115}$$

$$u_2 = \frac{1}{a + d^2}\left\{cd + \sqrt{a^2 + a(d^2 - c^2)}\right\}, \tag{10.116}$$

As the value of $f(u)$ are negative at $u = -1$ and $u = 1$, u_1 and u_2 are in the range $[-1, 1]$ (see Fig. 10.9). We will define u' as

$$u' = \frac{c}{d} = \frac{L_z}{L_3} = \cos\beta', \tag{10.117}$$

The meaning of β' is the initial tilt of the rotating x_3-axis of a top from the vertical z-axis.

(a) When $u' < u_1$,

$$\dot{\alpha} = \frac{c - du}{1 - u^2} = \frac{d(u' - u)}{1 - u^2} < 0.$$

The rotating axis x_3 always precesses to the negative α-direction nodding between colatitudes β_1 and β_2. Suppose a polar spherical coordinate system of radius r, longitude α, and colatitude β centering at the center of mass of the gyroscope. We will discuss the nutation of the gyroscope by tracing the locus

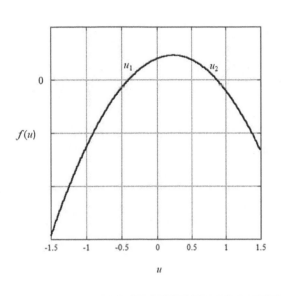

Figure 10.9: Dependence of $f(u)$ on u for the case of a gyroscope.

of the x_3-axis intersecting with the spherical surface of the spherical polar co-ordinates. The nutation is the combination of the nodding and the precession to the negative α-direction. The direction of the precession is negative every-where, so that the locus is tangent at β_1 and β_2 and likes sinusoidal curve as is shown in Fig. 10.10(a).

(b) When $u' = u_1$,

$$\dot{\alpha} = \frac{d(u' - u)}{1 - u^2} \begin{cases} = 0, & \text{at} \quad u = u_1 , \\ < 0, & \text{at} \quad u_1 < u \le u_2 . \end{cases}$$

The rotating axis x_3 precesses in the negative α-direction everywhere except at colatitude β_1 where it stops for a moment. While it nods between colat-itude β_1 and β_2. The locus drawn by the x_3-axis on the spherical surface makes cusps at β_1, and is tangent at β_2 (Fig. 10.10(b)).

(c) When $u_1 \le u' \le u_2$,

$$\dot{\alpha} = \frac{d(u' - u)}{1 - u^2} \begin{cases} > 0, & \text{at} \quad u_1 < u < u' , \\ = 0, & \text{at} \quad u = u' , \\ < 0, & \text{at} \quad u' < u < u_2 . \end{cases}$$

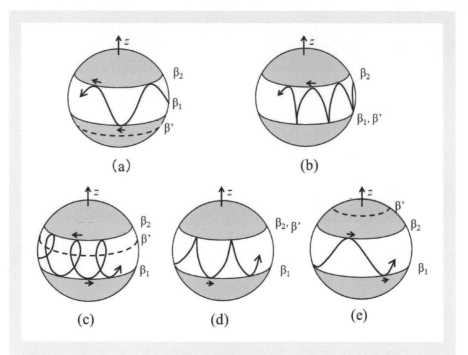

Figure 10.10: The possible locus shape drawn by the rotating axis on the spherical surface.

The rotating axis x_3, precesses to the positive α-direction between β_1 and β', stops at β' for a moment, and precesses to the negative α-direction between β' and β_2. While the x_3-axis nods up and down between β_1 and β_2. So, the locus of the x_3-axis makes loops as is shown in Fig. 10.10(c).

(d) When $u' = u_2$,

$$\dot{\alpha} = \frac{d(u' - u)}{1 - u^2} \begin{cases} > 0, & \text{at} \quad u_1 \leq u < u_2 , \\ = 0, & \text{at} \quad u = u_2 . \end{cases}$$

The rotating axis x_3 precesses to the positive α-direction everywhere except at colatitude β_2 where it stops for a moment. While it nods between colatitudes β_1 and β_2. Then the locus drawn by the x_3-axis on the spherical surface makes cusps at β_2, and is tangent at β_1 as is shown in Fig. 10.10(d).

(e) When $u' > u_2$,

$$\dot{\alpha} = \frac{c - du}{1 - u^2} = \frac{d(u' - u)}{1 - u^2} > 0.$$

The rotating axis x_3 always precesses to the positive α-direction nodding between colatitude β_1 and β_2. Then, the locus on the spherical surface is tangent at β_1 and β_2, and likes a sinusoidal curve as is shown in Fig. 10.10(e).

2. A top on a point support

For the case of a top on a point support, the rotational motion is permitted under the condition $-1 < u \leq 1$ and $f(u) \geq 0$. The roots of the cubic equation $f(u) = 0$ are classified to three cases: (1) two imaginary roots and a real root, (2) an equal root and a real root, and (3) three real roots.

(1) *The case of three real roots*
The behavior of $f(u)$ is shown in Fig. 10.11 when $f(u) = 0$ has tree real roots, u_1, u_2, and u_3 $(u_1 < u_2 < u_3)$.
Because

$$f(-1) = -(c+d)^2 < 0,$$
$$f(1) = -(c-d)^2 < 0,$$

it is obvious that u_1 and u_2 are in the range $[-1, 1]$ and $u_3 > 1$.

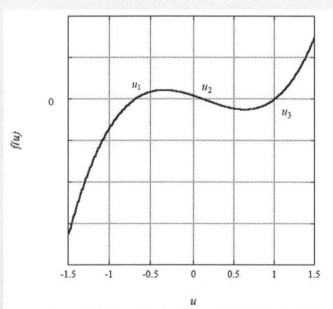

Figure 10.11: Dependence of $f(u)$ on u for the case that the cubic equation $f(u) = 0$ has three real roots.

(a) When $u' < u_1$,

$$\dot{\alpha} = \frac{d(u' - u)}{1 - u^2} < 0.$$

The rotating axis x_3 always precesses in the negative α-direction nodding between colatitude β_1 and β_2. Then the locus drawn by the x_3-axis on the spherical surface is tangent at β_1 and β_2, and likes a sinusoidal curve as is shown in Fig. 10.10(a).

(b) When $u' = u_1$,

$$\dot{\alpha} = \frac{d(u' - u)}{1 - u^2} \begin{cases} = 0, & \text{at} \quad u = u_1, \\ < 0, & \text{at} \quad u_1 < u \le u_2. \end{cases}$$

The rotating axis x_3 precesses in the negative α-direction everywhere except at colatitude β_1 where it stops for a moment. While it nods between colatitude β_1 and β_2. The locus drawn by the x_3-axis on the spherical surface makes cusps at β_1, and is tangent at β_2 (Fig. 10.10(b)).

(c) When $u_1 < u' < u_2$,

$$\dot{\alpha} = \frac{d(u' - u)}{1 - u^2} \begin{cases} > 0, & \text{at} \quad u_1 \le u < u', \\ = 0, & \text{at} \quad u = u', \\ < 0, & \text{at} \quad u' < u \le u_2. \end{cases}$$

The rotating axis x_3 precesses in the positive α-direction between colatitude β_1 and β', stops at β' for a moment, and precesses in the negative α-direction between colatitude β_1 and β'. While it nods between colatitude β_1 and β_2. The locus drawn by the x_3-axis on the spherical surface must have loops, and is tangent at colatitude β_1 and β_2 as is shown in Fig. 10.10(c).

(d) When $u' = u_2$,

$$\dot{\alpha} = \frac{d(u' - u)}{1 - u^2} \begin{cases} > 0, & \text{at} \quad u_1 \le u < u_2, \\ = 0, & \text{at} \quad u = u_2. \end{cases}$$

The rotating axis x_3 precesses in the positive α-direction everywhere except at colatitude β_2 where it stops for a moment. While it nods between colatitude β_1 and β_2. Then the locus drawn by the x_3-axis on the spherical surface makes cusps at colatitude β_2, and is tangent at β_1 as is shown in Fig. 10.10(d).

(e) When $u' > u_2$

$$\dot{\alpha} = \frac{d(u' - u)}{1 - u^2} > 0.$$

The rotating axis x_3 always precesses in the positive α-direction nodding between colatitude β_1 and β_2. So, the locus drawn by the x_3-axis on the spherical surface is tangent at colatitude β_1 and β_2, and likes a sinusoidal curve as is shown in Fig. 10.10(e).

(2) *The case of an equal root and a real root*
In this case, nodding does not occur because $\dot{\beta}$ is zero only at colatitude $\beta_1 = \cos^{-1} u_1$. The conditions that $f(u) = 0$ has an equal root are

$$f(u) = \frac{df(u)}{du} = 0, \quad \text{at} \quad u = u_1,$$

which become

$$a - bu_1 = \frac{(c - du_1)^2}{1 - u_1{}^2}, \tag{10.118}$$

and

$$b(1 - u_1{}^2) + 2u_1 \frac{(c - du_1)^2}{1 - u_1{}^2} - 2d(c - du_1) = 0.$$

Substituting (10.108) into the above equation, we find

$$2u_1 \dot{\alpha}_1^2 - 2d\dot{\alpha}_1 + b = 0. \tag{10.119}$$

The roots of (10.119) are

$$\dot{\alpha}_1 = \frac{1}{2u_1} \left\{ d \pm \sqrt{d^2 - 2bu_1} \right\}. \tag{10.120}$$

The condition that (10.120) has real roots is

$$d^2 - 2bu_1 = \left(\frac{L_3}{I_1'} \right)^2 - \frac{4Mgl}{I_1'} \cos \beta_1 \geq 0,$$

$$\omega_3 \geq \left[\frac{4I_1'Mgl}{I_3{}^2} \cos \beta_1 \right]^{1/2}. \tag{10.121}$$

Equation (10.121) is the condition that the regular precession is able to occur. If $\pi/2 \leq \beta_1 < \pi$, the precession occurs for non-zero value of ω_3. While if $0 \leq \beta_1 < \pi/2$, the precession occurs when ω_3 is larger than the critical value given by (10.121).
We will consider the high-speed rotation about the x_3-axis, namely

$$\omega_3 \gg \left[\frac{4I_1'Mgl}{I_3{}^2} \cos \beta_1 \right]^{1/2}. \tag{10.122}$$

Using (10.107), (10.122) is rewritten as

$$d^2 \gg 2bu_1$$

Expanding (10.120) in a Taylor series retaining a term of first order of small quantity, we obtain

$$\dot{\alpha}_1 \cong \frac{d}{2u_1}\left\{1 \pm \left(1 - \frac{bu_1}{d^2}\right)\right\}.$$

Thus, there are two kinds of precessions, the fast precession and the slow precession.

$$\dot{\alpha}_1 = \begin{cases} \dfrac{d}{u_1} = \dfrac{I_3\omega_3}{I_1{}'\cos\beta_1}, & \text{fast precession}, \\[2mm] \dfrac{b}{2d} = \dfrac{Mgl}{I_3\omega_3}, & \text{slow precession}. \end{cases} \qquad (10.123)$$

3. A top on a level surface

The rotational motion of a top on a level surface is permitted under the conditions that $0 < u \le 1$ and $f(u) \ge 0$.
Because

$$f(0) = a - c^2 > 0$$

usually, $f(u) = 0$ has one real root in the range $[0, 1]$. Therefore, only the slow and fast regular precession of (10.123) are possible for a top on a level surface.

Example 8. The slow precession of a top on a level surface is easily derived from the angular momentum equation (9.6).

Answer
We will consider the precession of a top whose rotational axis has the angle θ from the vertical axis. Let the moment of inertia of the top be I, the mass be M, the angular velocity be ω, the acceleration due to gravity be $\mathbf{g}$, the center of mass be G, the contact point of the axis and the level surface be P and the position vector from P to G be $\mathbf{l}$. The angular momentum of the top is

$$\mathbf{L} = I\boldsymbol{\omega}. \qquad (10.124)$$

The magnitude of the tangential torque of the precession becomes

$$\mathbf{N} = \mathbf{l} \times M\mathbf{g},$$
$$N = Mgl\sin\theta. \qquad (10.125)$$

The torque $\mathbf{N}$ obeys the right-hand rule together with $\mathbf{d}$ and $\mathbf{g}$, and is perpendicular to the angular momentum $\mathbf{L}$, so that $\mathbf{N}$ does not change the magnitude of $\mathbf{L}$ but only changes its direction. Suppose that the head of $\mathbf{L}$ rotates at the constant angular velocity $\boldsymbol{\Omega}$. Letting the increment of the angle be $\Delta\phi$ and the increment of $\mathbf{L}$ be $\Delta\mathbf{L}$ in a small-time increment Δt (Fig. 10.12), the magnitude of $\Delta\mathbf{L}$ becomes

$$\Delta L = L\sin\theta\Delta\phi. \qquad (10.126)$$

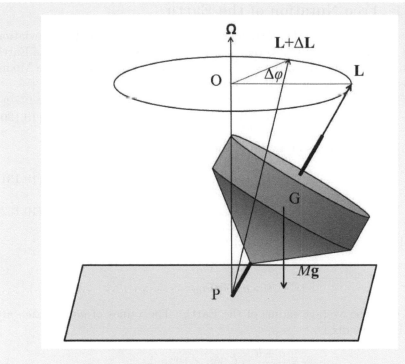

Figure 10.12: Precession of a top.

Dividing (10.126) through by Δt and in the limit $\Delta t \to 0$,

$$\lim_{\Delta t \to 0} \frac{\Delta L}{\Delta t} = L \sin \theta \lim_{\Delta t \to 0} \frac{\Delta \phi}{\Delta t},$$

$$\frac{dL}{dt} = L \sin \theta \Omega = N. \tag{10.127}$$

Substituting (10.125) into (10.127) and using (10.124) yields

$$\Omega = \frac{Mgl}{L} = \frac{Mgl}{I\omega}. \tag{10.128}$$

We obtain the period of the precession of the top,

$$T = \frac{2\pi}{\Omega} = \frac{2\pi I \omega}{Mgl}. \tag{10.129}$$

10.5   Free Nutation and Precession of the Earth

In this section, we will discuss the free nutation and the precession of the Earth using the results of the previous section.

10.5.1   Free Nutation of the Earth

The Earth is an *oblate spheroid* bulging around the equator, but the deviation from a sphere is very small so that we will discuss free rotation of the Earth ignoring the torque due to the gravitational force by the Sun and the Moon. Suppose a spheroid of mass M defined by

$$\frac{x_1{}^2}{a^2} + \frac{x_2{}^2}{a^2} + \frac{x_3{}^2}{b^2} = 1. \tag{10.130}$$

The principal moments of inertia in the x_1-, x_2-, x_3-components are

$$I_1 = I_2 = \frac{1}{5}M(a^2 + b^2), \tag{10.131}$$

$$I_3 = \frac{2}{5}Ma^2. \tag{10.132}$$

We will apply (10.131) and (10.132) to the Earth,

$$a = a_0 + \Delta a_0,$$
$$b = a_0 - \Delta a_0,$$

where a_0 is the average radius of the Earth. The values of a_0 and Δa_0 are found in Appendix D,

$$a_0 = \frac{a+b}{2} = 6.368 \times 10^6 \,[\mathrm{m}],$$

$$\Delta a_0 = \frac{a-b}{2} = 1.05 \times 10^4 \,[\mathrm{m}].$$

Supposing that the Earth is the rigid body of uniform density, the principal moments of inertia are

$$I_1 = I_2 = \frac{1}{5}M\{(a_0 + \Delta a_0)^2 + (a_0 - \Delta a_0)^2\} \cong \frac{2}{5}Ma_0{}^2 = I_0, \tag{10.133}$$

$$I_3 = \frac{2}{5}M(a_0 + \Delta a_0)^2 \cong I_0\left(1 + \frac{2\Delta a_0}{a_0}\right) = I_0 + \Delta I_0, \tag{10.134}$$

$$\frac{I_3 - I_1}{I_1} = \frac{\Delta I_0}{I_0} = \frac{2\Delta a_0}{a_0} = 3.30 \times 10^{-3}. \tag{10.135}$$

Applying (10.135) to (10.83), we can get the period of free nutation of the Earth.

$$T = \frac{2\pi}{\Omega'} = \frac{I_0}{\Delta I_0}\frac{2\pi}{\Omega} = \frac{a_0}{2\Delta a_0}\frac{2\pi}{\Omega} = 303\,[\mathrm{days}]. \tag{10.136}$$

The observed period of the Earth's nutation is 433 days.[2] The difference between the observed period and the calculated one may be due to the inhomogeneous density distribution and nonrigidity of the Earth. Thus, the fractional

---

[2]This nutation is known as the *Chandler wobble* which was found by American astronomer Seth Carlo Chandler in 1891.

difference of the principal moment of inertia of the Earth is obtained as

$$\frac{\Delta I_0}{I_0} = \frac{1}{433} = 2.31 \times 10^{-3} \tag{10.137}$$

due to the observed value. In the next section, we will calculate the period of the precession of the Earth according to the value of (10.137).

10.5.2 Precession of the Earth

As is discussed in the previous subsection, the Earth is an oblate spheroid bulging around the equator. Further, the obliquity of the Earth is 23.4° at present,[3] so that the gravitational forces due to the Sun and the Moon exert on the Earth's bulge to make its axis vertical to the ecliptic plane.

(1) Precession due to Solar Torque

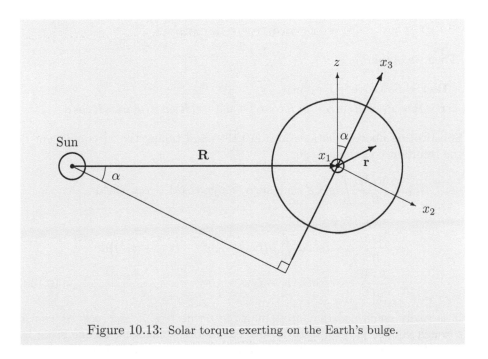

Figure 10.13: Solar torque exerting on the Earth's bulge.

We will discuss the precession due to solar torque. As shown in Fig. 10.13, the principal axes of the Earth are x_1, x_2, x_3, $\mathbf{R}$ is the position vector from the center of the Sun to the center of the Earth, $\mathbf{r}$ is the position vector from

---

[3]The obliquity of the Earth's axis varies between 22.18° and 24.5° at the period of 41,000 years.

the center of the Earth to an arbitrary point of the Earth's interior, z is the axis vertical to the ecliptic plane and α is the obliquity of the Earth's axis. The gravitational attraction of the Sun exerting on the volume element ΔV at $\mathbf{r}$ is given by

$$\Delta \mathbf{F} = -\frac{GM\rho\Delta V}{|\mathbf{R}+\mathbf{r}|^3}(\mathbf{R}+\mathbf{r}) = -\frac{GM\rho\Delta V}{R^3}\left(1 - \frac{3\mathbf{R}\cdot\mathbf{r}}{R^2}\right)(\mathbf{R}+\mathbf{r}), \quad (10.138)$$

where M is the mass of the Sun, ρ is the density at $\mathbf{r}$. The leading term of the right-hand side of (10.138) is the gravitational attraction for the Earth to orbit around the Sun, so that the torque of the Sun exerting on the Earth is

$$\mathbf{N}_S \cong \frac{3GM}{R^5}\int_V \rho(\mathbf{R}\cdot\mathbf{r})(\mathbf{r}\times\mathbf{R})dv. \quad (10.139)$$

From Fig. 10.14, we find

$$\mathbf{R} = R\cos\alpha\,\mathbf{e}_2 + R\sin\alpha\,\mathbf{e}_3,$$
$$\mathbf{r} = x_1\mathbf{e}_1 + x_2\mathbf{e}_2 + x_3\mathbf{e}_3.$$

Then we obtain

$$\mathbf{R}\cdot\mathbf{r} = Rx_2\cos\alpha + Rx_3\sin\alpha,$$
$$\mathbf{r}\times\mathbf{R} = \mathbf{e}_1(x_2R\sin\alpha - x_3R\cos\alpha) + \mathbf{e}_2(-x_1R\sin\alpha) + \mathbf{e}_3 x_1 R\cos\alpha.$$

Substituting above equations into (10.139) and taking the x_1-component of the resultant equation, we get

$$\begin{aligned}
N_{S1} &= \frac{3GM}{R^3}\int_V \rho(x_2{}^2\cos\alpha\sin\alpha - x_2 x_3\cos^2\alpha + x_2 x_3\sin^2\alpha \\
&\qquad - x_3{}^2\cos\alpha\sin\alpha)dv \\
&= \frac{3GM}{R^3}\cos\alpha\sin\alpha\int_V \rho\{(x_1{}^2 + x_2{}^2) - (x_3{}^2 + x_1{}^2)\}dv \\
&= \frac{3GM}{R^3}\cos\alpha\sin\alpha(I_3 - I_2). \quad (10.140)
\end{aligned}$$

It is easily confirmed that cross-multiple terms (e.g., $-x_2 x_3\cos^2\alpha$) vanish through volume integral in the above calculation. The solar torque given by (10.140) is maximum when the Earth is at a solstice (point A in Fig. 10.14). We should recalculate when the Earth is at position P and take the average over the Earth's orbit. For simplicity, we suppose that the Earth's orbit is a circle of radius R, the position of the Sun is O and the $\angle$AOP $= \theta$. Then vector $\mathbf{R}$ becomes

$$\mathbf{R} = R\sin\theta\,\mathbf{e}_1 + R\cos\theta\cos\alpha\,\mathbf{e}_2 + R\cos\theta\sin\alpha\,\mathbf{e}_3,$$

which yields

$$\mathbf{R} \cdot \mathbf{r} = R(x_1 \sin\theta + x_2 \cos\theta \cos\alpha + x_3 \cos\theta \sin\alpha),$$

$$\mathbf{r} \times \mathbf{R} = \mathbf{e}_1 R(x_2 \cos\theta \sin\alpha - x_3 \cos\theta \sin\alpha) + \mathbf{e}_2 R(x_3 \sin\theta - x_1 \cos\theta \sin\alpha)$$
$$+ \mathbf{e}_3 R(x_1 \cos\theta \cos\alpha - x_2 \sin\theta).$$

Substituting above equations into (10.139) and taking the x_1-component of the resultant equation, we get

$$\begin{aligned}
N_{S1} &= \frac{3GM}{R^3} \int_V \rho(x_1 x_2 \sin\theta \cos\theta \sin\alpha - x_1 x_3 \sin\theta \cos\theta \cos\alpha \\
&\quad + x_2{}^2 \cos^2\theta \sin\alpha \cos\alpha - x_2 x_3 \cos^2\theta \cos^2\alpha \\
&\quad + x_2 x_3 \cos^2\theta \sin^2\alpha - x_3{}^2 \cos^2\theta \sin\alpha \cos\alpha) dv \\
&= \frac{3GM}{R^3} \cos\alpha \sin\alpha \cos^2\theta \int_V \rho\{(x_1{}^2 + x_2{}^2) - (x_3{}^2 + x_1{}^2)\} dv \\
&= \frac{3GM}{R^3} \cos\alpha \sin\alpha \cos^2\theta (I_3 - I_2).
\end{aligned} \tag{10.141}$$

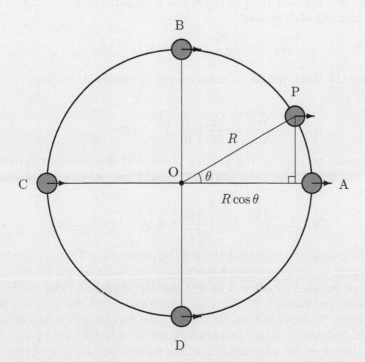

Figure 10.14: Torque due to the Sun at arbitrary positions along the Earth's orbit. Small arrows show the component parallel to the ecliptic plane of the angular velocity of the Earth's rotation. Plan view from the polar star.

Taking the average of (10.141) over the Earth's orbit, we obtain

$$
\overline{N}_{S1} = \frac{1}{2\pi} \int_0^{2\pi} N_{S1} d\theta = \frac{3GM}{2\pi R^3} \sin\alpha \cos\alpha (I_3 - I_2) \int_0^{2\pi} \frac{1 + \cos 2\theta}{2} d\theta
$$
$$
= \frac{3GM}{2R^3} \cos\alpha \sin\alpha (I_3 - I_2). \tag{10.142}
$$

Replacing $\Omega \to \Omega_{\rm SP}$, $L \to I_3\omega_3$, $\theta \to \alpha$, $N \to \overline{N}_{S1}$ in (10.127), we get the equation for the angular velocity of the precession due to solar torque.

$$
I_3\omega_3 \Omega_{\rm SP} = \frac{3GM}{2R^3} \cos\alpha (I_3 - I_2),
$$
$$
\Omega_{\rm SP} = \frac{3GM}{2R^3\omega_3} \cos\alpha \frac{I_3 - I_2}{I_3} = \frac{3GM}{2R^3\omega_3} \cos\alpha \frac{\Delta I_0}{I_0}. \tag{10.143}
$$

Substituting the value of (10.37) and solar values in Appendix C into (10.143), we find that $\Omega_{\rm SP} = 1.72 \times 10^{-12}\,{\rm s}^{-1}$. Thus, the period of the precession due to solar torque becomes $T_{\rm SP} = 1.16 \times 10^5$ years.

(2) Precession due to Lunar Torque

Next, we will consider the precession due to lunar torque. Substituting lunar values into (10.142), we find

$$
\overline{N}_{L1} = \frac{3GM_{\rm L}}{2R_{\rm L}^{\,3}} \cos\alpha \sin\alpha (I_3 - I_2). \tag{10.144}
$$

The ratio of lunar torque to solar torque is obtained dividing (10.144) by (10.142)

$$
\frac{\overline{N}_{L1}}{\overline{N}_{S1}} = \frac{M_{\rm L}}{M}\left(\frac{R}{R_{\rm L}}\right)^3 = 2.23. \tag{10.145}
$$

The angular velocity of the Earth's precession due to lunar torque is obtained substituting the value of (10.137) and lunar values in Appendix C into (10.143).

$$
\Omega_{\rm LP} = \frac{3GM_{\rm L}}{2R_{\rm L}^{\,3}\omega_3} \cos\alpha \frac{\Delta I_0}{I_0} = 3.83 \times 10^{-12}\,[{\rm s}^{-1}]. \tag{10.146}
$$

Then the period of the precession due to lunar torque is $T_{\rm SP} = 5.20 \times 10^4$ years. Nonlinear interaction between lunar torque and solar torque produces several periods of precession, referred to as *lunisolar precession*. One of the periods of lunisolar precession is 21,200 years, which is close to the observed period of 25,772 years. In 1920, Milutin Milanković, the Serbian physicist, proposed the so-called *Milanković theory* that glacial and interglacial ages repeat according to the change of the solar insolation due to the combination of three orbital elements of the Earth.[4]

---

[4]The first element is the precession, the second element is the changing of the *eccentricity* at periods of 95,000, 125,000, and 400,000 years, and the third element is the oscillation of the obliquity at a period of 41,000 years.

10.6 Problems

1. Suppose an annular cylinder of mass M, inner radius a, outer radius b and height d. Obtain the moment of inertia about the axis passing through the center of the rigid body and vertical to the circular plate.

2. Suppose an annular cylinder of mass M, inner radius a, outer radius b and height d. Obtain the moment of inertia about the axis passing through the center of the rigid body and vertical to the side.

3. Suppose a thin spherical shell of mass M and radius a. Obtain the moment of inertia about the axis passing through the center of the spherical shell.

4. Suppose a hollow sphere of mass M, inner radius a and outer radius b. Obtain the moment of inertia about the axis passing through the center of the hollow sphere. Confirm that the result coincides with that of the previous problem in the limit $b \to a$.

5. Suppose that a rigid sphere of radius b is place on a level surface and a rigid ball of radius a ($a \ll b$) is put on the top of the sphere T. At some moment, the ball begins to roll down from rest without sliding and departs from the sphere at point P. Letting the center of the sphere be O, obtain the angle $\angle$TOP.

6. When the Earth was born by collisions and mergers of planetoids, it was a sphere of uniform density, mass 5.974×10^{24} kg and radius 6.369×10^6 m. During 10^8 years since the Earth's birth, the gravitational differentiation occurred so that the heavier material subsided toward the center of the Earth and the lighter material rose up toward the Earth's surface. Thus, the Earth consists of the two-layer structure; the inner sphere is the core of radius 3.480×10^6 m and density 1.200×10^4 kg m^{-3} and the outer sphere is the mantle of density 4.256×10^3 kg m^{-3}.

   (1) Calculate the moment of inertia about the Earth's axis at the Earth's birth.

   (2) Calculate the moment of inertia about the Earth's axis after the gravitational differentiation[5]. Suppose that the Earth's radius did not change before and after the gravitational differentiation.

   (3) Obtain the fractional change of the angular velocity of the Earth's rotation before and after the gravitational differentiation.

7. Suppose an elliptic plate of mass M, semimajor axis a, semiminor axis b and thickness d. Calculate the moment of inertia of the plate about the axis passing through the center and vertical to the plate.

---

[5] The structure and the density distribution are simplified from Shimadu 1967.

8. Suppose that a cylinder of mass M, radius a and length l is rotating about a fixed axis passing through the center of the cylinder and parallel to the side. What is the minimum force vertical to the axis to ride across the difference $h(< a)$ between the two-level surfaces?

9. Suppose that a massless and inextensible string is rolled around a fixed pulley of radius a, mass M, and moment of inertia $I = Ma^2/2$. Attaching bob 1 of mass m_1 and bob 2 of mass m_2 $(m_1 > m_2)$ to both ends of the string, the bobs are kept at rest. At time $t = 0$ the bobs are released from rest, obtain the falling speed of bob 1 at time t using the equation of motion for the two bobs and the angular momentum equation for the fixed pulley. Suppose that there is no slip between the string and the fixed pulley, and let the magnitude of the acceleration due to gravity be g.

10. In Cartesian coordinates, three particles of mass m are set at (1,1,0), (−1,1,0), and (0,−1,1). They are connected by massless rigid rods with each other. Calculate the inertia tensor of the system.

11. In the previous problem, obtain the basis vectors of the principal axes and the principal moments of inertia.

12. Obtain the principal moments of inertia of a spheroid of mass M with the shape defined by

$$\frac{x^2}{a^2} + \frac{y^2}{a^2} + \frac{z^2}{b^2} \leq 1.$$

10.7 Reference

1. Shimadu Y.: *Evolution of the Earth*, Iwanami Shoten Publishers, Tokyo (1967), *in Japanese*.

Chapter 11

Orbital Motion of Planets

Johannes Kepler found three laws of the orbital motion of planets based on the enormous observational data of Tycho Brahe. Isaac Newton established the basis of classical mechanics and found the law of universal gravitation in which Kepler's three laws played a very important role. In this chapter, we will review the process classical mechanics was established and prove Kepler's three laws exactly. Further, we will discuss the universal gravitation exerted by bodies of finite extent, and the oceanic tides and tidal effects on the Earth–Moon system. We will discuss the general orbits due to a central force. As an application of the orbital motion, Rutherford scattering is discussed precisely.

11.1   The Law of Universal Gravitation

Newton considered that an apple falls to the ground owing to some attractive force exerted by the Earth and the same force exerts on the Moon. Orbital motion of the Moon around the Earth is the same that the Moon is continuously falling to the Earth. Newton calculated the falling distance of the Moon in 1 s, from which he obtained the acceleration due to gravity at the center of the Moon. He found that the ratio of the magnitude of the acceleration due to gravity at the center of the Moon to that of the Earth's surface is almost equal to the square ratio of the Earth's radius to the distance between the Earth and the Moon. We will review the calculation performed by Newton. Here we will use the following physical quantities,

$r = 3.84 \times 10^8$ m: the distance from the Earth to the Moon,

$a = 6.37 \times 10^6$ m: the radius of the Earth,

$T = 2.36 \times 10^6$ s: the orbital period of the Moon.

At first, we will calculate the distance that the Moon falls to the Earth in 1 s. Let the center of the Earth be O, the position of the Moon at time $t(s)$ be A and the position of the Moon at $t + 1(s)$ be B. Suppose that the Moon would move to point C at $t + 1(s)$ if it traveled at the constant velocity (Fig. 11.1).

DOI: 10.1201/9781003310068-11

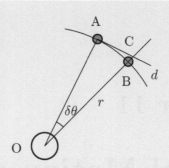

Figure 11.1: The fall of the Moon to the Earth.

Let $\angle AOB = \delta\theta$, and we will regard that a line segment $\overline{AC}$ is nearly equal to an arc $\overset{\frown}{AB}$. As $\triangle OAC$ is a right-angled triangle,

$$(r + d)^2 = r^2 + (r\delta\theta)^2,$$

$$d = r\left\{\sqrt{1 + \delta\theta^2} - 1\right\} \simeq \frac{1}{2}r(\delta\theta)^2. \tag{11.1}$$

holds due to the *Pythagorean theorem*. Next, we will obtain $\delta\theta$. The angular velocity of the orbital motion of the moon is

$$\omega = \frac{2\pi}{2.36 \times 10^6} = 2.66 \times 10^{-6}\,[\text{s}^{-1}].$$

Therefore,

$$\delta\theta = \omega \times 1.00 = 2.66 \times 10^{-6}. \tag{11.2}$$

Substituting (11.2) into (11.1), we find

$$d = \frac{1}{2}r(\delta\theta)^2 = 1.36 \times 10^{-3}\,[\text{m}].$$

Letting the magnitude of the acceleration due to gravity at the center of the moon be g_M, the falling distance of the moon toward the Earth during $1\,\text{s}$ is

$$d = \frac{1}{2}g_\text{M} \times 1.00^2\,[\text{m}].$$

We obtain

$$g_\text{M} = \frac{2d}{1.00^2} = 2.72 \times 10^{-3}\,[\text{m s}^{-2}].$$

The ratio of the value of the acceleration due to gravity at the center of the moon to that at the Earth's surface is

$$\frac{g_\text{M}}{g_0} = \frac{2.72 \times 10^{-3}}{9.80} = 2.78 \times 10^{-4}.$$

The square ratio of the Earth's radius to the distance from the Earth to the moon is

$$\left(\frac{a}{r}\right)^2 = \left(\frac{6.37 \times 10^6}{3.84 \times 10^8}\right)^2 = 2.75 \times 10^{-4}.$$

Thus we find that both values are almost equal. Before the above calculation, Newton showed that there exists an attractive force between the Sun and a planet which is inversely proportional to the square of the orbital radius, supposing a circular orbit, using Newton's second law, and Kepler's third law. Newton considered that the force acts not only between two celestial bodies but also between any two bodies in the universe, and he proposed the law of universal gravitation that an attractive force exerts between two bodies proportional to the product of their masses and inversely proportional to the square of the distance between them. The gravitational force that body A of mass M exerts on body B of mass m is

$$\mathbf{F} = -G\frac{Mm}{r^2}\left(\frac{\mathbf{r}}{r}\right), \tag{11.3}$$

where

$$G = (6.67259 \pm 0.00030) \times 10^{-11}\,[\mathrm{m^3\,kg^{-1}\,s^{-2}}]$$

is the gravitational constant and $\mathbf{r}$ is the position vector from body A to body B (Fig. 11.2).

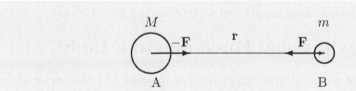

Figure 11.2: Universal gravitation.

The First Cosmic Speed
An artificial satellite of mass m is orbiting around the Earth at a constant speed v just above the Earth's surface. Ignoring the air resistance and letting the Earth's radius be a, the equation of motion for the artificial satellite is

$$m\frac{v^2}{a} = mg_0,$$

$$v = \sqrt{ag_0} = 7.89 \times 10^3\,[\mathrm{m\,s^{-1}}], \tag{11.4}$$

where g_0 is the magnitude of the acceleration due to gravity at the Earth's surface. The speed given by (11.4) is called the *first cosmic speed*. The orbital period of the artificial satellite is

$$T = \frac{2\pi a}{v} = 2\pi \sqrt{\frac{a}{g_0}} = 5.06 \times 10^3 \, [\text{s}].$$

Example 1. An artificial satellite which is orbiting around the Earth in the equatorial plane at the same angular velocity as that of the Earth's rotation seems to be at rest from an observer on the Earth, so it is called the *geosynchronous satellite*. Calculate the altitude h of the geosynchronous satellite supposing that the mass of the Earth is 5.97×10^{24} kg, the Earth's radius is 6.37×10^6 m and the gravitational constant is $G = 6.67 \times 10^{-11} \, \text{m}^3 \, \text{kg}^{-1} \, \text{s}^{-2}$.

Answer
The angular velocity of the Earth is

$$\Omega = \frac{2\pi}{24 \times 60 \times 60} = 7.27 \times 10^{-5} \, [\text{s}^{-1}]. \tag{11.5}$$

Letting the radius of the Earth be a, the mass of the Earth be M and the mass of the satellite be m, the equation of motion of the satellite is

$$m(a + h)\Omega^2 = G \frac{mM}{(a + h)^2},$$

$$h = \left(\frac{GM}{\Omega^2} \right)^{1/3} - a = 3.59 \times 10^7 \, [\text{m}]. \tag{11.6}$$

Thus, the altitude of the geosynchronous satellite is 3.59×10^7 m.

11.2   Gravitational Force due to a Body

Newton's law of universal gravitation regards the Sun and planets as a particle (a point mass). This approximation is reasonable because the radii of the Sun and planets are negligibly small compared with the distance between them. However, when we discuss the gravitational force between an apple and the Earth, we cannot regard the Earth as a particle. In this section we will discuss the gravitational force exerted by bodies of finite extent.

Example 2. Suppose a uniform sphere of mass M and radius a. Obtain the potential of the universal gravitation and the gravitational force due to the sphere on a particle of mass m at point Q with the distance $z(> a)$ from the center of the sphere O. Let the gravitational constant be G.

Answer
Let's take coordinates as shown in Fig. 11.3. Taking an arbitrary point $P(r, \phi, \theta)$ on the spherical shell of radius r, and the distance from point P

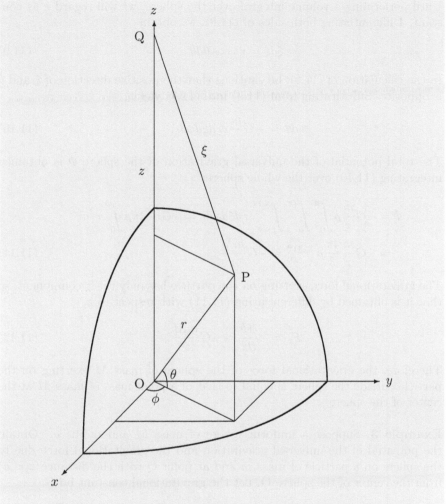

Figure 11.3: The universal gravitation of the uniform sphere.

to point Q as ξ, the potential of the universal gravitation $d\Phi$ of the volume element $dv = r^2 \cos\theta dr d\phi d\theta$ at P on the particle at Q is

$$d\Phi = -G\frac{m}{\xi}\rho r^2 \cos\theta dr d\phi d\theta, \qquad (11.7)$$

where $\rho = 3M/4\pi a^3$ is the density of the sphere. From the *cosine formula*, we find

$$\xi^2 = z^2 + r^2 - 2rz\sin\theta . \qquad (11.8)$$

Until performing a volume integral over the sphere, we will regard r as constant. Differentiating both sides of (11.8), we obtain

$$\xi d\xi = rz \cos\theta d\theta. \tag{11.9}$$

In the calculation of (11.9), be cautious that the positive direction of ξ and θ is opposite. Substituting from (11.9) into (11.7) yields,

$$d\Phi = -G\frac{m}{z}\rho r d\xi d\phi dr. \tag{11.10}$$

The total potential of the universal gravitation of the sphere Φ is obtained integrating (11.10) over the whole sphere,

$$\Phi = -G\frac{m}{z}\rho \int_0^a \int_0^{2\pi} \int_{z-r}^{z+r} r d\xi d\phi dr = -G\frac{m}{z}4\pi\rho \int_0^a r^2 dr$$

$$= -G\frac{m}{z}\frac{4\pi}{3}\rho\left[r^3\right]_0^a = -G\frac{mM}{z}. \tag{11.11}$$

The gravitational force exerting on the particle has only the z-component, so that it is obtained by differentiating (11.11) with respect to z,

$$F_z = -\frac{\partial\Phi}{\partial z} = -G\frac{mM}{z^2}. \tag{11.12}$$

Therefore, the gravitational force of the sphere of mass M exerting on the particle outside the sphere is equal to that of a point mass of mass M at the center of the sphere.

Example 3. Suppose a uniform sphere of mass M and radius a. Obtain the potential of the universal gravitation and the gravitational force due to the sphere on a particle of mass m and at point Q with the distance $z(< a)$ from the center of the sphere O. Let the gravitational constant be G.

Answer
Let's take coordinates as shown in Fig. 11.4. Taking an arbitrary point P(r, ϕ, θ) on the spherical shell of radius r, and the distance from point P to point Q as ξ, the potential of the universal gravitation $d\Phi$ of the volume element $dv = r^2 \cos\theta dr d\phi d\theta$ at point P on the particle at point Q is

$$d\Phi = -G\frac{m}{\xi}\rho r^2 \cos\theta dr d\phi d\theta, \tag{11.13}$$

where $\rho = 3M/4\pi a^3$ is the density of the sphere. Until performing volume integration over the sphere, we will regard r as constant. Substituting from (11.9) into (11.13),

$$d\Phi = -G\frac{m}{z}\rho r d\xi d\phi dr. \tag{11.14}$$

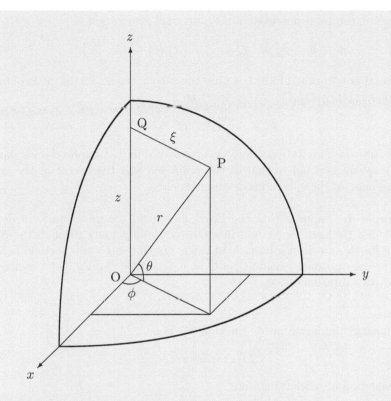

Figure 11.4: The universal gravitation of the uniform sphere exerting on a particle in the sphere.

Separating the total potential of the sphere Φ to the inner part $\Phi_i(r \leq a_i)$ and the outer part $\Phi_o(a_i \leq r \leq a)$, the potential of the inner part is

$$\Phi_i = -G\frac{m}{z}\rho \int_0^{a_i} \int_0^{2\pi} \int_{z-r}^{z+r} r d\xi d\phi dr = -G\frac{m}{z}4\pi\rho \int_0^{a_i} r^2 dr$$

$$= -G\frac{m}{z}\frac{4\pi}{3}\rho \left[r^3\right]_0^{a_i} = -G\frac{mM_i}{z}, \tag{11.15}$$

where M_i is the mass of the inner sphere of radius a_i. Next, the potential of the outer part is

$$\Phi_o = -G\frac{m}{z}\rho \int_{a_i}^a \int_0^{2\pi} \int_{r-z}^{z+r} r d\xi d\phi dr = -Gm4\pi\rho \int_{a_i}^a r dr$$

$$= -Gm4\pi\rho\frac{1}{2}\left[r^2\right]_{a_i}^a = -\frac{3}{2}Gm\left(\frac{M}{a} - \frac{M_i}{a_i}\right). \tag{11.16}$$

Therefore, the total potential of the universal gravitation is

$$\Phi = \Phi_i + \Phi_o = -G\frac{mM_i}{z} - \frac{3}{2}Gm\left(\frac{M}{a} - \frac{M_i}{a_i}\right). \tag{11.17}$$

The total gravitational force has only the z-component, so that it is obtained by differentiating (11.17) with respect to z

$$F_z = -\frac{\partial \Phi}{\partial z} = -G\frac{mM_i}{z^2}. \tag{11.18}$$

Therefore, the gravitational force exerted by a uniform sphere on a particle in the sphere is equal to that of the point mass at the center of the sphere having mass of the sphere inside the particle.

Example 4. Suppose that the Earth is a rigid sphere of uniform density $\rho = 5.52 \times 10^3 \, \mathrm{kg\,m^{-3}}$. Let's drill a straight hole passing through the center of the Earth and drop a particle of mass m. How does the particle behave? Let the gravitational constant be $G = 6.67 \times 10^{-11} \, \mathrm{m^3\,kg^{-1}\,s^{-2}}$ and ignore the Earth's rotation.

Answer
The mass of the Earth inside the radius r is,

$$M(r) = \rho\frac{4}{3}\pi r^3.$$

The equation of motion becomes

$$m\frac{d^2r}{dt^2} = -\frac{Gm}{r^2}\frac{4}{3}\rho\pi r^3. \tag{11.19}$$

Assuming the exponential type solution as

$$r \propto \exp\left(\tilde{\imath}\lambda t\right), \tag{11.20}$$

where $\tilde{\imath}$ is the imaginary unit. Substituting (11.20) into (11.19), we get

$$-\lambda^2 + \frac{4}{3}\rho\pi G = 0,$$

$$\lambda = \pm\sqrt{\frac{4}{3}\rho\pi G} = \pm\omega. \tag{11.21}$$

The general solution of (11.19) is

$$r = \alpha\exp\left(\tilde{\imath}\omega t\right) + \beta\exp\left(-\tilde{\imath}\omega t\right).$$

Taking the real part of r, we get

$$\Re\{r\} = \alpha_r\cos\left(\omega t\right) - \alpha_i\sin\left(\omega t\right) + \beta_r\cos\left(\omega t\right) + \beta_i\sin\left(\omega t\right)$$
$$= A\cos\left(\omega t - \theta_0\right). \tag{11.22}$$

The particle oscillates with the period

$$T = \frac{2\pi}{\omega} = 5.06 \times 10^3 \, [\mathrm{s}].$$

11.3 Universal Gravitation and Gravity

Let the acceleration due to Earth's gravitation at height z be g^*, the equation of motion of a particle of mass m is

$$my^* = G\frac{Mm}{(a+z)^2},$$

$$g^* = \frac{GM}{(a+z)^2},\tag{11.23}$$

where a is the Earth's radius and M is the mass of the Earth. The acceleration due to gravity at the Earth's surface g_0^* is obtained by putting $z = 0$ in (11.23).

$$g_0^* = \frac{GM}{a^2}.\tag{11.24}$$

Using (11.23) and (11.24),

$$g^* = \frac{GM}{(a+z)^2} = \frac{GM}{a^2}\left(1+\frac{z}{a}\right)^{-2} = g_0^*\left(1+\frac{z}{a}\right)^2.\tag{11.25}$$

The Earth is rotating at the angular velocity $\boldsymbol{\Omega}$, so that the centrifugal force exerts on anyone on the Earth. The gravity of the Earth $m\mathbf{g}$ is the vector sum of the gravitational force $m\mathbf{g}^*$ and the centrifugal force $m\Omega^2\mathbf{R}$ ($\mathbf{R}$ is the position vector from the rotational axis to the particle). Therefore, we find

$$\mathbf{g} = \mathbf{g}^* + \Omega^2\mathbf{R}.\tag{11.26}$$

As is clear from Fig. 11.5, the gravity is perpendicular to the spherical surface only at both poles and the equator. If the Earth were a perfect sphere, the gravity would have an equatorward component parallel to the surface. Thus, the Earth deforms to the rotational spheroid until the equatorward component of the gravity vanishes. As a consequence, the equatorial radius of the Earth is about 21 km larger than the polar radius.

The Mass of the Earth
We can calculate the mass of the Earth M using (11.24). Letting the Earth's radius be a and the magnitude of the acceleration due to gravity at the Earth's surface be g_0^*,

$$g_0^* = \frac{GM}{a^2},$$

$$M = \frac{g_0^* a^2}{G} = 5.97 \times 10^{24}\,[\text{kg}].\tag{11.27}$$

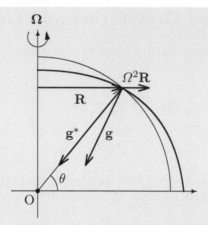

Figure 11.5: The gravitational force $\mathbf{g}^*$ and the Earth's gravity exerting on a particle of unit mass. The bold solid line represents the Earth's surface (a rotational spheroid) and the thin solid line represents the sphere.

11.4   Oceanic Tides

Oceanic tides occur about twice a day. This phenomenon is well documented by the calculation of the gravitational force of the Moon and the Sun exerting on a parcel of sea water of unit mass at an arbitrary point. The depth of the sea is shallow compared to the Earth's radius so that an arbitrary point in the sea may be regarded to be at the Earth's surface. Coordinates are so chosen as shown in Fig. 11.6. Namely, we will take the origin O at the center of the Earth, the z-axis as the extension line from the center of the moon O' to O, the x-axis as an arbitrary direction perpendicular to the z-axis and the y-axis so as to obey the right-hand rule. Letting an arbitrary point in the sea be P(x, y, z), the distance between O' and O be R, the distance between O' and P be r, the potential of the universal gravitation Φ for a parcel of sea water of unit mass at point P is,

$$
\begin{aligned}
\Phi &= -\frac{GM}{r} = -\frac{GM}{\sqrt{(R+z)^2 + x^2 + y^2}} \\
&= -\frac{GM}{R}\left[\left(1+\frac{z}{R}\right)^2 + \left(\frac{x}{R}\right)^2 + \left(\frac{y}{R}\right)^2\right]^{-1/2} \\
&= -\frac{GM}{R}\left(1+\varepsilon\right)^{-1/2},
\end{aligned}
\tag{11.28}
$$

where

$$
\varepsilon = \left(\frac{x}{R}\right)^2 + \left(\frac{y}{R}\right)^2 + \left(\frac{z}{R}\right)^2 + \frac{2z}{R}.
$$

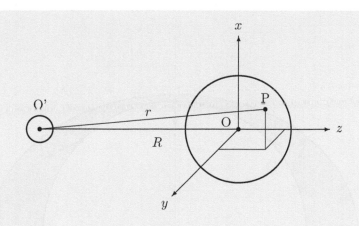

Figure 11.6: Coordinates of the Earth–Moon system.

Expanding (11.28) in a Taylor series and neglecting all terms of order ε^3 and higher, we obtain

$$\Phi = -\frac{GM}{R} + \frac{GM}{R^2}z - \frac{GM}{2R^3}(2z^2 - x^2 - y^2). \tag{11.29}$$

The x-, y-, z-components of the gravitational force exerting on a parcel of sea water of unit mass are obtained by partial differentiation of (11.28) with respect to x, y, z.

$$F_x = -\frac{\partial \Phi}{\partial x} = -\frac{GM}{R^3}x, \tag{11.30}$$

$$F_y = -\frac{\partial \Phi}{\partial y} = -\frac{GM}{R^3}y, \tag{11.31}$$

$$F_z = -\frac{\partial \Phi}{\partial z} = -\frac{GM}{R^2} + \frac{2GM}{R^3}z. \tag{11.32}$$

The first term of the right-hand side of (11.32) is the attractive force of the moon for the Earth to orbit about the center of mass of the Earth–Moon system,[1] so that the force does not contribute to drive the sea water relative to the Earth. Tidal force exerting on a sea water parcel is shown in the x-z plane in Fig. 11.7.

11.5   The Effect of Oceanic Tides

There is a lag of 2.5 hours between the time of crossing a meridian of the moon and the time of flood tides at the meridian. This is due to the tidal friction

---

[1] The center of mass of the Earth–Moon system is 4.67×10^6 m from the center of the Earth to the Moon (the radius of the Earth is about 6.37×10^6 m).

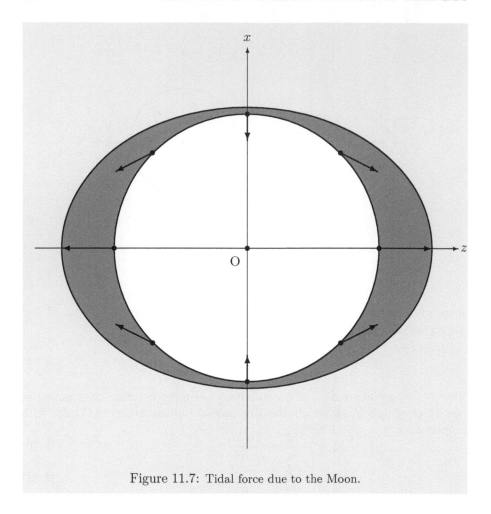

Figure 11.7: Tidal force due to the Moon.

between the solid Earth and the sea water, which causes the decrease of the angular momentum of the Earth's rotation. We will consider the effect of the tidal friction on the Moon. The angular momentum of the Earth–Moon system is conserved, so that the angular momentum of the orbital motion of the Moon increases while the angular momentum of the Earth's rotation decreases. Let's ignore the angular momentum of the rotation of the Moon compared with the angular momentum of the orbital motion. Let the mass of the Earth be M, the moment of inertia of the Earth be I, the mass of the Moon be m. At time t, let the angular velocity of the Earth be Ω, the orbital radius of the Moon be r and the orbital speed be v. After a small-time increment Δt, $\Omega \rightarrow \Omega + \Delta\Omega$, $r \rightarrow r + \Delta r$ and $v \rightarrow v + \Delta v$. We obtain the following equation owing to the law of angular momentum conservation,

$$I\Omega + mvr = I(\Omega + \Delta\Omega) + m(v + \Delta v)(r + \Delta r),$$
$$0 = I\Delta\Omega + mv\Delta r + mr\Delta v. \tag{11.33}$$

The equation of motion of the Moon is

$$m\frac{v^2}{r} = G\frac{mM}{r^2},$$
$$v^2 = \frac{GM}{r}. \tag{11.34}$$

Applying (11.34) at $t + \Delta t$, we obtain

$$(v + \Delta v)^2 = \frac{GM}{r + \Delta r}.$$

Expanding the above equation in a Taylor series and neglecting all terms of order Δ^2 and higher, we get

$$v^2 + 2v\Delta v = \frac{GM}{r}\left(1 - \frac{\Delta r}{r}\right). \tag{11.35}$$

Subtracting (11.34) from (11.35) and using (11.34), we find

$$\Delta v = -\frac{GM}{2r^2v}\Delta r = -\frac{v}{2r}\Delta r. \tag{11.36}$$

Substituting from (11.36) into (11.33) yields,

$$\Delta r = -\frac{2I}{mv}\Delta\Omega. \tag{11.37}$$

Substituting (11.37) into (11.36), we obtain

$$\Delta v = \frac{I}{mr}\Delta\Omega. \tag{11.38}$$

As $\Delta\Omega < 0$, $\Delta r > 0$ and $\Delta v < 0$. Therefore, the Moon retreats from the Earth and the orbital speed decreases. About 400 million years ago, in the Devonian period of the Paleozoic era, one year was 400 days, which was proved by the analyses of fossils of corals and stromatolites. Based on this value, we calculate Δr and Δv per year. Physical quantities necessary for the calculation are as follows:

$$\begin{cases} M = 5.97 \times 10^{24}\,\text{kg} : \text{the mass of the Earth,} \\ I = 8.07 \times 10^{37}\,\text{kg m}^2 : \text{the moment of inertia of the Earth,} \\ m = 7.35 \times 10^{22}\,\text{kg} : \text{the mass of the Moon,} \\ r = 3.84 \times 10^8\,\text{m} : \text{the distance from the Earth to the Moon,} \\ v = 1.02 \times 10^3\,\text{m s}^{-1} : \text{the orbital speed of the Moon.} \end{cases}$$

Substituting above values[2] into (11.37) and (11.38), we obtain $\Delta r = 3.75 \times 10^{-2}$ m, $\Delta v = -5.00 \times 10^{-8}$ m s^{-1}. Namely, the Moon is retreating from the Earth by 3.75 cm a year. Using (11.34) and (11.37), we can estimate the orbital radius of the Moon in *Devonian period* to be 3.70×10^{8} m. Calculation is left to readers in the problem of Section 11.11.

Problem 1. In spite that no seas ever existed on Mercury and the Moon, why do the orbital periods and their rotational periods synchronize?

11.6   Orbital Motion of Planets and Kepler's Three Laws

Tycho Brahe (Fig. 11.8), a Danish nobleman and astronomer, made accurate and comprehensive astronomical observations and left vast records. Johannes Kepler (Fig. 11.9), a German mathematician and astronomer, inherited the records of Tycho Brahe and analyzed them and found three laws concerning the orbital motion of planets. Later, the three laws were referred to as *Kepler's three laws*.

1. *The first law*: The orbit of a planet is an ellipse with the Sun at one focus.

2. *The second law*: A planet's radius vector sweeps out the equal areas in equal time intervals, i.e., the area speed of the planet's radius vector is constant.

3. *The third law*: The square of the orbital period of a planet is proportional to the cube of the semimajor axis of the elliptic orbit.

Before Kepler, orbits of celestial bodies were considered to be circles in both geocentrism and heliocentrism. Kepler found that the observational data of Tycho Brahe were reasonably understood assuming the planetary orbits were ellipses. Kepler's three laws were very important findings and became the forerunners of the law of universal gravitation and Newtonian mechanics.

11.7   Proof of Kepler's Three Laws

A planet is orbiting around the Sun due to the centripetal force, universal gravitation, between the Sun and the planet, so that the equation of motion for the planet is adequately described in plane polar coordinates. Using the acceleration terms in plane polar coordinates of (2.39), the r- and θ-components

---

[2]The moment of inertia of the Earth is after Shimadu 1967.

⋆⋆⋆⋆⋆⋆⋆⋆⋆⋆⋆⋆⋆⋆⋆⋆ Tycho Brahe (1546–1601) ⋆⋆⋆⋆⋆⋆⋆⋆⋆⋆⋆⋆⋆⋆⋆⋆

Figure 11.8: Tycho Brahe was a Danish nobleman, astronomer, and astrologer. Receiving financial support from Danish king, Frederick II, Brahe built the Uraniborg astronomical observatory and the Stjerneborg astronomical observatory on Island Hven (belonging to modern-day Sweden) and observed the motion of celestial bodies from 1576 to 1596. In his era telescopes had not yet been invented, so he made observations using his own naked eyes and measuring equipment. After the disagreements with the new Danish king, Christian IV, in 1597, the king of the Holy Roman Empire Rudolf II invited him to Prague in Czechoslovakia, and Brahe moved to Prague with his observational instruments and records and built a new observatory at Benátky nad Jizerou. In spite of the fact that his observational records were inconsistent with the geocentric theory, Brahe could not abandon it and advocated the modified geocentric theory, named geo-heliocentric theory, that the Sun and Moon orbited around the Earth and other celestial bodies orbited around the Sun. After Brahe's death, the vast and accurate observational records were inherited by Johannes Kepler and were brought to fruition as Kepler's three laws.

⋆⋆⋆

of the equation of motion become

$$m\frac{d^2r}{dt^2} - mr\left(\frac{d\theta}{dt}\right)^2 = -G\frac{mM}{r^2}, \tag{11.39}$$

★★★★★★★★★★★★★★ Johannes Kepler (1571–1630) ★★★★★★★★★★★★★★★

Figure 11.9: Johannes Kepler was a German mathematician, astronomer, and astrologer. He was born on December 27, 1571, at Weil der Stadt in southern Germany. He entered the University of Tübingen in 1587 and studied philosophy and theology. After graduating from the university, he gave lectures on mathematics and astronomy at the Protestant School in Graz from 1594 to 1599. In 1600, he was invited by Tycho Brahe to Prague. Two days after the death of Brahe on October 24, 1601, Kepler was appointed his successor as the imperial mathematician. Thus Kepler inherited Brahe's observational data and analyzed them during the following 11 years. In 1609, he published *Astronomia Nova* in which he described Kepler's first and second laws. After Emperor Rudolf II died in 1612, Kepler moved to Linz as a teacher of mathematics at the district school. In 1619, he issued Kepler's third law. Kepler moved to Ulm, at that time a Free Imperial City, in 1626 and then to Regensburg (Fig. 11.10) in Bavaria in 1628 and died of illness in 1630. Kepler made clear the orbital motion of planets, analyzing Brahe's vast data thanks to his outstanding ability of mathematics. Kepler's work was one of the foundations of Newton's law of universal gravitation and Newtonian mechanics.

★★★

$$2m\frac{dr}{dt}\frac{d\theta}{dt} + mr\frac{d^2\theta}{dt^2} = 0. \tag{11.40}$$

Transforming (11.40), we get

Figure 11.10: Regensburg, nowadays in the Free State of Bavaria, the place of Kepler's death, is one of Germany oldest and most beautiful towns, founded by the Romans in 179 A.D., and situated at the northernmost point of the river Danube. Regensburg became a Free Imperial City in 1245 and was seat to the Perpetual Diet of the Holy Roman Empire from 1663 to 1806. Since 2006 the historic city center of Regensburg has been a UNESCO World Heritage site. The Danube flows right through the town and the famous Stone Bridge built in 1146 remains medieval feature. The Gothic St. Peter's Cathedral at the center of the photo, which was begun to construct in 1273, has twin towers with a height of 105 m. Kepler's house, now the Kepler Museum, is at the right side of the photo and is facing the first street from the other side of the river.

$$\frac{1}{r}\frac{d}{dt}\left(r^2\frac{d\theta}{dt}\right) = 0,$$

$$r^2\frac{d\theta}{dt} = h(\text{const.}). \tag{11.41}$$

From (11.41),

$$\frac{d\theta}{dt} = \frac{h}{r^2}. \tag{11.42}$$

Substituting (11.42) into (11.39), we find

$$\frac{d^2r}{dt^2} - \frac{h^2}{r^3} = -\frac{GM}{r^2}. \tag{11.43}$$

Let the position of the Sun be focus F_1, the position of the planet at time t be P, the position of the planet at $t + \delta t$ be Q and $\angle PF_1Q = \delta\theta$ as illustrated in Fig. 11.11. As $\overline{PQ} \cong r\delta\theta$, the area δS swept out by the planet's radius vector in a time increment δt is

$$\delta S \cong \frac{1}{2}r^2\delta\theta.$$

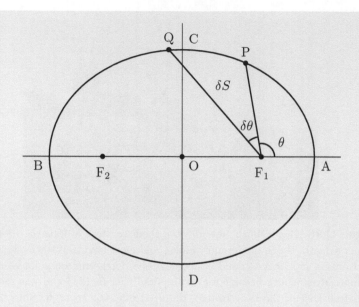

Figure 11.11: Orbit of a planet and areal speed of the radius vector.

Dividing the above equation through by δt and taking the limit $\delta t \to 0$,

$$\lim_{\delta t \to 0} \frac{\delta S}{\delta t} = \frac{1}{2} r^2 \lim_{\delta t \to 0} \frac{\delta \theta}{\delta t},$$

$$\frac{dS}{dt} = \frac{1}{2} r^2 \frac{d\theta}{dt} = \frac{1}{2} h, \tag{11.44}$$

with the aid of (11.41). Thus *Kepler's second law* is proved.

Using (11.42), we will transform the independent variable t to θ as

$$\frac{d}{dt} = \frac{d\theta}{dt} \frac{d}{d\theta} = \frac{h}{r^2} \frac{d}{d\theta}. \tag{11.45}$$

Applying (11.45) to (11.43), we obtain

$$\frac{h}{r^2} \frac{d}{d\theta} \left(\frac{h}{r^2} \frac{dr}{d\theta} \right) - \frac{h^2}{r^3} = -\frac{GM}{r^2}. \tag{11.46}$$

Putting $\xi = 1/r$, (11.46) becomes

$$\frac{d^2 \xi}{d\theta^2} + \xi = \frac{GM}{h^2}. \tag{11.47}$$

The general solution of (11.47) is

$$\xi = \frac{GM}{h^2} + A \cos (\theta - \theta_0) = \frac{1}{r},$$

$$r = \frac{\dfrac{h^2}{GM}}{1 + \dfrac{h^2 A}{GM}\cos(\theta - \theta_0)} = \frac{\eta}{1 + \varepsilon \cos(\theta - \theta_0)}, \qquad (11.48)$$

where

$$\eta = \frac{h^2}{GM}, \qquad \varepsilon = \frac{Ah^2}{GM}. \qquad (11.49)$$

Equation (11.48) represents a *conic section*; a hyperbola when $\varepsilon > 1$, a parabola when $\varepsilon = 1$, and an ellipse when $\varepsilon < 1$. Planetary orbits should remain a finite range so that ε must be smaller than unity and then planetary orbits must be ellipses. Thus *Kepler's first law* is proved.

Measuring θ from the line segment $\overline{F_1 A}$, namely, putting $\theta_0 = 0$, (11.48) becomes

$$r = \frac{\eta}{1 + \varepsilon \cos \theta}. \qquad (11.50)$$

We will consider an elliptic orbit shown in Fig. 11.11. Let's take the origin as O, the semimajor axis as the x-axis, a focus and the position of the Sun as F_1, another focus as F_2, the intersections of the ellipse and the x-axis as A and B, the intersections of the ellipse and the y-axis as C and D and the position of a planet as $P(r, \theta)$. Letting the semimajor axis be a and the semiminor axis be b, we find from (11.50)

$$\overline{F_1 A} = \frac{\eta}{1 + \varepsilon}, \qquad \overline{F_1 B} = \frac{\eta}{1 - \varepsilon},$$

$$\overline{AB} = 2a = \frac{\eta}{1 + \varepsilon} + \frac{\eta}{1 - \varepsilon} = \frac{2\eta}{1 - \varepsilon^2},$$

$$a = \frac{\eta}{1 - \varepsilon^2}. \qquad (11.51)$$

When the planet comes to C,

$$\overline{OF_1} = -r \cos \theta = \overline{OA} - \overline{F_1 A}$$

$$= a - \frac{\eta}{1 + \varepsilon} = \frac{\eta}{1 - \varepsilon^2} - \frac{\eta}{1 + \varepsilon} = \frac{\varepsilon \eta}{1 - \varepsilon^2},$$

$$-r \cos \theta = \frac{\varepsilon \eta}{1 - \varepsilon^2}. \qquad (11.52)$$

From (11.50) and (11.52), we get

$$\overline{F_1 C} = r = \eta - r \varepsilon \cos \theta = \eta + \frac{\varepsilon^2 \eta}{1 - \varepsilon^2} = \frac{\eta}{1 - \varepsilon^2},$$

$$b^2 = \overline{OC}^2 = \overline{F_1 C}^2 - \overline{OF_1}^2 = \frac{\eta^2}{1 - \varepsilon^2},$$

$$b = \frac{\eta}{\sqrt{1 - \varepsilon^2}} = (\eta a)^{1/2}. \qquad (11.53)$$

The area of the ellipse is obtained using (11.51) and (11.53),

$$S = \pi ab = \pi \eta^{1/2} a^{3/2}. \tag{11.54}$$

Then the orbital period T is obtained by dividing (11.54) through by the area speed given by (11.44), we obtain

$$T = \frac{S}{dS/dt} = \frac{2\pi \eta^{1/2} a^{3/2}}{h}. \tag{11.55}$$

Taking the square of (11.55) and using (11.49), we find

$$T^2 = \frac{4\pi^2 \eta}{h^2} a^3 = \frac{4\pi^2}{GM} a^3. \tag{11.56}$$

Thus, *Kepler's third law* is proved.

Problem 2. Transform the equation of an ellipse in plane polar coordinates

$$r = \frac{\eta}{1 + \varepsilon \cos \theta}$$

to the equation in two-dimensional Cartesian coordinates

$$\frac{x^2}{a^2} + \frac{y^2}{b^2} = 1.$$

Problem 3. Show that the area of an ellipse of the semimajor axis a and the semiminor axis b is given by πab.

11.8   Escape Velocity

The work W necessary for moving a particle of mass m at distance r from the center of a sphere of mass M to infinity against the gravitational attraction is

$$W = \int_r^\infty G\frac{Mm}{r^2} dr$$
$$= GMm \left[-\frac{1}{r} \right]_r^\infty = G\frac{Mm}{r}.$$

The difference of the potential energy at infinity due to the universal gravitation of the sphere and that at the distance r is equal to the work W. Then, taking the potential energy at infinity as reference, the potential energy at r is

$$U_\infty - U(r) = W = G\frac{Mm}{r},$$
$$U(r) = -W = -G\frac{Mm}{r}. \tag{11.57}$$

Example 5. The escape velocity (the second cosmic speed)
Suppose an antiaircraft gun which can project a shell at high speed. What is
the initial speed v_0 for the shell to be able to escape from the gravitational
attraction of the Earth? Let the Earth's radius be $a = 6.37 \times 10^6$ m, the
mass of the Earth be $M = 5.97 \times 10^{24}$ kg and the gravitational constant be
$G = 6.67 \times 10^{-11}$ m^3 kg^{-1} s^{-2} and ignore the air resistance.

Answer
Taking the potential energy at infinity as reference, we find from (11.57) that
the potential energy possessed by a particle of mass m at distance r from the
center of the Earth is

$$U(r) = -G\frac{Mm}{r}.$$

Letting the speed of the shell at infinity be v_∞, the mechanical energy at the
Earth's surface and at infinity are equal, so that

$$-G\frac{Mm}{a} + \frac{1}{2}mv_0{}^2 = 0 + \frac{1}{2}mv_\infty{}^2 > 0,$$

$$v_0 > \sqrt{\frac{2GM}{a}} = 1.12 \times 10^4 \, [\mathrm{m\,s^{-1}}].$$

It may seem unrealistic projecting a shell at high speed to escape the Earth's
gravity, but the escape velocity is an important physical quantity in geophysics. It is the measure whether a planet can hold its atmosphere or not.
Planets of small mass such as Mercury and Mars lost most of their atmosphere
which they had possessed at their birth.

11.9 General Orbits due to a Central Force

In this section, we will discuss the orbital motion of an object due to a central
force. Suppose that the potential energy due to a central force is

$$U(r) = -\frac{\alpha}{r}. \tag{11.58}$$

The sum of the kinetic energy and the potential energy of the orbital motion
is

$$\frac{1}{2}m\left[\left(\frac{dr}{dt}\right)^2 + r^2\left(\frac{d\theta}{dt}\right)^2\right] - \frac{\alpha}{r} = E. \tag{11.59}$$

The conservation of the angular momentum is

$$r^2\frac{d\theta}{dt} = h(\text{const.}).$$

Thus

$$\frac{d\theta}{dt} = \frac{h}{r^2}.$$

(11.60)

Using (11.60), we find the following relation

$$\frac{dr}{dt} = \frac{dr}{d\theta}\frac{d\theta}{dt} = \frac{h}{r^2}\frac{dr}{d\theta}.$$

(11.61)

Substituting (11.60) and (11.61) into (11.59), we obtain

$$\frac{\pm h \, dr}{r^2 \sqrt{\frac{2}{m}\left(E + \frac{\alpha}{r} - \frac{mh^2}{2r^2}\right)}} = d\theta.$$

(11.62)

Putting $\xi = 1/r$, (11.62) becomes

$$\frac{\mp d\xi}{\sqrt{\frac{2E}{mh^2} + \frac{\alpha^2}{m^2 h^4} - \left(\xi - \frac{\alpha}{mh^2}\right)^2}} = d\theta.$$

(11.63)

Integrating (11.63), we get

$$\mp \cos^{-1} \frac{\xi - \alpha/mh^2}{\sqrt{\frac{2E}{mh^2} + \frac{\alpha^2}{m^2 h^4}}} = \theta - \theta_0.$$

Remembering $r = 1/\xi$, we find

$$r = \frac{mh^2/\alpha}{1 + \sqrt{1 + \frac{2mEh^2}{\alpha^2}} \cos(\theta - \theta_0)}.$$

(11.64)

Here, we will define η and ε as

$$\eta = \frac{mh^2}{\alpha},$$

(11.65)

$$\varepsilon = \sqrt{1 + 2mEh^2/\alpha^2},$$

(11.66)

which are called the *semi latus rectum* and the *eccentricity*, respectively. Then (11.64) becomes

$$r = \frac{\eta}{1 + \varepsilon \cos(\theta - \theta_0)}.$$

(11.67)

Equation (11.67) represents a conic section. A conic section is defined as the ensemble of point P whose distance from the fixed-point F and the distance from the reference line $\overline{\text{LM}}$ have the constant ratio ε (Fig. 11.12). Namely,

$$\frac{\overline{\text{PR}}}{\overline{\text{PF}}} = \frac{1}{\varepsilon}, \quad \overline{\text{PR}} = \frac{r}{\varepsilon}, \quad \overline{\text{QS}} = \frac{\eta}{\varepsilon}.$$

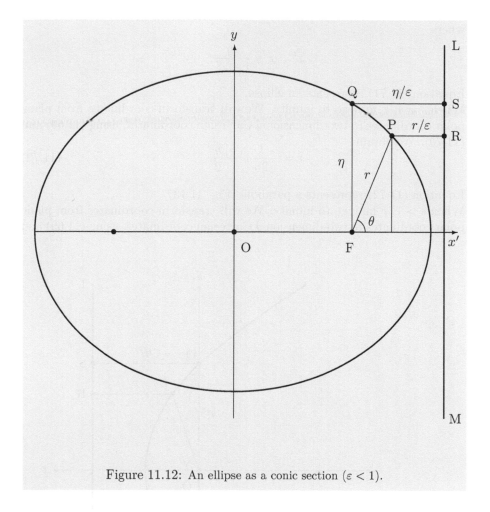

Figure 11.12: An ellipse as a conic section ($\varepsilon < 1$).

Thus, we find

$$r = \frac{\eta}{1 + \varepsilon \cos \theta}.\tag{11.68}$$

Equation (11.68) is the same as (11.67) when $\theta_0 = 0$.
When $\varepsilon < 1$, r remains finite. We will transform coordinates from plane polar coordinates to two-dimensional Cartesian coordinates, putting

$$x = r \cos \theta,\tag{11.69}$$
$$y = r \sin \theta.\tag{11.70}$$

We obtain

$$\frac{x'^2}{\eta^2/(1 - \varepsilon^2)^2} + \frac{y^2}{\eta^2/(1 - \varepsilon^2)} = 1,\tag{11.71}$$

where

$$x' = x - \frac{\varepsilon\eta}{1 - \varepsilon^2}.$$

Equation (11.71) represents an ellipse.

When $\epsilon = 1$, r may go to infinity. We will transform coordinates from plane polar coordinates to two-dimensional Cartesian coordinates, using (11.69) and (11.70). We obtain

$$x = -\frac{1}{2\eta}y^2 + \frac{\eta}{2}. \qquad (11.72)$$

Equation (11.72) represents a parabola (Fig. 11.13).

When $\epsilon > 1$, r may go to infinity. We will transform coordinates from plane polar coordinates to two-dimensional Cartesian coordinates, using (11.69) and

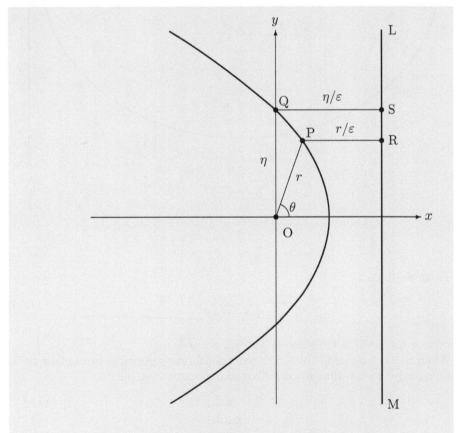

Figure 11.13: A parabola as a conic section ($\varepsilon = 1$).

(11.70). Then we get

$$\frac{x'^2}{\eta^2/(\varepsilon^2-1)^2} - \frac{y^2}{\eta^2/(\varepsilon^2-1)} = 1, \qquad (11.73)$$

where

$$x' = x - \frac{\varepsilon\eta}{\varepsilon^2-1}.$$

Equation (11.72) represents a hyperbola (Fig. 11.14). When x' and y go to infinity, they approach to asymptotes

$$y = \pm\sqrt{\varepsilon^2-1}x'. \qquad (11.74)$$

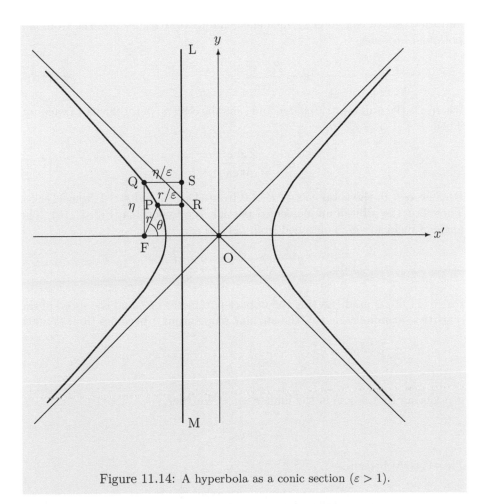

Figure 11.14: A hyperbola as a conic section ($\varepsilon > 1$).

Letting the angle between an asymptote and the x-axis be ψ, we find

$$\tan \psi = \frac{y}{x'} = \pm\sqrt{\varepsilon^2 - 1}. \tag{11.75}$$

11.10   Rutherford Scattering

From 1908 to 1913, Geiger and Marsden performed a series of experiments investigating the spatial extent of atomic nuclei under the supervision of Professor Ernest Rutherford. They injected a beam of α-particles into thin gold foil and measured the scattering angle of α-particles.

In 1911, Rutherford calculated the differential scattering cross section of α-particles assuming that positive charges concentrate in spherical nuclei based on classical mechanics. Supposing that the positive charge of an atom is Ze and the positive charge of an incidence charged particle is $Z'e$, the Coulomb potential becomes

$$\Phi = \frac{ZZ'e^2}{4\pi\varepsilon_0 r} = -\frac{\alpha}{r}, \tag{11.76}$$

where e is the unit electric charge and ε_0 is the electric permittivity of vacuum. Then,

$$\alpha = -\frac{ZZ'e^2}{4\pi\varepsilon_0}. \tag{11.77}$$

Because $\alpha < 0$, the total energy $E > 0$ from (11.59), and $\varepsilon > 1$ from (11.66). Therefore, the orbit of an incident α-particle is a hyperbola (Fig. 11.15). The angle of incidence ψ is obtained from (11.75) as,

$$\tan \psi = \frac{y}{x'} = \sqrt{\varepsilon^2 - 1} = \frac{\sqrt{2mEh}}{|\alpha|}, \tag{11.78}$$

where (11.66) is used. Letting the impact parameter be b and the speed of the α-particle at infinity be v_∞, the angular momentum h becomes from (11.60),

$$h = r^2 \frac{d\psi}{dt} = bv_\infty,$$

considering that $\theta = \psi$ in the limit $r \to \infty$, Further,

$$E = \frac{1}{2}mv_\infty{}^2.$$

Thus (11.78) becomes

$$\tan \psi = \sqrt{\varepsilon^2 - 1} = \frac{mbv_\infty{}^2}{|\alpha|}. \tag{11.79}$$

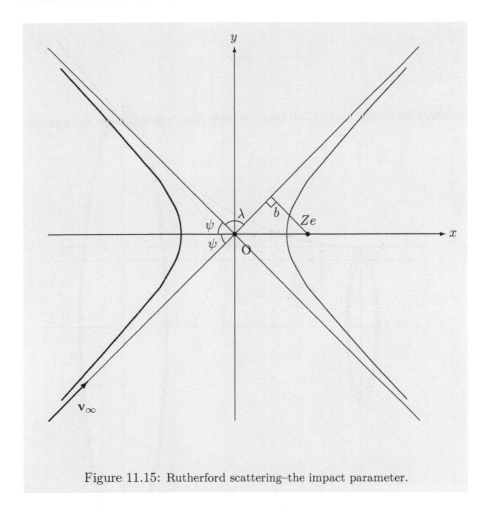

Figure 11.15: Rutherford scattering–the impact parameter.

Letting the scattering angle be λ, the relationship between λ and ψ is obtained referring to Fig. 11.16,

$$\psi = \frac{\pi}{2} - \frac{\lambda}{2}. \tag{11.80}$$

From (11.79) and (11.80), we find

$$\cot \frac{\lambda}{2} = \sqrt{\varepsilon^2 - 1} = \frac{mbv_\infty^2}{|\alpha|}. \tag{11.81}$$

The infinitesimal area dS of an annulus of radius b and $b - db$ is

$$dS = 2\pi b |db| = \pi \left(\frac{|\alpha|}{mv_\infty^2} \right)^2 \frac{\cos \lambda/2}{\sin^3 \lambda/2} d\lambda. \tag{11.82}$$

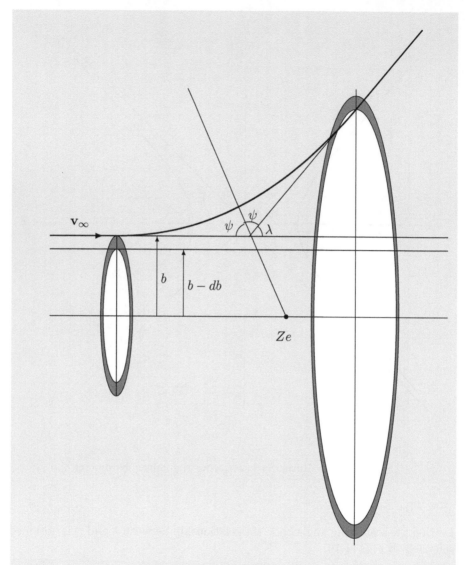

Figure 11.16: Rutherford scattering–the differential scattering cross section.

The infinitesimal solid angle $d\Omega$ confined between λ and $\lambda + d\lambda$ is

$$d\Omega = 2\pi \sin \lambda d\lambda. \tag{11.83}$$

From (11.82) and (11.83), we get the differential scattering cross section,

$$\frac{dS}{d\Omega} = \left(\frac{|\alpha|}{2mv_\infty{}^2}\right)^2 \frac{1}{\sin^4\lambda/2} = \left(\frac{ZZ'e^2}{8\pi\varepsilon_0 mv_\infty{}^2}\right)^2 \frac{1}{\sin^4\lambda/2}. \tag{11.84}$$

Suppose that the number of incidents α-particles in a unit time to the scattering cross section A is N, the number of scattered α-particles into a solid angle between λ and $\lambda + d\lambda$ is

$$dN = \frac{N}{A} dS = \frac{N}{A} \left(\frac{ZZ'e^2}{8\pi\varepsilon_0 mv_\infty{}^2} \right)^2 \frac{1}{\sin^4 \lambda/2} d\Omega. \tag{11.85}$$

Equation (11.85) is referred to as the *Rutherford scattering formula*. In the scattering experiment, the distance of the closest approach of an α-particle to the atomic nucleus is obtained using (11.68),

$$r_{\min} = \frac{\eta}{1+\varepsilon} = \frac{1}{1+\varepsilon} \frac{mh^2}{|\alpha|}. \tag{11.86}$$

The relation between λ and ε is obtained from (11.81) as

$$\varepsilon = \frac{1}{\sin \lambda/2}. \tag{11.87}$$

Substituting (11.86) into (11.85), we get

$$r_{\min} = \frac{mh^2}{|\alpha|} \frac{\sin \lambda/2}{1 + \sin \lambda/2}. \tag{11.88}$$

Values of physical quantities of the right-hand side of (11.87) are obtained by scattering experiments, so that the approximate spatial extent of an atomic nucleus can be calculated. Rutherford found the size of an atomic nucleus to be 10^{-14} m, which was far smaller than 10^{-10} m believed in those days.

11.11 Problems

1. The Moon is orbiting around the Earth at the period $T = 2.73 \times 10$ days. What is the orbital radius of the Moon, supposing that the orbit is a circle? Let the mass of the Earth be $M = 5.97 \times 10^{24}$ kg, and the gravitational constant be $G = 6.67 \times 10^{-11}$ m^3 kg^{-1} s^{-2}.

2. Obtain the gravitational attraction of a disk exerting on a particle of mass m at distance z along the axis passing the center of the disk and vertical to it. Let the radius of the disk be a, the area density be σ and the gravitational constant be G.

3. Obtain the gravitational attraction of an infinite plate exerting on a particle of mass m at distance z from the plate. Let the area density of the plate be σ and the gravitational constant be G.

4. Obtain the potential of the universal gravitation and the gravitational attraction of a spherical shell of radius a and mass M exerting on a particle of mass m at distance $z(z > a)$ from the center of the spherical shell. Let the gravitational constant be G.

5. Obtain the potential of the universal gravitation and the gravitational attraction of a spherical shell of radius a and mass M exerting on a particle of mass m at distance $z (z < a)$ from the center of the spherical shell. Let the gravitational constant be G.

6. Obtain the orbital radius of the Moon in the Devonian period by solving the following problems.

    (1) Calculate the period of the Earth's rotation using the fact that one year had 400 days in the Devonian period.

    (2) Obtain the differential equation concerning r and Ω, eliminating v from (11.34) and (11.37).

    (3) Calculate the orbital radius of the Moon in the Devonian period by integrating the above differential equation from the Devonian period to the present.

7. Using the law of the universal gravitation, prove Kepler's third law that the square of the orbital period of a planet is proportional to the cube of the orbital radius. Let the orbit of the planet be a perfect circle.

8. An ellipse is defined as an ensemble of points whose sum of the distance from two fixed points (focuses) is constant. Show that the equation of an ellipse becomes

$$\frac{x^2}{a^2} + \frac{y^2}{b^2} = 1, \qquad a > b,$$

   in two-dimensional Cartesian coordinates. Let the position of focuses be $(c, 0)$ and $(-c, 0)$.

9. An artificial satellite, made of small two rigid balls of mass m connected by a massless rigid rod of length $2l$, is orbiting around the Earth. Discuss the posture stability of the satellite, letting the distance from the Earth's center to the center of mass of the satellite be r, the mass of the Earth be M, and the coefficient of the universal gravity be G.

10. In the previous problem, the artificial satellite is oscillating around its equilibrium position in small amplitude. Obtain the period of the oscillatory motion.

11. When the Earth was born by collisions and mergers of planetoids, it was a sphere of uniform density ρ, mass M, and radius a. During about 10^8 years after the Earth's birth, the gravitational differentiation occurred, so that the heavier material subsided toward the center of the Earth and the lighter material rose up toward the surface. Then the Earth consisted of the inner sphere of radius a_1 and density ρ_1 and the outer sphere of density ρ_2. How much potential energy is released by the gravitational differentiation? Suppose that the Earth's radius was unchanged through the process.

12. Letting $M = 5.974 \times 10^{24}$ kg, $a = 6.369 \times 10^6$ m, $a_1 = 3.480 \times 10^6$ m, $\rho = 5.520 \times 10^3$ kg m^{-3}, $\rho_1 = 1.200 \times 10^4$ kg m^{-3}, $\rho_2 = 4.256 \times 10^3$ kg m^{-3} in the previous problem, calculate potential energy released through the gravitational differentiation.

13. Following the result of the previous problem, calculate the temperature increment through the gravitational differentiation. Suppose that all released potential energy was used for heating the Earth. Let the mass of the Earth be $M_E = 5.97 \times 10^{24}$ kg and the mean specific heat of the Earth's material be $c = 6.50 \times 10^2$ J kg^{-1} K^{-1}.

11.12 Reference

1. Shimadu Y.: *Evolution of the Earth*, Iwanami Shoten Publishers, Tokyo (1967), *in Japanese*.

Chapter 12

Introduction to Geophysical Fluid Dynamics

Gases and liquids are called fluids collectively. A fluid consists of many molecules, which move about randomly and collide with each other. In fluid dynamics, we ignore molecular behavior of a fluid and treat it as a continuous medium, or *continuum*. A small fluid element with an infinitesimal volume but containing a large number of molecules is referred to as a *fluid parcel*, which is the analogous concept of a particle or a point mass in Newtonian mechanics. Physical quantities (velocity, pressure, density, and temperature) are defined for a fluid parcel.

The fundamental equations in fluid dynamics are the momentum equations, thermodynamic energy equation, the continuity equation, and the equation of state. The dependent variables are velocity $\mathbf{v} = (u, v, w)$, pressure p, density ρ, and temperature T. In this book, we will adopt the *Boussinesq approximation*, by which we can treat the dynamics of the Earth's atmosphere, oceans and laboratory fluid systems comprehensively. In this chapter, we will introduce fundamental equations necessary for studying geophysical fluid dynamics.

12.1 Individual Rate of Change and Local Rate of Change

Fluid motions are governed by three fundamental conservation laws of mass, momentum and energy. The conservation laws are derived by considering the budgets of mass, momentum, and energy for an infinitesimal control volume in the fluid. Two types of control volume are used in fluid dynamics, one is the

DOI: 10.1201/9781003310068-12

Lagrangian control volume and the other is the *Eulerian control volume*. The former consists of an infinitesimal mass of tagged fluid parcels, and moves about following the fluid motion, changing its volume and shape. In the Lagrangian frame of reference, a physical quantity is measured for a tagged fluid parcel and the changing rate of it is called the *individual change*. The Lagrangian formulation is useful for deriving conservation laws since they can be stated most simply for a particular control volume. The latter does not change its position and shape while the fluid parcels consisting of it change in time. In the Eulerian frame of reference, a physical quantity is measured at the fixed position and the changing rate of it is called the *local change*. The Eulerian formulation is convenient for solving problems mathematically, since physical laws are described in a set of partial differential equations. Initially in this chapter, we will show the relationship between the Lagrangian formulation and the Eulerian one.

As shown in Fig. 12.1, suppose that a fluid parcel is located at $\mathbf{r}(t) = (x(t), y(t), z(t))$ at time t and has a physical quantity $\Phi(t)$, which is a function of time and space as

$$\Phi(t) = \Phi(t, x(t), y(t), z(t)).$$

In a small time increment δt, the parcel moves to position $\mathbf{r}(t) + \delta \mathbf{r} = (x(t) + \delta x, y(t) + \delta y, z(t) + \delta z)$, and $\Phi(t)$ becomes $\Phi(t + \delta t)$. Expanding $\Phi(t + \delta t)$ in

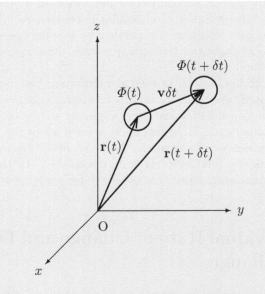

Figure 12.1: The individual rate of change of a fluid parcel.

a Taylor series, we get

$$\Phi(t + \delta t) = \Phi(t) + \frac{\partial \Phi}{\partial t}\delta t + \frac{\partial \Phi}{\partial x}\delta x + \frac{\partial \Phi}{\partial y}\delta y + \frac{\partial \Phi}{\partial z}\delta z,$$

neglecting all terms of order δ^2 and higher. Translating the first term of the right-hand side to the left-hand side, and dividing the resultant equation through by δt, we find

$$\frac{\Phi(t + \delta t) - \Phi(t)}{\delta t} = \frac{\partial \Phi}{\partial t} + \frac{\partial \Phi}{\partial x}\frac{\delta x}{\delta t} + \frac{\partial \Phi}{\partial y}\frac{\delta y}{\delta t} + \frac{\partial \Phi}{\partial z}\frac{\delta z}{\delta t}.$$

Taking the limit $\delta t \to 0$, we obtain

$$\begin{aligned}
\frac{d\Phi}{dt} &= \frac{\partial \Phi}{\partial t} + \frac{\partial \Phi}{\partial x}\frac{dx}{dt} + \frac{\partial \Phi}{\partial y}\frac{dy}{dt} + \frac{\partial \Phi}{\partial z}\frac{dz}{dt} \\
&= \frac{\partial \Phi}{\partial t} + \left(u\frac{\partial \Phi}{\partial x} + v\frac{\partial \Phi}{\partial y} + w\frac{\partial \Phi}{\partial z}\right).
\end{aligned} \tag{12.1}$$

Using the vector operator yields,

$$\frac{\partial \Phi}{\partial t} = \frac{d\Phi}{dt} - \mathbf{v} \cdot \boldsymbol{\nabla}\Phi. \tag{12.2}$$

$d\Phi/dt$ is the individual rate of change of Φ following the fluid motion and is called the *total derivative* or the *substantial derivative*. While $\partial\Phi/\partial t$ represents the local rate of change of Φ at a fixed position, which is referred to as the *local derivative*. The second term of the right-hand side $-\mathbf{v} \cdot \boldsymbol{\nabla}\Phi$ is called the *advection term*. For instance, supposing the air temperature T as a physical quantity, the individual time change occurs by radiation, thermal conduction and latent heat release.

12.2   The Continuity Equation

We will derive the continuity equation, i.e., the mass conservation law in a fluid system. We consider a rectangular infinitesimal volume element $\delta V = \delta x \delta y \delta z$ centered at (x, y, z) in a Cartesian coordinate system as shown in Fig. 12.2.

Let the surface perpendicular to the x-axis and at $x - \delta x/2$ be A and at $x + \delta x/2$ be B. The mass inflow per unit time through surface A is

$$\rho(x - \frac{\delta x}{2})u(x - \frac{\delta x}{2})\delta y\delta z = \left[\rho u - \frac{\partial}{\partial x}(\rho u)\frac{\delta x}{2}\right]\delta y\delta z,$$

where use is made of a Taylor series expansion, neglecting all terms of order δx^2 and higher. While the mass outflow per unit time through surface B is obtained similarly,

$$\rho(x + \frac{\delta x}{2})u(x + \frac{\delta x}{2})\delta y\delta z = \left[\rho u + \frac{\partial}{\partial x}(\rho u)\frac{\delta x}{2}\right]\delta y\delta z.$$

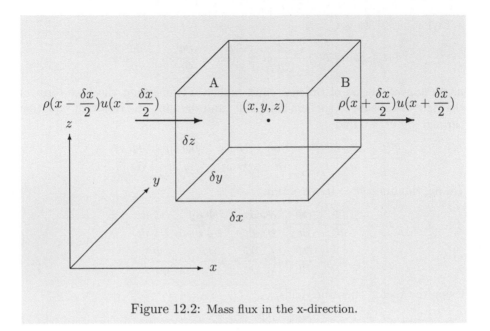

Figure 12.2: Mass flux in the x-direction.

Thus the net mass inflow into the volume element per unit time is

$$\left[\rho u - \frac{\partial}{\partial x}(\rho u)\frac{\delta x}{2}\right]\delta y \delta z - \left[\rho u + \frac{\partial}{\partial x}(\rho u)\frac{\delta x}{2}\right]\delta y \delta z = -\frac{\partial}{\partial x}(\rho u)\delta x \delta y \delta z.$$

Similar discussions hold for the y- and z-directions, and the total mass inflow into the volume element is

$$-\left[\frac{\partial}{\partial x}(\rho u) + \frac{\partial}{\partial y}(\rho v) + \frac{\partial}{\partial z}(\rho w)\right]\delta x \delta y \delta z = -\left[\boldsymbol{\nabla} \cdot (\rho \mathbf{v})\right]\delta x \delta y \delta z.$$

The net rate of mass inflow into the fixed volume should be equal to the accumulation rate of mass in it, so that

$$\frac{\partial \rho}{\partial t}\delta x \delta y \delta z = -\left[\boldsymbol{\nabla} \cdot (\rho \mathbf{v})\right]\delta x \delta y \delta z,$$

$$\frac{\partial \rho}{\partial t} + \boldsymbol{\nabla} \cdot (\rho \mathbf{v}) = 0. \tag{12.3}$$

Equation (12.3) is the *mass divergent form* of the continuity equation, and the alternative form of *velocity divergence form* is

$$\frac{1}{\rho}\frac{d\rho}{dt} + \boldsymbol{\nabla} \cdot \mathbf{v} = 0, \tag{12.4}$$

where (12.1) is used. For the incompressible fluid,[1] (12.4) becomes the simplest form as

$$\nabla \cdot \mathbf{v} = 0. \tag{12.5}$$

12.3 Forces Exerting on Fluid Parcels

In this section, we will discuss forces exerting on fluid parcels to describe momentum equations for fluids. These forces are classified into *body forces* and *surface forces*. The body forces exert on the center of mass of a fluid parcel, whose magnitude is proportional to the mass of the parcel. The surface forces act across the surface of the fluid parcel, whose magnitude has nothing to do with the mass of the parcel. The examples of the former are the gravitational force and the Coriolis force, and examples of the latter are the pressure gradient force and the viscous force.

12.3.1 The Pressure Gradient Force

The pressure is defined as the force per unit area exerting normally on a surface of a body. In the fluid interior, the pressure is caused by the momentum transfer per unit time and unit area on some material surface owing to random molecular motions. Suppose a rectangular infinitesimal volume element $\delta V = \delta x \delta y \delta z$ centered at (x, y, z) in a fluid as shown in Fig. 12.3. Letting the surface perpendicular to the x-axis and at $x - \delta x/2$ be A and the surface at $x + \delta x/2$ be B, the pressure at (x, y, z) be $p(x, y, z)$, and the density be $\rho(x, y, z)$, the pressure at the surface A is expressed in a Taylor series expansion as

$$p\left(x - \frac{\delta x}{2}\right) = p(x) - \frac{\partial p}{\partial x}\frac{\delta x}{2}, \tag{12.6}$$

neglecting all terms of order δx^2 and higher. Thus the x-component of the force exerting on the volume element at surface A is

$$F_{xA} = \left\{p(x) - \frac{\partial p}{\partial x}\frac{\delta x}{2}\right\}\delta y \delta z. \tag{12.7}$$

Similarly, the x-component of the force exerting on the volume element at surface B is

$$F_{xB} = \left\{p(x) + \frac{\partial p}{\partial x}\frac{\delta x}{2}\right\}\delta y \delta z. \tag{12.8}$$

---

[1]Fluid whose parcel does not change its density following the motion is called incompressible fluid. Namely, the individual rate of change of the density is zero $d\rho/dt = 0$. When the advective velocity is small compared with the acoustic speed, the above equation is approximated $\partial \rho/\partial t = 0$.

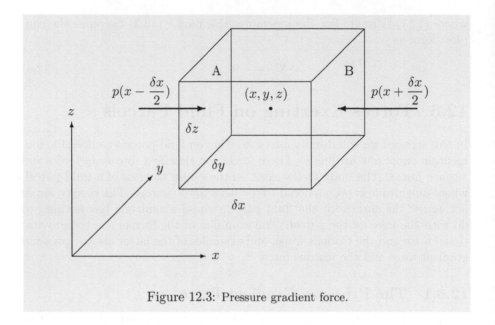

Figure 12.3: Pressure gradient force.

Therefore, the net x-component of the force exerting on the volume element is

$$F_x = -\frac{\partial p}{\partial x}\delta x \delta y \delta z. \tag{12.9}$$

The mass of the volume element is $m = \rho \delta x \delta y \delta z$, so that the x-component of the force per unit mass is

$$\frac{F_x}{m} = -\frac{1}{\rho}\frac{\partial p}{\partial x}. \tag{12.10}$$

Similarly, the y- and z-components of the force are obtained as

$$\frac{F_y}{m} = -\frac{1}{\rho}\frac{\partial p}{\partial y}, \tag{12.11}$$

$$\frac{F_z}{m} = -\frac{1}{\rho}\frac{\partial p}{\partial z}. \tag{12.12}$$

The vector form of the force due to pressure is

$$\frac{\mathbf{F}_p}{m} = -\frac{1}{\rho}\left(\frac{\partial p}{\partial x}\mathbf{e}_x + \frac{\partial p}{\partial y}\mathbf{e}_y + \frac{\partial p}{\partial z}\mathbf{e}_z\right) = -\frac{1}{\rho}\nabla p. \tag{12.13}$$

The force due to pressure takes the form of the gradient of the pressure, so it is called the pressure gradient force.

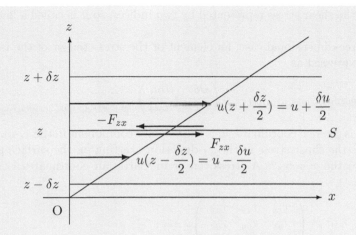

Figure 12.4: Two-dimensional shear flow.

12.3.2 Viscous Force

We assume the two-dimensional flow in the x-direction changing its magnitude in the z-direction as illustrated in Fig. 12.4. Suppose that the distance of adjacent two fluid layers is δz. If the fluid is viscous, the forces act between the fluid layers. This type of viscosity is caused by the momentum transfer of the basic flow owing to random molecular motions, and it is called the molecular viscosity. When the momentum transfer of the basic flow occurs owing to eddy motions, the viscosity is referred to as the eddy viscosity which is far larger in magnitude than the molecular viscosity. The viscous force is expressed in the indexed form as F_{zx}, in which the index z means that the force acts on the surface perpendicular to the z-axis and the index x means that the force acts in the x-direction. The force F_{zx} is proportional to the velocity difference between two fluid layers δu and the area of the interface of two-layers S and inversely proportional to the distance of adjacent fluid layers δz, namely

$$F_{zx} = \mu \frac{S \delta u}{\delta z},$$

where μ is the constant of proportionality called the *dynamic viscosity coefficient*. Dividing both sides by S and taking the limit $\delta z \to 0$, we find

$$\frac{F_{zx}}{S} = \tau_{zx} = \mu \frac{du}{dz}, \tag{12.14}$$

which is the shear stress represented by two indices, so it is called a 2nd-order tensor.[2]

In three-dimensional case, an element of the stress tensor of the isotropic fluid is expressed as

$$\tau_{ij} = \mu \left(\frac{\partial v_i}{\partial x_j} + \frac{\partial v_j}{\partial x_i} \right), \tag{12.15}$$

where x_i is the i-th coordinate, u_i is the velocity component in the x_i-direction, and τ_{ij} is the shear stress in the x_j-direction exerting on the surface perpendicular to the x_i-axis.[3] A stress tensor in Cartesian coordinates is written as

$$\tilde{\sigma} = \begin{pmatrix} \tau_{xx} & \tau_{xy} & \tau_{xz} \\ \tau_{yx} & \tau_{yy} & \tau_{yz} \\ \tau_{zx} & \tau_{zy} & \tau_{zz} \end{pmatrix}.$$

$$= \mu \begin{pmatrix} 2\dfrac{\partial u}{\partial x} & \dfrac{\partial u}{\partial y} + \dfrac{\partial v}{\partial x} & \dfrac{\partial u}{\partial z} + \dfrac{\partial w}{\partial x} \\ \dfrac{\partial u}{\partial y} + \dfrac{\partial v}{\partial x} & 2\dfrac{\partial v}{\partial y} & \dfrac{\partial w}{\partial y} + \dfrac{\partial v}{\partial z} \\ \dfrac{\partial u}{\partial z} + \dfrac{\partial w}{\partial x} & \dfrac{\partial w}{\partial y} + \dfrac{\partial v}{\partial z} & 2\dfrac{\partial w}{\partial z} \end{pmatrix}. \tag{12.16}$$

Referring Fig. 12.5, we will consider the net stress acting on a rectangular infinitesimal volume element $\delta V = \delta x \delta y \delta z$ centered at (x, y, z). Let the surface perpendicular to the z-axis and at $z - \delta z/2$ be C and at $z + \delta z/2$ be D, the stress at (x, y, z) be $\tau(x, y, z)$, and the density be $\rho(x, y, z)$. The stress on surface C is expressed in a Taylor series expansion as

$$\tau(z - \frac{\delta z}{2}) = \tau(z) - \frac{\partial \tau}{\partial z} \frac{\delta z}{2},$$

neglecting all terms of order δz^2 and higher. The stress exerted by the lower fluid on the volume element across surface C is in the negative x-direction, so that the x-component of the viscous force at surface C is

$$F_{zxC} = - \left\{ \tau_{zx}(z) - \frac{\partial \tau_{zx}}{\partial z} \frac{\delta z}{2} \right\} \delta x \delta y.$$

Similarly the x-component of the viscous force at surface D is

$$F_{zxD} = \left\{ \tau_{zx}(z) + \frac{\partial \tau_{zx}}{\partial z} \frac{\delta z}{2} \right\} \delta x \delta y.$$

---

[2]A scalar is a single number and is called a 0th-order tensor, and a vector is represented by one index and is called a 1st-order tensor.

[3]The deduction of (12.15) is out of the scope of this book. Readers who want to learn in detail, refer to textbooks of mechanics of deformable bodies and fluid dynamics (e.g., Batchelor, 1983).

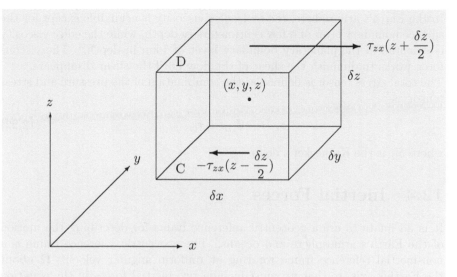

Figure 12.5: The stress of the x-component due to the z-directional shear.

The net x-component of the viscous force per unit mass is

$$\frac{F_{zxC} + F_{zxD}}{m} = \frac{1}{\rho \delta x \delta y \delta z} \frac{\partial \tau_{zx}}{\partial z} \delta x \delta y \delta z = \frac{1}{\rho} \frac{\partial \tau_{zx}}{\partial z}. \tag{12.17}$$

Similar discussion holds for surfaces perpendicular to the x- and y-axes, so that

$$\begin{aligned}
\frac{F_x}{m} &= \frac{1}{\rho} \left(\frac{\partial \tau_{xx}}{\partial x} + \frac{\partial \tau_{yx}}{\partial y} + \frac{\partial \tau_{zx}}{\partial z} \right) \\
&= \nu \nabla^2 u + \nu \frac{\partial}{\partial x} \left(\frac{\partial u}{\partial x} + \frac{\partial v}{\partial y} + \frac{\partial w}{\partial z} \right) \\
&= \nu \nabla^2 u, \tag{12.18}
\end{aligned}$$

where $\nu = \mu/\rho$ is referred to as the *kinematic viscosity coefficient*. The viscous forces in the y- and z-directions are also written as

$$\frac{F_y}{m} = \nu \nabla^2 v, \tag{12.19}$$

$$\frac{F_z}{m} = \nu \nabla^2 w. \tag{12.20}$$

The viscous force in the vector form is

$$\frac{\mathbf{F}_s}{m} = \nu \nabla^2 \mathbf{v}. \tag{12.21}$$

In the Earth's atmosphere, the molecular viscosity is negligible except for the surface boundary layer of a few centimeters in depth, while the eddy viscosity is essential in the planetary boundary layer of 1 km in depth.[4] The viscous force works to diminish the shear of the flow until the shear disappears.

The total stress tensor is defined as the combination of the pressure and stress tensor as

$$\sigma_{ij} = -\delta_{ij} p + \tau_{ij}, \tag{12.22}$$

where δ_{ij} is the Kronecker's delta.

12.4 Inertial Forces

It is adequate to use a geocentric reference frame for describing the motion of the Earth's atmosphere and oceans. The geocentric reference frame is a non-inertial reference frame rotating at uniform angular velocity $\mathbf{\Omega}$ about the Earth's axis, so that we must include two inertial forces in the equation of motion. One is the centrifugal force and the other is the Coriolis force. The former is combined with the universal gravitation and is included in the effective gravity as

$$\mathbf{g} = \mathbf{g}^* + \Omega^2 \mathbf{R} . \tag{12.23}$$

The latter is shown as follows, applying (4.20) for unit mass

$$\mathbf{F}_{\mathrm{Co}} = -2\mathbf{\Omega} \times \mathbf{v} = -2 \begin{vmatrix} \mathbf{e}_x & \mathbf{e}_y & \mathbf{e}_z \\ 0 & \Omega\cos\theta & \Omega\sin\theta \\ u & v & w \end{vmatrix},$$

$$= \mathbf{e}_x(2v\Omega\sin\theta - 2w\Omega\cos\theta) + \mathbf{e}_y(-2u\Omega\sin\theta) + \mathbf{e}_z(2u\Omega\cos\theta). \tag{12.24}$$

12.5 The Momentum Equations

In the absolute reference frame, the equation of motion of a parcel of unit mass moving at velocity $\mathbf{v}$ relative to the Earth is written

$$\frac{d_a \mathbf{v}_a}{dt} = -\frac{1}{\rho}\nabla p + \mathbf{g} + \nu\nabla^2\mathbf{v}, \tag{12.25}$$

using forces considered in the previous section.

Then, we can obtain the equation of motion in geocentric coordinate system (r, ϕ, θ),

$$\frac{d\mathbf{v}}{dt} = -2\mathbf{\Omega} \times \mathbf{v} - \frac{1}{\rho}\nabla p + \mathbf{g} + \nu\nabla^2\mathbf{v}, \tag{12.26}$$

---

[4]The planetary boundary layer or the Ekman boundary layer will be discussed in Chapter 13.

where use is made of (4.17), and the centrifugal force $\Omega^2\mathbf{R}$ is included in the effective gravity $\mathbf{g}$.

We will set local Cartesian coordinates (x, y, z) at point $P(r, \phi, \theta)$, taking the x-, y-, z-axes as ϕ-, θ-, r-directions, respectively. Namely

$$dx = r\cos\theta d\phi, \quad dy = r d\theta, \quad dz = d(r - a), \tag{12.27}$$

where a is the Earth's radius. Letting basis vectors be $(\mathbf{e}_x, \mathbf{e}_y, \mathbf{e}_z)$ and velocity components be (u, v, w), the velocity is designated

$$\mathbf{v} = u\mathbf{e}_x + v\mathbf{e}_y + w\mathbf{e}_z,$$

where

$$u = r\cos\theta\frac{d\phi}{dt}, \quad v = r\frac{d\theta}{dt}, \quad w = \frac{dr}{dt}, \tag{12.28}$$

We can replace r by a in (12.28) in fairly good approximation, because the depth of the Earth's atmosphere and oceans are negligibly small compared with the Earth's radius. Taking into account the acceleration terms due to the Earth's curvature, we obtain the component equations of (12.26),

$$\frac{du}{dt} - \frac{uv\tan\theta}{a} + \frac{uw}{a} = -\frac{1}{\rho}\frac{\partial p}{\partial x} + 2\Omega v\sin\theta - 2\Omega w\cos\theta + \nu\nabla^2 u, \tag{12.29}$$

$$\frac{dv}{dt} + \frac{u^2\tan\theta}{a} + \frac{vw}{a} = -\frac{1}{\rho}\frac{\partial p}{\partial y} - 2\Omega u\sin\theta + \nu\nabla^2 v, \tag{12.30}$$

$$\frac{dw}{dt} - \frac{u^2 + v^2}{a} = -\frac{1}{\rho}\frac{\partial p}{\partial z} - g + 2\Omega u\cos\theta + \nu\nabla^2 w, \tag{12.31}$$

These equations are the fluid dynamics version of Newton's second law, and are referred to as the *Navier–Stokes equations*.

12.6 Simplified Coordinate Systems

The acceleration terms due to the Earth's curvature in (12.29), (12.30), and (12.31) are sufficiently small in magnitude compared with other terms. We often use the simplified coordinate systems neglecting the curvature effect of the Earth and retaining the dynamic effect of the Earth's rotation. They are coordinate systems called the f-plane approximation and the β-plane approximation.

12.6.1 The f-plane Approximation

Let us place a tangent plane at latitude θ_0, taking the x-axis eastward, the y-axis northward and the z-axis upward (Fig. 12.6). As for the boundary, it is usual to put rigid walls at $y = \pm D$ and the cyclic condition in the x-direction, by which we can represent the round Earth. The Coriolis parameter f is constant through the channel as,

$$f_0 = 2\Omega \sin\theta_0. \tag{12.32}$$

The f-plane approximation is applied for the phenomena in which the vertical component of the angular velocity of the Earth's rotation is crucial but its latitudinal dependence is not necessarily important. For instance, the midlatitude depressions in the atmosphere and the inertial oscillations in the oceans or great lakes (Section 14.1) are well formulated in the f-plane approximation.

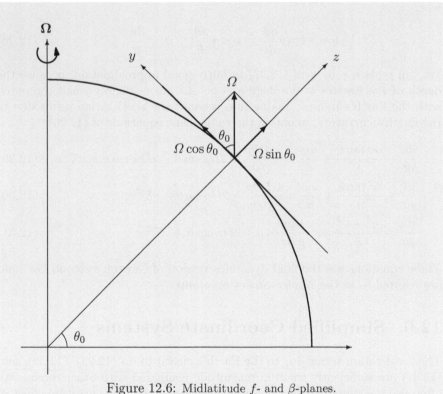

Figure 12.6: Midlatitude f- and β-planes.

12.6.2   The β-plane Approximation

We will place a tangent plane at latitude θ_0, taking the x-axis eastward, the y-axis northward, and the z-axis upward. As for the boundary, it is usual to put rigid walls at $y = \pm D$ and the cyclic condition in the x-direction.

We consider the latitudinal dependence of the Coriolis parameter f, which is expanded in a Taylor series neglecting all terms of order y^2 and higher as

$$f = f_0 + \left(\frac{df}{dy}\right)_{\theta_0} y = f_0 + \left(\frac{df}{ad\theta}\right)_{\theta_0} y$$
$$= f_0 + \beta y, \tag{12.33}$$

where

$$\beta = \left(\frac{df}{ad\theta}\right)_{\theta_0} = \frac{2\Omega\cos\theta_0}{a}. \tag{12.34}$$

Conventionally β is used as the coefficient of y, so that this simplified coordinate system is called the β-plane approximation, and the dynamic effect of the variation of f with latitude is called the β-effect. The magnitude of β is $\cong 10^{-11}\,\mathrm{m}^{-1}\,\mathrm{s}^{-1}$, so the β-effect becomes essential for large-scale motions in geophysical fluids, e.g., Rossby waves (Section 14.2) in the Earth's atmosphere and the meandering of the Gulf Stream in the Atlantic Ocean.

12.7 The Boussinesq Approximation

The Boussinesq approximation treats the density of fluids as constant ρ_0 except when it appears in terms multiplied by the acceleration due to gravity. Of course, the density change due to the pressure change is ignored. Namely, Boussinesq approximation treats fluids as incompressible fluids, so that the continuity equation and the viscous term in the Navier–Stokes equations turn out simple forms such as (12.5) and (12.21). The advantage of employing the Boussinesq approximation is that we can use the common Navier–Stokes equations, the continuity equation and the thermodynamic energy equation for the Earth's atmosphere, oceans and fluid systems in laboratories.
Here, we will consider how the buoyancy force is expressed in the Boussinesq approximation. The gravity force exerting on a unit volume of fluids becomes

$$\rho\mathbf{g} = (\rho_0 + \delta\rho)\mathbf{g}. \tag{12.35}$$

We will show in a later section, the first term of the right-hand side balances the vertical component of the basic portion of pressure gradient force and the buoyancy force is the second term of the right-hand side $\delta\rho\mathbf{g}$. The buoyancy is written using temperature deviation as

$$\delta\rho\mathbf{g} = (\rho - \rho_0)\mathbf{g} = -\rho_0\alpha(T - T_0)\mathbf{g}, \tag{12.36}$$

where α is the coefficient of thermal expansion, T is the temperature, and T_0 is the average temperature.

12.8   Scale Analysis

The conventional technique to non-dimensionalize independent variables and dependent variables by their characteristic physical values of each variable is called the *scale analysis* or *scaling*. In the process of the scaling, several nondimensional parameters appear, which are useful to evaluate the contribution of each term in the governing equations. If the nondimensional parameters between the different fluid systems are equal, it is said that the *dynamic similarity law* holds and the solutions of governing equations are commonly valid in both systems. This similarity law guarantees the validity of the numerical simulations and laboratory simulations for the phenomena in the Earth's atmosphere and oceans.

Supposing that the characteristic values of the horizontal scale is L, the height scale is H, the horizontal velocity scale is U, the vertical velocity scale is W, the vertical pressure fluctuation scale is ΔP and the horizontal pressure fluctuation scale is δP, the variables are written as

$$x = Lx^*, \quad y = Ly^*, \quad z = Hz^*, \quad u = Uu^*,$$
$$v = Uv^*, \quad w = Ww^*, \quad t = \tfrac{L}{U}t^*, \quad p = \delta P(\text{or } \Delta P)p^*,$$

where the variables with asterisks are nondimensional variables of the order of unity. At first, we will nondimensionalize the continuity equation (12.5).

$$\frac{U}{L}\left(\frac{\partial u^*}{\partial x^*} + \frac{\partial v^*}{\partial y^*}\right) + \frac{W}{H}\frac{\partial w^*}{\partial z^*} = 0. \tag{12.37}$$

From 12.37 we find

$$W \cong \frac{H}{L}U. \tag{12.38}$$

Many disturbances in geophysical fluids have the sinusoidal structure in the x- and y-directions, so two terms of the horizontal divergence tend to cancel each other. Thus the horizontal divergence reduces to around one tenth, therefore

$$W \cong 10^{-1}\frac{H}{L}U. \tag{12.39}$$

Next, let us nondimensionalize the Navier–Stokes equations (12.29), (12.30), and (12.31).

$$R_o\left(\frac{\partial u^*}{\partial t^*} + u^*\frac{\partial u^*}{\partial x^*} + v^*\frac{\partial u^*}{\partial y^*} + w^*\frac{\partial u^*}{\partial z^*}\right) - \frac{L}{a}R_o\left(\tan\theta u^*v^* - A_s u^*w^*\right)$$
$$= -\frac{\delta P}{\rho_0 U^2}R_o\frac{\partial p^*}{\partial x^*} + \left(v^* - A_s\frac{w^*}{\tan\theta}\right)$$
$$+ \frac{1}{T_a^{1/2}}\left(\frac{\partial^2}{\partial x^{*2}} + \frac{\partial^2}{\partial y^{*2}} + \frac{1}{A_s^2}\frac{\partial^2}{\partial z^{*2}}\right)u^*, \tag{12.40}$$

$$R_\text{o} \left(\frac{\partial v^*}{\partial t^*} + u^* \frac{\partial v^*}{\partial x^*} + v^* \frac{\partial v^*}{\partial y^*} + w^* \frac{\partial v^*}{\partial z^*} \right) + \frac{L}{a} R_\text{o} \left(\tan \theta u^{*2} + A_\text{s} v^* w^* \right)$$

$$= -\frac{\delta P}{\rho_0 U^2} R_\text{o} \frac{\partial p^*}{\partial y^*} - u^*$$

$$+ \frac{1}{T_\text{a}^{1/2}} \left(\frac{\partial^2}{\partial x^{*2}} + \frac{\partial^2}{\partial y^{*2}} + \frac{1}{A_\text{s}^2} \frac{\partial^2}{\partial z^{*2}} \right) v^*, \tag{12.41}$$

$$A_\text{s} R_\text{o} \left(\frac{\partial w^*}{\partial t^*} + u^* \frac{\partial w^*}{\partial x^*} + v^* \frac{\partial w^*}{\partial y^*} + w^* \frac{\partial w^*}{\partial z^*} \right) - \frac{L}{a} R_\text{o} \left(u^{*2} + v^{*2} \right)$$

$$= -\frac{R_\text{o}}{A_\text{s}} \frac{\Delta P}{\rho_0 U^2} \frac{\partial p^*}{\partial z^*} - \frac{L}{U^2} R_\text{o} g + \frac{1}{\tan \theta} u^*$$

$$+ \frac{A_\text{s}}{T_\text{a}^{1/2}} \left(\frac{\partial^2}{\partial x^{*2}} + \frac{\partial^2}{\partial y^{*2}} + \frac{1}{A_\text{s}^2} \frac{\partial^2}{\partial z^{*2}} \right) w^*. \tag{12.42}$$

Three nondimensional parameters appear in the above equations. The *Rossby number*

$$R_\text{o} = \frac{U}{2\Omega \sin \theta L} = \frac{U}{fL} \tag{12.43}$$

is the ratio of the speed of the phenomena to the speed of the Earth's rotation. The *Taylor number*

$$T_\text{a} = \frac{f^2 L^4}{\nu^2} \tag{12.44}$$

is the square ratio of the Coriolis force to the viscous force. And the *aspect ratio*

$$A_\text{s} = \frac{H}{L} \tag{12.45}$$

is the ratio of the vertical scale to the horizontal scale of the phenomena.

12.9 Basic Balance Equations

In this section, four important balance equations will be derived which are valid in the large-scale motions in the Earth's atmosphere and oceans. The characteristic motions are synoptic waves[5] and Rossby waves in the atmosphere, and oceanic currents represented by the Gulf Stream, the Kuroshio Current and the Antarctic Circumpolar Current.

Let us evaluate the magnitude of terms of the Navier–Stokes equations (12.40) and (12.42) based on the following characteristic scales of variables for synoptic-scale motions in the Earth's atmosphere:

---

[5] Waves in the Earth's atmosphere concerning the weather change with horizontal scale of 2,000–3,000 km.

$L \cong 10^6\,\mathrm{m}$ horizontal scale

$H \cong 10^4\,\mathrm{m}$ vertical scale

$a \cong 10^7\,\mathrm{m}$ Earth's radius

$f \cong 10^{-4}\,\mathrm{s}^{-1}$ Coriolis parameter

$U \cong 10\,\mathrm{m\,s}^{-1}$ horizontal velocity scale

$W \cong 10^{-2}\,\mathrm{m\,s}^{-1}$ vertical velocity scale

$\delta P \cong 10^3\,\mathrm{Pa}$ horizontal pressure fluctuation scale

$\Delta P \cong 10^5\,\mathrm{Pa}$ vertical pressure scale

$\nu \cong 10\,\mathrm{m^2\,s}^{-1}$ eddy viscosity coefficient

$\rho_0 \cong 1\,\mathrm{kg\,m}^{-3}$ mean density.

12.9.1   The Hydrostatic Equation

Let us evaluate the magnitude of terms of the vertical Navier–Stokes equation (12.42).

$$R_\mathrm{o} \cong 10^{-1}$$

$$A_\mathrm{s} R_\mathrm{o} \cong 10^{-3}$$

$$\frac{L}{a} R_\mathrm{o} \cong 10^{-2}$$

$$\frac{R_\mathrm{o}\Delta P}{A_\mathrm{s}\rho_0 U^2} \cong 10^4$$

$$\frac{L}{U^2} R_\mathrm{o} g \cong 10^4$$

$$\frac{A_\mathrm{s}}{T_\mathrm{a}^{1/2}} \cong 10^{-9}$$

As two terms, the pressure gradient force and the gravity force, are dominant in (12.42), (12.31) becomes

$$\frac{\partial p}{\partial z} = -\rho g. \tag{12.46}$$

Equation (12.46) is referred to as the *hydrostatic equation*. This equation is also obtained, supposing the static balance of forces exerting on an infinitesimal circular column of fluid as shown in Fig. 12.7. Three forces acting on the fluid column are balanced as follows:

$$p(z)S - p(z + \delta z)S - Mg = 0, \tag{12.47}$$

where S is the cross section and $M = \rho S \delta z$ is the mass of the fluid column. Expanding the second term of the left-hand side of (12.47) in a Taylor series, neglecting all terms of order δz^2 and higher, and dividing the resultant

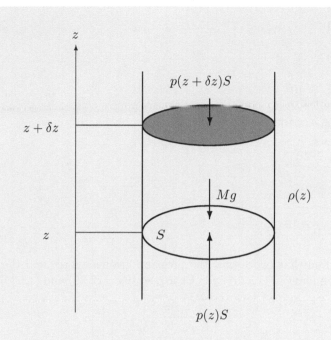

Figure 12.7: The hydrostatic equilibrium.

equation through by M, we find

$$\frac{dp}{dz} = -\rho g. \tag{12.48}$$

In the Boussinesq approximation, the change of the density is taken into account only when the density is multiplied by the acceleration due to gravity **g**, otherwise it is treated as the constant value ρ_0.

Problem 1. The global average of the mean sea level pressure of the atmosphere is 1013.25 hPa. Calculate the global average of the height[6] of the atmosphere, supposing the density of the atmosphere is constant ($\rho = 1.00\,\mathrm{kg\,m^{-3}}$). Let the magnitude of the acceleration due to gravity be $9.80\,\mathrm{m\,s^{-2}}$.

12.9.2 The Geostrophic Approximation

We will evaluate the magnitude of terms of the horizontal Navier–Stokes equations (12.40) and (12.41).

---

[6]The global average of the tropopause height is about 1.0×10^4 m, under which 75% of the atmospheric mass exists.

$$R_o \cong 10^{-1}$$

$$\frac{L}{a} \cong 10^{-1}$$

$$\frac{\delta P}{\rho_0 U^2} R_o \cong 10^0$$

$$\frac{L}{a} R_o \tan\theta \cong 10^{-2}$$

$$\frac{L}{a} R_o A_s \cong 10^{-4}$$

$$\frac{1}{T_a^{1/2}} \cong 10^{-7}$$

$$\frac{1}{A_s^2 T_a^{1/2}} \cong 10^{-3}$$

We can find that two terms, the pressure gradient force and the Coriolis force, are dominant in (12.39) and (12.40), so that (12.29) and (12.30) become

$$fv_g = \frac{1}{\rho_0}\frac{\partial p}{\partial x} \tag{12.49}$$

$$fu_g = -\frac{1}{\rho_0}\frac{\partial p}{\partial y}. \tag{12.50}$$

Equations (12.49) and (12.50) are called the *geostrophic equations* and the velocity subscripted by g is called the *geostrophic flow*. In the Northern Hemisphere, the geostrophic flow travels parallel to the isobars with the lower pressure to the left-hand side. In the midlatitude, the difference between the actual flow and the geostrophic flow is within 10%.

12.9.3   The Quasi-Geostrophic Approximation

When we retain terms an order of magnitude smaller than the leading terms in the horizontal Navier–Stokes equations, we obtain the set of prognostic equations as follows:

$$\frac{du}{dt} = f(v - v_g), \tag{12.51}$$

$$\frac{dv}{dt} = -f(u - u_g), \tag{12.52}$$

where the pressure gradient forces are replaced by the geostrophic flows. Equations (12.51) and (12.52) are called the *quasi-geostrophic equations*, which are prediction equations of the flow field. The synoptic-scale flow field evolves due to the small difference between the actual flow and the geostrophic flow.

12.9.4 The Thermal Flow Balance

The geostrophic flow has the vertical shear in the presence of the horizontal temperature gradient or the horizontal density gradient. Differentiating (12.49) and (12.50) with respect to z, and applying the hydrostatic equation (12.48), we find

$$\frac{\partial v_g}{\partial z} = \frac{1}{f\rho_0} \frac{\partial}{\partial x}\left(\frac{\partial p}{\partial z}\right) = -\frac{g}{f\rho_0}\frac{\partial \rho}{\partial x}, \tag{12.53}$$

$$\frac{\partial u_g}{\partial z} = -\frac{1}{f\rho_0} \frac{\partial}{\partial y}\left(\frac{\partial p}{\partial z}\right) = \frac{g}{f\rho_0}\frac{\partial \rho}{\partial y}, \tag{12.54}$$

or in the vector form

$$\frac{\partial \mathbf{v}_g}{\partial z} = -\frac{g}{f\rho_0}\mathbf{e}_z \times \nabla\rho. \tag{12.55}$$

Equation (12.55) is called the *thermal flow balance*, which states that in the Northern Hemisphere the vertical shear of the geostrophic flow is positive when the density gradient is positive to the left-side from the direction of the geostrophic flow.

12.10 Circulation and Vorticity

Deforming Newton's second law, we derived the angular momentum equation for the rotational motion of a particle or a many-particle system. The equivalence of the angular momentum equation in fluid dynamics is the *vorticity equation* derived from the Navier–Stokes equations. For this preparation, we will introduce circulation and vorticity in this section.

12.10.1 Circulation

Circulation is a scalar quantity and a macroscopic measure of the rotational motion of fluids in a finite area. The circulation C is defined about a close contour in a fluid as

$$C = \oint \mathbf{v}\cdot dl, \tag{12.56}$$

where $\mathbf{v}$ is the velocity of a fluid parcel and dl is the displacement vector locally tangent to the contour. By convention the line integral is performed counterclockwise along the contour. Differentiating (12.56) with respect to time t, we obtain

$$\frac{dC}{dt} = \oint \frac{d\mathbf{v}}{dt}\cdot dl + \oint \mathbf{v}\cdot\frac{d}{dt}dl, \tag{12.57}$$

The second term of the right-hand side becomes

$$\oint \mathbf{v} \cdot d\left(\frac{d\mathbf{l}}{dt}\right) = \oint \mathbf{v} \cdot d\mathbf{v}$$

$$= \frac{1}{2} \oint d(\mathbf{v} \cdot \mathbf{v}) = \frac{1}{2} \oint d|\mathbf{v}|^2 = 0.$$

Thus (12.57) becomes

$$\frac{dC}{dt} = \oint \frac{d\mathbf{v}}{dt} \cdot d\mathbf{l}. \tag{12.58}$$

From (12.26), the Navier–Stokes equation for inviscid fluids in the inertial reference frame is

$$\frac{d\mathbf{v}}{dt} = -\frac{1}{\rho}\boldsymbol{\nabla}p + \boldsymbol{\nabla}\varPhi, \tag{12.59}$$

where the acceleration due to gravity $\mathbf{g}$ is replaced by the potential $\varPhi$. Taking a scalar product of (12.59) and $d\mathbf{l}$, and integrating along the closed contour, we get

$$\oint \frac{d\mathbf{v}}{dt} \cdot d\mathbf{l} = \oint \left(-\frac{1}{\rho}\boldsymbol{\nabla}p + \boldsymbol{\nabla}\varPhi\right) \cdot d\mathbf{l}$$

$$= \int_S \left\{\frac{1}{\rho^2}\left(\boldsymbol{\nabla}\rho \times \boldsymbol{\nabla}p\right) \cdot \mathbf{n}dS\right\}, \tag{12.60}$$

where S is the surface enclosed by the closed contour and $\mathbf{n}dS$ is the area element vector whose direction and the counterclockwise sense integration obey the *right-hand screw rule*. In the process of deducing (12.60), use is made of *Stokes' theorem*[7] and the following vector identity

$$\boldsymbol{\nabla} \times \boldsymbol{\nabla}\varPhi = 0.$$

From (12.58), (12.59), and (12.60), we obtain

$$\frac{dC}{dt} = \int_S \frac{1}{\rho^2}\left(\boldsymbol{\nabla}\rho \times \boldsymbol{\nabla}p\right) \cdot \mathbf{n}dS, \tag{12.61}$$

The right-hand side of (12.61) is called the *solenoid term*. For a *barotropic fluid* whose density is a function only of pressure, the solenoid term is zero, so that the circulation is conserved following the motion. This conservation law is called *Kelvin's circulation theorem*,[8] which is the fluid dynamics analog of the angular momentum conservation for rigid bodies.

---

[7] $\oint \mathbf{V} \cdot d\mathbf{l} = \int_S (\boldsymbol{\nabla} \times \mathbf{V}) \cdot \mathbf{n}dS$, where $\mathbf{V}$ is an arbitrary vector.

[8] Lord Kelvin appeared previously in the profile of James Joule in Chapter 5.

12.10.2 Absolute circulation

Suppose a closed contour at latitude θ in an inviscid fluid whose area normal to the Earth surface is S. Integrating (12.25) around the contour, we obtain

$$\oint \frac{d\mathbf{v}_a}{dt} \cdot dl = \oint \left(-\frac{1}{\rho} \nabla p + \nabla \Phi \right) \cdot dl,$$

$$\frac{dC_a}{dt} = \int_S \left\{ \frac{1}{\rho^2} (\nabla \rho \times \nabla p) \cdot d\mathbf{S} \right\}, \tag{12.62}$$

where C_a is called the absolute circulation. In the barotropic fluid, the solenoid term $\nabla \rho \times \nabla p$ is zero, so that C_a is conserved following the motion.

$$C_a = \oint \mathbf{v}_a \cdot dl = \oint (\mathbf{v} + \Omega \times \mathbf{r}) \cdot dl$$

$$= C + \int_S \nabla \times \Omega \times \mathbf{r} \cdot d\mathbf{S} = C + 2\Omega A_e, \tag{12.63}$$

where A is the area encompassed by the closed contour and $A_e = A \sin \theta$ is the projection of A to the equatorial plane.

Problem 2. Prove the following relation.

$$\nabla \times \Omega \times \mathbf{r} = 2\Omega,$$

12.10.3 Vorticity

Applying Stokes' theorem to (12.56), we find

$$C = \oint \mathbf{v} \cdot dl = \int_S (\nabla \times \mathbf{v}) \cdot \mathbf{n} dS. \tag{12.64}$$

Dividing (12.64) through by the area S, and taking the limit $S \to 0$, we get

$$\lim_{S \to 0} \frac{C}{S} = \lim_{S \to 0} \frac{1}{S} \int_S (\nabla \times \mathbf{v}) \cdot \mathbf{n} dS = (\nabla \times \mathbf{v}) \cdot \mathbf{n} = \omega \cdot \mathbf{n}, \tag{12.65}$$

where

$$\omega = \nabla \times \mathbf{v} \tag{12.66}$$

is called the vorticity, which is defined at any point in the fluid and is the microscopic measure of the rotational motion of fluid. In Cartesian coordinates, the vorticity is written as

$$\omega = \begin{pmatrix} \mathbf{e}_x & \mathbf{e}_y & \mathbf{e}_z \\ \dfrac{\partial}{\partial x} & \dfrac{\partial}{\partial y} & \dfrac{\partial}{\partial z} \\ u & v & w \end{pmatrix}$$

$$= \mathbf{e}_x \left(\frac{\partial w}{\partial y} - \frac{\partial v}{\partial z} \right) + \mathbf{e}_y \left(\frac{\partial u}{\partial z} - \frac{\partial w}{\partial x} \right) + \mathbf{e}_z \left(\frac{\partial v}{\partial x} - \frac{\partial u}{\partial y} \right)$$

$$= (\xi, \eta, \zeta). \tag{12.67}$$

Example 1. Suppose an infinitesimal rectangle in the x-y plane as shown in Fig. 12.8. Let the position of the center of the rectangle be (x, y) and its sides be δx and δy. Calculate the circulation along the closed contour of the rectangle and obtain the vorticity.

Answer
Calculating the circulation along the side of the rectangle, we get

$$\delta C = u(y - \frac{\delta y}{2})\delta x + v(x + \frac{\delta x}{2})\delta y - u(y + \frac{\delta y}{2})\delta x - v(x - \frac{\delta x}{2})\delta y,$$

Expanding the above equation in a Taylor series neglecting all terms of order

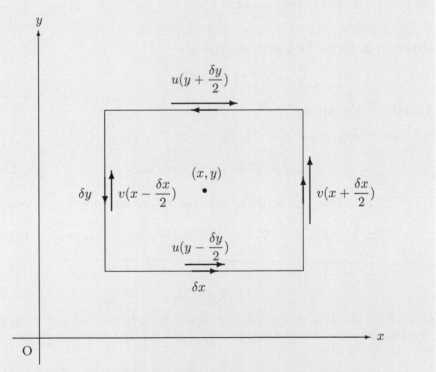

Figure 12.8: Circulation and vorticity for an infinitesimal rectangle in the x-y plane.

δt^2 and higher, we get the circulation

$$\delta C = \left(\frac{\partial v}{\partial x} - \frac{\partial u}{\partial y}\right) \delta x \delta y. \tag{12.68}$$

Dividing (12.68) through by the area $\delta S = \delta x \delta y$, we obtain the z-component vorticity

$$\zeta = \frac{\partial v}{\partial x} - \frac{\partial u}{\partial y}.$$

12.10.4 The Quasi-Geostrophic Vorticity Equation

We will derive the vorticity equation for a quasi-geostrophic flow. Differentiating (12.52) with respect to x and (12.51) with respect to y and subtracting the latter from the former, we find

$$\frac{d(\zeta + f)}{dt} = -\left(\frac{\partial u}{\partial x} + \frac{\partial v}{\partial y}\right)(\zeta + f) = \frac{\partial w}{\partial z}(\zeta + f). \tag{12.69}$$

The case of uniform fluid depth

Suppose a fluid with uniform depth H on a midlatitude β-plane. Integrating (12.69) with respect to z from the bottom $z = h_B$ to the top $z = h_T$,

$$H\frac{d(\zeta + f)}{dt} = [w(h_T) - w(h_B)](\zeta + f)$$

$$= \left(\frac{dh_T}{dt} - \frac{dh_B}{dt}\right)(\zeta + f) = \frac{dH}{dt}(\zeta + f) = 0,$$

$$\frac{d\eta}{dt} = \frac{d(\zeta + f)}{dt} = 0. \tag{12.70}$$

We call $\eta = \zeta + f$ the *absolute vorticity*, and (12.68) is the conservation law of absolute vorticity.

The case of variable fluid depth

Suppose a fluid with variable depth H on a midlatitude β-plane. Integrating (12.67) with respect to z from the bottom $z = h_B$ to the top $z = h_T$, we get

$$H\frac{d(\zeta + f)}{dt} = \frac{dH}{dt}(\zeta + f),$$

$$\frac{d}{dt}\left(\frac{\zeta + f}{H}\right) = 0. \tag{12.71}$$

We call $(\zeta + f)/H$ the *absolute potential vorticity*, and (12.71) is the conservation law of the absolute potential vorticity.

Let us consider a flow in a channel flow with the constant Coriolis parameter f. The top and bottom boundaries are flat, and a semi-spherical obstacle is set at the bottom. The basic flow is uniform and directed to the positive x-direction. Suppose that a cylindrical fluid column is advected by the basic flow from the upstream of the obstacle as shown in Fig. 12.9. When the column comes to the obstacle, the fluid column begins to shrink so that it acquires the negative relative vorticity due to the conservation of the absolute potential vorticity. Then the fluid column turns to the negative y-direction. After passing through the top of the obstacle, the fluid column begins to stretch and the relative vorticity approaches zero until it passes through the obstacle. After passing the obstacle, the fluid column is steered by the uniform flow.

12.10.5   Ertel Potentialvorticity

In a stably stratified baroclinic fluid, suppose a fluid column with a cross section δS confined between a constant density surfaces $\rho_1 = \rho$ and $\rho_2 = \rho - \delta\rho$ (Fig. 12.10). Let the distance of two layers at the fluid column be δz and its

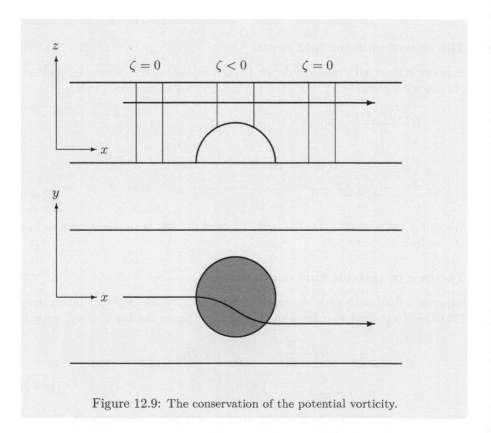

Figure 12.9: The conservation of the potential vorticity.

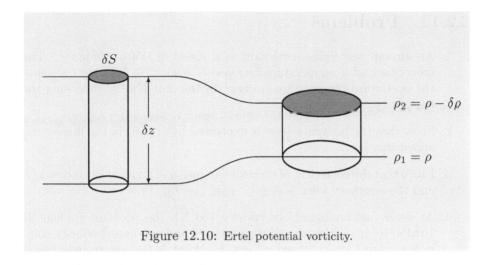

Figure 12.10: Ertel potential vorticity.

absolute circulation be C_a. Integrating (12.62) from ρ_1 to ρ_2, we obtain

$$(\rho_2 - \rho_1)\frac{dC_a}{dt} = -\delta\rho\frac{dC_a}{dt} = 0, \qquad (12.72)$$

In the limit $\delta S \to 0$, we find

$$C_a = (\zeta + f)\,\delta S. \qquad (12.73)$$

Therefore,

$$-\delta\rho\delta S\,(\zeta + f) = \text{const.} \qquad (12.74)$$

While, the mass of the fluid column δM is conserved following the motion.

$$\delta M = \rho\delta S\delta z = \text{const.} \qquad (12.75)$$

Dividing (12.74) by (12.75) for each side, we obtain

$$(\zeta + f)\left(-\frac{1}{\rho}\frac{\delta\rho}{\delta z}\right) = \text{const.}$$

Taking the limit $\delta z \to 0$, we get

$$(\zeta + f)\left(-\frac{1}{\rho}\frac{\partial\rho}{\partial z}\right) = \text{const.} \qquad (12.76)$$

$(\zeta + f)\left(-\dfrac{1}{\rho}\dfrac{\partial\rho}{\partial z}\right)$ is the Ertel potential vorticity in the Boussinesq approximation.

12.11   Problems

1. An aircraft was flying northward at a speed of $1.50 \times 10^2 \, \mathrm{m\,s^{-1}}$. The crew observed a temperature decrease of $6.00 \times 10^{-4} \, \mathrm{K\,s^{-1}}$. Calculate the northward temperature gradient in the unit $\mathrm{K\,m^{-1}}$, supposing the temperature to be stationary.

2. Show that the buoyancy force is expressed by (12.36) in the Boussinesq approximation.

3. Prove that the coefficient of thermal expansion of an ideal gas is $\alpha = 1/T$ and the buoyancy force is $\delta\rho\mathbf{g} = -\rho_0\mathbf{g}(T - T_0)/T_0$.

4. At some meteorological observatory ($30°\mathrm{N}$), the pressure gradient is $10 \, \mathrm{hPa}/10^6 \, \mathrm{m}$ northward and the eastward and northward velocity components are $13.0 \, \mathrm{m\,s^{-1}}$ and $2.5 \, \mathrm{m\,s^{-1}}$. What is the geostrophic wind? Next, calculate the wind after one hour. Let the density of the air be $\rho_0 = 1.20 \, \mathrm{kg\,m^{-3}}$.

5. What is the vorticity of a fluid in the *rigid body rotation* at the angular velocity $\mathbf{\Omega}$?

12.12   Reference

1. Batchelor G. K.: *An Introduction to Fluid Dynamics*, Cambridge University Press, New York (1983).

Chapter 13

Phenomena in Geophysical Fluids: Part I

In this chapter, we will present several dynamic phenomena in the Earth's atmosphere, oceans, and laboratory fluid systems. They are the Taylor–Proudman column, the Ekman flow, the Kelvin–Helmholtz instability, the Rayleigh–Bénard convection, and the Taylor vortices. Except for the Rayleigh–Bénard convection and the Kelvin–Helmholtz instability, the topics are concerned with the mechanics of rotating fluids.

13.1   The Taylor–Proudman Theorem

The Taylor–Proudman theorem or the Taylor–Proudman column shows most prominently the dynamic characteristics of rotating fluids, namely, the degeneracy of the freedom of fluid motion from three to two.

For inviscid fluids, the Navier–Stokes equation (12.25) becomes

$$\frac{d\mathbf{v}}{dt} = -2\mathbf{\Omega} \times \mathbf{v} - \frac{1}{\rho}\boldsymbol{\nabla}p - \boldsymbol{\nabla}\Phi, \tag{13.1}$$

where

$$-\boldsymbol{\nabla}\Phi = \mathbf{g}.$$

Supposing that the fluid motion is steady and the angular velocity of the relative motion is sufficiently small compared with the angular velocity of the system $\mathbf{\Omega}$ (the Rossby number is far smaller than unity), (13.1) becomes,

$$0 = -2\mathbf{\Omega} \times \mathbf{v} - \frac{1}{\rho}\boldsymbol{\nabla}p - \boldsymbol{\nabla}\Phi, \tag{13.2}$$

DOI: 10.1201/9781003310068-13

★★★★★★★★★★★★★★★★ G.I. Taylor   (1886–1975) ★★★★★★★★★★★★★★★★

Figure 13.1: A portrait of Sir Geoffrey Ingram Taylor in his young days at Trinity College, Cambridge (©Trinity College, Cambridge. Used with permission.). G.I. Taylor was a British physicist and mathematician. He was born in St. John's Wood, Middlesex, on March 7, 1886. He read mathematics at Trinity College, Cambridge and became a Reader in Dynamical Meteorology at Trinity College. At that time, Taylor studied the turbulent motion in the atmosphere and published *Turbulent motion in fluids*. During World War I he worked at the Royal Aircraft Factory, and returned to Trinity College after the war and conducted laboratory experiments on the Taylor–Proudman column. In 1923, he was appointed to a Royal Society research professorship as a Yarrow Research Professor, which enabled him to concentrate research work. He made great contributions to fluid mechanics and solid mechanics, e.g., Taylor vortices in Couette flow, Rayleigh–Taylor instability, Taylor dispersion, Taylor's frozen turbulence hypothesis, the deformation of crystalline material, and so on. During World War II, he participated in the Manhattan Project and studied implosion instability problems. In 1944, Taylor received a knighthood and was awarded the Copley medal from the Royal Society. He passed away in Cambridge, Cambridgeshire, on June 27, 1975.

★★

Taking the curl of (13.2), we get

$$\boldsymbol{\Omega} \cdot \boldsymbol{\nabla} \mathbf{v} = 0,$$
$$\frac{\partial \mathbf{v}}{\partial z} = 0, \tag{13.3}$$

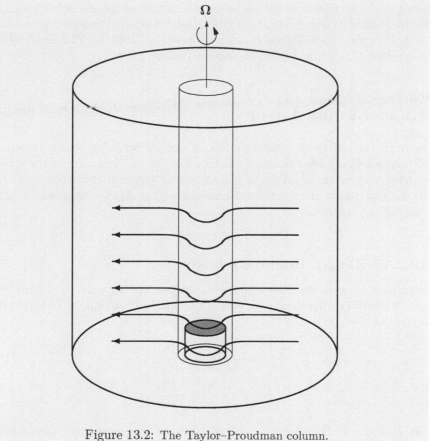

Figure 13.2: The Taylor–Proudman column.

where the following vector identitics are used.

$$\nabla \times \nabla \Phi = 0,$$
$$\nabla \times (\mathbf{A} \times \mathbf{B}) = \mathbf{A}(\nabla \cdot \mathbf{B}) - \mathbf{B}(\nabla \cdot \mathbf{A}) - (\mathbf{A} \cdot \nabla)\mathbf{B} + (\mathbf{B} \cdot \nabla)\mathbf{A}.$$

This equation was first derived by S. Hough[1] (1895), and later presented by J. Proudman in 1916. In 1917, G.I. Taylor (Fig. 13.1) performed laboratory experimonts using a rotating fluid in a cylindrical vessel with a moving obstacle at the bottom as illustrated in Fig. 13.2. A semi-spherical obstacle was attached to a screw crossing the base of the rotating cylinder and was moved slowly by the screw. A relative flow was caused and the flow was visualized

---

[1]Sydney Samuel Hough (1870–1923) was a British applied mathematician and astronomer. He obtained the solution of the Laplace tidal equation which were eigenfunctions named later Hough functions.

using dye. The flow from the bottom to the height of the obstacle avoided the obstacle without riding over it. Further, the flow above the obstacle avoided it as if there were a solid circular column above the obstacle. This fluid column was referred to as the Taylor–Proudman column.

13.2 Ekman Layer

Near the boundary of rotating fluids, a special boundary layer named the *Ekman boundary layer* develops. Let us discuss the dynamics of the Ekman boundary layer limiting the discussion to the Northern Hemisphere ($f > 0$). As for the case of the Southern Hemisphere ($f < 0$), we will cover it in the chapter end problem.

13.2.1 Ekman Boundary Layer

Suppose a viscous, stationary flow whose departure from the rigid body rotation is small (i.e., the Rossby number is far smaller than unity). The governing equations are

$$-fv = -\frac{1}{\rho}\frac{\partial p}{\partial x} + \nu\frac{\partial^2 u}{\partial z^2}, \tag{13.4}$$

$$fu = -\frac{1}{\rho}\frac{\partial p}{\partial y} + \nu\frac{\partial^2 v}{\partial z^2}. \tag{13.5}$$

We may choose the x-axis as the direction of the geostrophic flow without losing generality. Then we have

$$fu_g = -\frac{1}{\rho}\frac{\partial p}{\partial y}, \tag{13.6}$$

$$0 = -\frac{1}{\rho}\frac{\partial p}{\partial x} . \tag{13.7}$$

Expressing the pressure gradient force in terms of the geostrophic flow, we find

$$-fv = \nu\frac{\partial^2 u}{\partial z^2}, \tag{13.8}$$

$$f(u - u_g) = \nu\frac{\partial^2 v}{\partial z^2}. \tag{13.9}$$

The boundary conditions are

$$\begin{cases} u = 0, & v = 0, & \text{at} & z = 0 , \\ u \to u_g, & v \to 0, & \text{as} & z \to \infty . \end{cases} \tag{13.10}$$

We will define the complex velocity as,

$$w = u + \tilde{i}v, \tag{13.11}$$

where $\tilde{i}$ is the imaginary unit. Multiplying (13.9) by the imaginary unit $\tilde{i}$ and adding the resultant equation to (13.8), we obtain

$$\frac{d^2 w}{dz^2} - \tilde{i}\frac{f}{\nu}(w - u_g) = 0. \tag{13.12}$$

Defining $w' = w - u_g$, (13.12) becomes

$$\frac{d^2 w'}{dz^2} - \tilde{i}\frac{f}{\nu}w' = 0. \tag{13.13}$$

Assuming an exponential type solution $w' \propto \exp(\eta z)$ and substitute it into (13.13), we get

$$\eta^2 = (\eta_r{}^2 - \eta_i{}^2) + \tilde{i}2\eta_r\eta_i = \tilde{i}\frac{f}{\nu},$$

$$\eta_r = \eta_i = \pm\sqrt{\frac{f}{2\nu}} \equiv \pm\gamma.$$

The general solution of (13.13) is

$$w' = \alpha\exp\left[\gamma(1 + \tilde{i})z\right] + \beta\exp\left[-\gamma(1 + \tilde{i})z\right], \tag{13.14}$$

where $\alpha = \alpha_r + \tilde{i}\alpha_i$ and $\beta = \beta_r + \tilde{i}\beta_i$ are complex coefficients. Remembering $w' = u - u_g + \tilde{i}v$, we find the general solutions in the component form,

$$u = u_g + [\alpha_r\exp(\gamma z) + \beta_r\exp(-\gamma z)]\cos\gamma z$$
$$- [\alpha_i\exp(\gamma z) + \beta_i\exp(-\gamma z)]\sin\gamma z, \tag{13.15}$$
$$v = [\alpha_i\exp(\gamma z) + \beta_i\exp(-\gamma z)]\cos\gamma z$$
$$+ [\alpha_r\exp(\gamma z) - \beta_r\exp(-\gamma z)]\sin\gamma z. \tag{13.16}$$

Applying the boundary conditions

$$\begin{cases} u = 0, \quad v = 0, \quad \text{at} \quad z = 0, \\ u \to u_g, \quad v \to 0, \quad \text{as} \quad z \to \infty, \end{cases} \tag{13.17}$$

We obtain the particular solutions,

$$u = u_g\left[1 - \exp(-\gamma z)\cos\gamma z\right], \tag{13.18}$$
$$v = u_g\exp(-\gamma z)\sin\gamma z. \tag{13.19}$$

In the Northern Hemisphere, the velocity rotates clockwise with height and the trajectory of the head of the velocity vectors is referred to as the *Ekman*

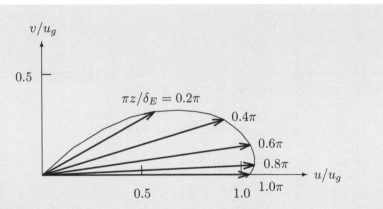

Figure 13.3: Hodograph of the Ekman flow. Velocity vectors are shown at normalized height $\pi z/\delta_E = 0.2\pi$, 0.4π, 0.6π, 0.8π, 1.0π.

spiral (Fig. 13.3). The flow is called the *Ekman flow*, which is parallel and equal to the geostrophic flow at height

$$\delta_E = \frac{\pi}{\gamma} = \pi\sqrt{\frac{2\nu}{f}}. \tag{13.20}$$

We call δ_E the *Ekman depth*. The viscous layer is referred to as the Ekman boundary layer or the planetary boundary layer.

Example 1. In the midlatitude, the depth of the atmospheric Ekman layer is 1.0×10^3 m. Obtain the mean eddy viscosity in the atmospheric boundary layer.

Answer
From (13.20), we find

$$\nu = \frac{f\delta_E{}^2}{2\pi^2} = \frac{(1.0 \times 10^3)^2 \times 1.0 \times 10^{-4}}{2 \times 3.14^2} = 5.0\,[\text{m}^2\,\text{s}^{-1}].$$

Problem 1. Calculate the mass flux in the Ekman layer due to the Ekman flow expressed by (13.18) and (13.19), supposing that density ρ is constant.

13.2.2 Oceanic Ekman Layer

Fridtjof Nansen, a Norwegian oceanographer and explorer, found that icebergs drifted 20° to 40° right from the direction of the prevailing winds during his expedition of the Fram in the Arctic Ocean from 1893 to 1896. He consulted about this phenomenon with Vilhelm Bjerknes, who was a professor of applied

★★★★★★★★★★★★★★ Vagn W. Ekman (1874–1954) ★★★★★★★★★★★★★★★

Figure 13.4: Vagn Walfrid Ekman was a Swedish oceanographer. He was born on May 3, 1874 at Stockholm. He studied physics at the University of Uppsala, and was interested in oceanography and fluid dynamics after taking the lectures of V. Bjerknes. Bjerknes invited Ekman to the University of Stockholm and inspired him to the problem posed by Nansen. Ekman obtained the solution, which came to be known as the famous Ekman spiral, and presented it as his doctoral thesis in 1902. From 1910 to 1939, he worked as a professor of mechanics and mathematical physics at the University of Lund, and studied theoretical and experimental oceanography. He died on March 9, 1954, in Gostad, Sweden.
★★★

mechanics and mathematical physics at the University of Stockholm. Bjerknes invited Vagn W. Ekman (Fig. 13.4), a doctoral student at the University of Uppsala, and inspired him to the problem posed by Nansen. Ekman solved the problem and presented it as his doctoral thesis in 1902.

Let us obtain the Ekman flow driven by wind stress. Taking the x-axis as the direction of the prevailing wind, the z-axis vertically upward and the y-axis so as to obey the right-hand rule. The governing equations are

$$-fv = \nu \frac{\partial^2 u}{\partial z^2}, \qquad (13.21)$$

$$fu = \nu \frac{\partial^2 v}{\partial z^2}. \qquad (13.22)$$

The boundary conditions are

$$\left\{ \begin{array}{ll} \rho_0 \nu \dfrac{\partial u}{\partial z} = \tau, & \dfrac{\partial v}{\partial z} = 0, \quad \text{at} \quad z = 0, \\ u \to 0, & v \to 0, \quad \text{as} \quad z \to -\infty. \end{array} \right. \qquad (13.23)$$

We will define the complex velocity as

$$w = u + \tilde{\imath} v. \qquad (13.24)$$

Multiplying (13.22) by the imaginary unit $\tilde{i}$ and adding the resultant equation to (13.21), we get

$$\frac{d^2w}{dz^2} - \tilde{i}\frac{f}{\nu}w = 0. \tag{13.25}$$

Supposing an exponential type solution $w \propto \exp \lambda z$ and substituting it into (13.26), we obtain

$$\lambda_r = \lambda_i = \pm\sqrt{\frac{f}{2\nu}} = \pm\gamma. \tag{13.26}$$

The general solution is

$$w = \alpha \exp\left(\gamma z\right)(\cos\gamma z + \tilde{i}\sin\gamma z) + \beta\exp\left(-\gamma z\right)(\cos\gamma z - \tilde{i}\sin\gamma z). \tag{13.27}$$

Separating w into the real and imaginary parts,

$$\begin{aligned}
u &= [\alpha_r \exp\left(\gamma z\right) + \beta_r \exp\left(-\gamma z\right)]\cos\gamma z \\
&\quad -[\alpha_i \exp\left(\gamma z\right) - \beta_i \exp\left(-\gamma z\right)]\sin\gamma z,
\end{aligned} \tag{13.28}$$

$$\begin{aligned}
v &= [\alpha_i \exp\left(\gamma z\right) + \beta_i \exp\left(-\gamma z\right)]\cos\gamma z \\
&\quad +[\alpha_r \exp\left(\gamma z\right) - \beta_r \exp\left(-\gamma z\right)]\sin\gamma z.
\end{aligned} \tag{13.29}$$

Applying the boundary conditions (13.23) to the general solutions, we get the particular solutions

$$u = \frac{\tau}{2\rho_0\gamma\nu}(\cos\gamma z + \sin\gamma z)\exp\left(\gamma z\right), \tag{13.30}$$

$$v = \frac{\tau}{2\rho_0\gamma\nu}(-\cos\gamma z + \sin\gamma z)\exp\left(\gamma z\right). \tag{13.31}$$

The flow is illustrated by a hodograph in Fig. 13.5.

Problem 2. Calculate the mass flux in the oceanic Ekman layer due to the Ekman flow expressed by (13.30) and (13.31). From the result, confirm the observed phenomenon by Nansen that icebergs drift 20° to 40° right from the direction of the prevailing wind.

13.2.3 Convergent Ekman Flow and Spin Down

1. The Convergent Ekman Flow

Let us consider the Ekman layer which develops in a fluid contained in a cylindrical vessel rotating at constant angular velocity Ω ($\Omega > 0$). Suppose that the fluid is in a rigid body rotation with angular velocity ω ($|\omega| \ll \Omega$) relative to the vessel. In this case, cylindrical coordinates (r, θ, z) are

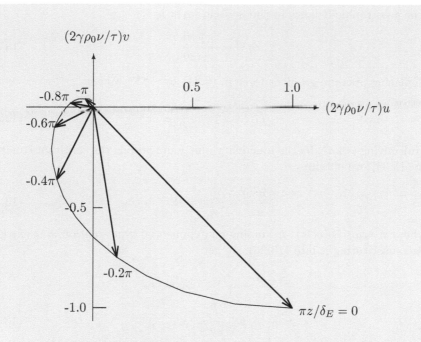

Figure 13.5: Hodograph of the oceanic Ekman flow. Velocity vectors are shown at normalized depth $\pi z/\delta_E = -0.2\pi$, -0.4π , -0.6π , -0.8π , -1.0π .

adequate, whose velocity components are (u, v, w). We find the azimuthal velocity V and the fluid depth H as

$$V = r\omega, \tag{13.32}$$

$$H = H_0 + \frac{\omega^2 r^2}{2g}, \tag{13.33}$$

where H_0 is the fluid depth at $r = 0$, and g is the magnitude of the acceleration due to gravity[2].

(1) *Ekman layer near the bottom*
The governing equations in the viscous boundary layer near the bottom are

$$-fv = -\frac{1}{\rho}\frac{\partial p}{\partial r} + \nu\frac{\partial^2 u}{\partial z^2}, \tag{13.34}$$

$$fu = \nu\frac{\partial^2 v}{\partial z^2}. \tag{13.35}$$

---

[2]The shape of the liquid surface is called a circular paraboloid

The geostrophic balance in the inviscid layer is

$$-fV = -\frac{1}{\rho}\frac{\partial p}{\partial r}, \tag{13.36}$$

Replacing pressure gradient force in (3.34) by $-fV$, we get

$$-f(v - V) = \nu\frac{\partial^2 u}{\partial z^2}, \tag{13.37}$$

Multiplying (13.37) by the imaginary unit $\tilde{i}$ and adding the resultant equation to (13.35), we obtain

$$\frac{d^2 w}{dz^2} - \tilde{i}\frac{f}{\nu}w = 0, \tag{13.38}$$

where $w = u + \tilde{i}(v - V)$. Assuming an exponential type solution $w \propto \exp(\lambda z)$ and substituting it into (13.38), we get

$$\lambda^2 = (\lambda_r{}^2 - \lambda_i{}^2) + \tilde{i}2\lambda_r\lambda_i = \tilde{i}\frac{f}{\nu},$$

$$\lambda_r = \lambda_i = \pm\sqrt{\frac{f}{2\nu}} = \pm\gamma. \tag{13.39}$$

The general solution is

$$w = \alpha\exp[\gamma(1 + \tilde{i})z] + \beta\exp[-\gamma(1 + \tilde{i})z], \tag{13.40}$$

where $\alpha = \alpha_r + \tilde{i}\alpha_i$ and $\beta = \beta_r + \tilde{i}\beta_i$ are complex coefficients. Remembering that $w = u + \tilde{i}(v - V)$, we find the general solutions,

$$u = \exp(\gamma z)(\alpha_r\cos\gamma z - \alpha_i\sin\gamma z)$$
$$+ \exp(-\gamma z)(\beta_r\cos\gamma z + \beta_i\sin\gamma z), \tag{13.41}$$
$$v = V + \exp(\gamma z)(\alpha_i\cos\gamma z + \alpha_r\sin\gamma z)$$
$$+ \exp(-\gamma z)(\beta_i\cos\gamma z - \beta_r\sin\gamma z). \tag{13.42}$$

The boundary conditions are

$$\begin{cases} u = 0, & v = 0, & \text{at} \quad z = 0, \\ u \to 0, & v \to V, & \text{as} \quad z \to \infty, \end{cases} \tag{13.43}$$

Applying the boundary conditions (13.43) to (13.41) and (13.42), we obtain

$$u = -V\exp(-\gamma z)\sin\gamma z = -rw\exp(-\gamma z)\sin\gamma z, \tag{13.44}$$
$$v = V\{1 - \exp(-\gamma z)\cos\gamma z\} = rw\{1 - \exp(-\gamma z)\cos\gamma z\}. \tag{13.45}$$

The horizontal divergence of the Ekman flow is

$$\mathbf{\nabla}_{\mathrm{H}} \cdot \mathbf{v} = \frac{1}{r}\frac{\partial(ru)}{\partial r} + \frac{\partial v}{r\partial\theta} = -2w\exp(-\gamma z)\sin\gamma z. \tag{13.46}$$

Integrating the continuity equation from $z = 0$ to $z = \delta_E$, we obtain the vertical velocity at the top of the Ekman layer.

$$\int_0^{\delta_E} \frac{\partial w}{\partial z} dz = -\int_0^{\delta_E} \mathbf{\nabla}_H \cdot \mathbf{v} dz = \frac{\omega}{\gamma} \{1 + \exp(-\pi)\},$$

$$w(\delta_E) - w(0) = \frac{\omega}{\gamma} \{1 + \exp(-\pi)\},$$

$$w(\delta_E) = \frac{\omega}{\gamma} \{1 + \exp(-\pi)\}. \tag{13.47}$$

When $\omega > 0$ ($\omega < 0$), upward (downward) motion proportional to the magnitude of the vorticity is caused at the top of the Ekman layer. This phenomenon is called the *Ekman pumping* or *Ekman suction*, which is very important mechanism for the growth of tropical cyclones, and the decay of the midlatitude depressions. Remembering that 2ω is the vorticity of the rigid body rotation, we obtain the general law for Ekman pumping replacing 2ω to ζ in (13.47).

$$w(\delta_E) = \frac{\zeta}{2\gamma} \{1 + \exp(-\pi)\}. \tag{13.48}$$

(2) *Ekman layer of the sea surface under the atmospheric vortices*
Let's take $z = 0$ at the sea surface. The governing equations near the liquid surface are

$$-fv = \nu \frac{\partial^2 u}{\partial z^2}, \tag{13.49}$$

$$fu = \nu \frac{\partial^2 v}{\partial z^2}. \tag{13.50}$$

Multiplying (13.49) by the imaginary unit $\tilde{\imath}$ and adding the resultant equation to (13.50), we obtain

$$\frac{d^2 w}{dz^2} - \tilde{\imath} \frac{f}{\nu} w = 0, \tag{13.51}$$

where $w = u + \tilde{\imath} v$. Assuming an exponential type solution $w \propto \exp(\lambda z)$ and substituting it into (13.51), we get

$$\lambda^2 = (\lambda_r{}^2 - \lambda_i{}^2) + \tilde{\imath} 2\lambda_r \lambda_i = \tilde{\imath} \frac{f}{\nu},$$

$$\lambda_r = \lambda_i = \pm\sqrt{\frac{f}{2\nu}} = \pm\gamma.$$

The general solution of (13.51) is

$$w = \alpha \exp[\gamma(1 + \tilde{\imath})z] + \beta \exp[-\gamma(1 + \tilde{\imath})z], \tag{13.52}$$

where $\alpha = \alpha_r + \tilde{\imath}\alpha_i$ and $\beta = \beta_r + \tilde{\imath}\beta_i$ are complex coefficients. Remembering that $w = u + \tilde{\imath}v$, we find the general solutions of (13.49) and (13.50).

$$
\begin{aligned}
u &= \exp\left(\gamma z\right)\left(\alpha_r \cos\gamma z - \alpha_i \sin\gamma z\right) \\
&\quad + \exp\left(-\gamma z\right)\left(\beta_r \cos\gamma z + \beta_i \sin\gamma z\right), \quad\quad (13.53) \\
v &= \exp\left(\gamma z\right)\left(\alpha_i \cos\gamma z + \alpha_r \sin\gamma z\right) \\
&\quad + \exp\left(-\gamma z\right)\left(\beta_i \cos\gamma z - \beta_r \sin\gamma z\right). \quad\quad (13.54)
\end{aligned}
$$

The boundary conditions are

$$
\begin{cases}
u = 0, & v = V = r\omega, \quad \text{at} \quad z = 0, \\
u \to 0, & v \to 0, \quad \text{as} \quad z \to -\infty,
\end{cases} \quad\quad (13.55)
$$

Applying the boundary conditions (13.55) to (13.53) and (13.54), we obtain

$$
u = -V \exp\left(\gamma z\right) \sin\gamma z = -r\omega \exp\left(\gamma z\right) \sin\gamma z, \quad\quad (13.56)
$$
$$
v = V \exp\left(\gamma z\right) \cos\gamma z = r\omega \exp\left(\gamma z\right) \cos\gamma z. \quad\quad (13.57)
$$

The horizontal divergence of the Ekman flow is

$$
\nabla_H \cdot \mathbf{v} = \frac{1}{r}\frac{\partial(ru)}{\partial r} + \frac{\partial v}{r\partial\theta} = -2\omega \exp\left(\gamma z\right) \sin\gamma z. \quad\quad (13.58)
$$

Integrating the continuity equation from $z = -\delta_E$ to $z = 0$, we get the vertical velocity at the bottom of the Ekman layer.

$$
\int_{-\delta_E}^{0} \frac{\partial w}{\partial z} dz = -\int_{-\delta_E}^{0} \nabla_H \cdot \mathbf{v} dz = -\frac{\omega}{\gamma}\{1 + \exp\left(-\pi\right)\},
$$
$$
w(0) - w(-\delta_E) = -\frac{\omega}{\gamma}\{1 + \exp\left(-\pi\right)\},
$$
$$
w(-\delta_E) = \frac{\omega}{\gamma}\{1 + \exp\left(-\pi\right)\}. \quad\quad (13.59)
$$

When $\omega > 0$ ($\omega < 0$), upward (downward) motion proportional to the magnitude of vorticity is caused at the top ($z = -\delta_E$) of the surface Ekman layer. This mechanism is observed for typhoons to induce upwelling of deep-sea water, which cools the sea surface temperature and depresses typhoons.

2. Spin Down

When $\zeta \ll f$ and f is constant, the quasi-geostrophic vorticity equation (12.66) is written as

$$
\frac{d\zeta}{dt} = f\frac{\partial w}{\partial z}. \quad\quad (13.60)
$$

Integrating (13.60) from the top of the Ekman layer to the top of the inviscid layer, we get

$$(H - \delta_E)\frac{d\zeta}{dt} = f[w(H) - w(\delta_E)] = -\frac{f}{2\gamma}\zeta,$$

$$\frac{d\zeta}{dt} = -\frac{f}{2\gamma H}\zeta, \tag{13.61}$$

where $\delta_E \ll H$ is taken into account. Solving (13.61) with the initial condition that $\zeta = \zeta_0$ at $t = 0$, we find

$$\zeta = \zeta_0 \exp\left(-\frac{f}{2\gamma H}t\right) = \zeta_0 \exp\left[-\left(\frac{f\nu}{2H^2}\right)^{1/2}t\right]. \tag{13.62}$$

The cyclonic vorticity of the inviscid layer causes upward motion in the Ekman boundary layer, which transports fluid having smaller vorticity to the inviscid interior and decreases its vorticity very rapidly. From the viewpoint of the angular momentum conservation of a rigid body, the expansion of the cross section of vortex tube in the inviscid layer owing to the Ekman suction corresponds to the increase of the moment of inertia, which causes the decrease of the angular velocity. The e-folding time $\tau = \sqrt{2/f\nu}H$ is called the *spin-down time*.

Problem 3. What is the spin-down time of the water with the depth of 1.0×10^{-1} m in the cylindrical vessel rotating at angular velocity $1.0\,\mathrm{s}^{-1}$? Suppose that the kinematic viscosity of the water is $1.0 \times 10^{-6}\,\mathrm{m^2\,s^{-1}}$. Next, calculate the characteristic diffusive time scale $\tau_d = H^2/\nu$ for the same water in the vessel.

13.3 Kelvin–Helmholtz Instability

Three sections from here treat small-scale phenomena on which the vertical component of the angular velocity of the system does not exert any effect. Then, we can ignore the Coriolis terms in the governing equations.

First of all, we will discuss the instability of the stratified shear flow known as the *Kelvin–Helmholtz instability*.

13.3.1 A Two-Layer Model

We will consider a two-layer fluid system confined to a channel as illustrated in Fig. 13.6. The channel has infinite extent in the x-direction, width D in the y-direction and depth H in the z-direction. Suppose that the zonal velocity, the density and the depth of the lower layer are U_1, ρ_1, and $H/2$, and those of the upper layer are U_2, ρ_2, and $H/2$ ($\rho_1 > \rho_2$, $U_1 < U_2$). The shear flow has

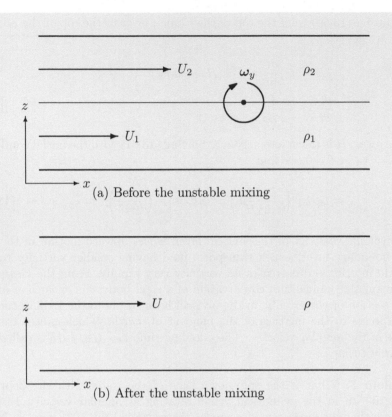

Figure 13.6: The Kelvin–Helmholtz instability in two-layer fluid. The upper panel is before the unstable mixing and the lower panel is after the unstable mixing.

the vorticity in the y-direction so that it has the tendency to turn over the system, while the stable stratification has the tendency to keep the system as it is. We will find the condition that the overturning or the instability takes place. Suppose that the mixing of fluids and the momentum completely occurs after the unstable overturning. As the result, the system becomes the final state whose density and zonal velocity are ρ and U, respectively. In order to evaluate the momentum and the mechanical energy of the system in finite range, we will consider the finite region of length L in the x-direction.

From the mass conservation law, we get

$$\rho_1 \frac{H}{2} DL + \rho_2 \frac{H}{2} DL = \rho HDL,$$

$$\rho = \frac{1}{2}(\rho_1 + \rho_2). \tag{13.63}$$

The law of momentum conservation yields

$$\rho_1 \frac{H}{2} DLU_1 + \rho_2 \frac{H}{2} DLU_2 = \rho HDLU,$$
$$U = \frac{\rho_1 U_1 + \rho_2 U_2}{2\rho}. \tag{13.64}$$

Substituting (13.63) into (13.64), we obtain

$$U = \frac{\rho_1 U_1 + \rho_2 U_2}{\rho_1 + \rho_2}. \tag{13.65}$$

Before the unstable mixing, the potential energy of the system is

$$\Phi_i = \rho_1 g \frac{H}{2} DL \frac{H}{4} + \rho_2 g \frac{H}{2} DL \frac{3H}{4}$$
$$= gDLH^2 \left(\frac{1}{8}\rho_1 + \frac{3}{8}\rho_2 \right), \tag{13.66}$$

and the kinetic energy is

$$K_i = \frac{1}{2}\rho_1 DL \frac{H}{2}U_1{}^2 + \frac{1}{2}\rho_2 DL \frac{H}{2}U_2{}^2$$
$$= \frac{1}{4}DLH(\rho_1 U_1{}^2 + \rho_2 U_2{}^2). \tag{13.67}$$

After the unstable mixing, the potential energy becomes

$$\Phi_f = \rho g DLH \frac{H}{2} = \frac{1}{4}(\rho_1 + \rho_2)gDLH^2, \tag{13.68}$$

and the kinetic energy is

$$K_f = \frac{1}{2}\rho DLHU^2 = \frac{1}{4}\frac{(\rho_1 U_1 + \rho_2 U_2)^2}{\rho_1 + \rho_2}DLH. \tag{13.69}$$

The difference of the mechanical energy before and after the unstable mixing is

$$E_i - E_f = (\Phi_i + K_i) - (\Phi_f + K_f),$$
$$= \frac{1}{8}(\rho_2 - \rho_1)gDLH^2 + \frac{\rho_1\rho_2}{4(\rho_1 + \rho_2)}DLH(U_2 - U_1)^2. \tag{13.70}$$

The necessary condition for unstable overturning to occur is $E_i - E_f > 0$, Assuming $\rho_1 \sim \rho_2 \simeq \rho$ and denoting $\delta\rho = \rho_1 - \rho_2$, the necessary condition becomes

$$\frac{\delta\rho g H}{\rho} < (U_2 - U_1)^2. \tag{13.71}$$

This unstable overturning is called the *shearing instability* or the *Kelvin–Helmholtz instability*, because it was first discussed by Hermann von Helmholtz (1868) using the two-layer model.

13.3.2 A Continuously Stratified Model

Let us consider a two-dimensional, inviscid and stably stratified fluid with vertical shear in a channel with width D and depth H. Taking the x-axis in the zonal flow direction and the z-axis vertically upward, the governing equations are

$$\frac{\partial u}{\partial t} + u\frac{\partial u}{\partial x} + w\frac{\partial u}{\partial z} = -\frac{1}{\rho_0}\frac{\partial p}{\partial x}, \tag{13.72}$$

$$\frac{\partial w}{\partial t} + u\frac{\partial w}{\partial x} + w\frac{\partial w}{\partial z} = -\frac{1}{\rho_0}\frac{\partial p}{\partial z} - \frac{\rho g}{\rho_0}, \tag{13.73}$$

$$\frac{\partial u}{\partial x} + \frac{\partial w}{\partial z} = 0, \tag{13.74}$$

$$\frac{\partial \rho}{\partial t} + u\frac{\partial \rho}{\partial x} + w\frac{\partial \rho}{\partial z} = 0, \tag{13.75}$$

where (u, w) are the velocity components in the x- and z-directions, p is the pressure, and ρ is the density. Dividing field variables into basic state portions and perturbation portions,

$$u = \bar{u}(z) + u', \quad w = w', \quad p = \bar{p}(z) + p', \quad \rho = \bar{\rho}(z) + \rho'. \tag{13.76}$$

Substituting from (13.76) into (13.72)–(13.75), and linearizing the resultant equations, we find

$$\frac{\partial u'}{\partial t} + \bar{u}\frac{\partial u'}{\partial x} + w'\frac{d\bar{u}}{dz} = -\frac{1}{\rho_0}\frac{\partial p'}{\partial x}, \tag{13.77}$$

$$\frac{\partial w'}{\partial t} + \bar{u}\frac{\partial w'}{\partial x} = -\frac{1}{\rho_0}\frac{\partial p'}{\partial z} - \frac{\rho' g}{\rho_0}, \tag{13.78}$$

$$\frac{\partial u'}{\partial x} + \frac{\partial w'}{\partial z} = 0, \tag{13.79}$$

$$\frac{\partial \rho'}{\partial t} + \bar{u}\frac{\partial \rho'}{\partial x} + w'\frac{d\bar{\rho}}{dz} = 0. \tag{13.80}$$

We define a stream function ψ' as

$$u' = -\frac{\partial \psi'}{\partial z}, \quad w' = \frac{\partial \psi'}{\partial x}. \tag{13.81}$$

Subtracting $\partial(13.77)/\partial z$ from $\partial(13.78)/\partial x$, we obtain a vorticity equation

$$\left(\frac{\partial}{\partial t} + \bar{u}\frac{\partial}{\partial x}\right)\left(\frac{\partial^2}{\partial x^2} + \frac{\partial^2}{\partial z^2}\right)\psi' - \frac{d^2\bar{u}}{dz^2}\frac{\partial \psi'}{\partial x} = -\frac{g}{\rho_0}\frac{\partial \rho'}{\partial x}. \tag{13.82}$$

Operating $(\partial/\partial t + \bar{u}\partial/\partial x)$ on (13.82) yields

$$\left(\frac{\partial}{\partial t} + \bar{u}\frac{\partial}{\partial x}\right)^2 \nabla_{\mathrm{H}}{}^2\psi' - \frac{d^2\bar{u}}{dz^2}\left(\frac{\partial}{\partial t} + \bar{u}\frac{\partial}{\partial x}\right)\frac{\partial \psi'}{\partial x}$$

$$= -\frac{g}{\rho_0}\frac{\partial}{\partial x}\left(\frac{\partial}{\partial t} + \bar{u}\frac{\partial}{\partial x}\right)\rho'. \tag{13.83}$$

Operating $\partial/\partial x$ on (13.80), we get

$$\frac{\partial}{\partial x}\left(\frac{\partial}{\partial t}+\bar{u}\frac{\partial}{\partial x}\right)\rho' = -\frac{d\bar{\rho}}{dz}\frac{\partial^2\psi'}{\partial x^2}. \tag{13.84}$$

Substituting from (13.84) into (13.83), we obtain

$$\left(\frac{\partial}{\partial t}+\bar{u}\frac{\partial}{\partial x}\right)^2 \nabla_{\mathrm{H}}{}^2\psi' - \frac{d^2\bar{u}}{dz^2}\left(\frac{\partial}{\partial t}+\bar{u}\frac{\partial}{\partial x}\right)\frac{\partial\psi'}{\partial x} = \frac{g}{\rho_0}\frac{d\bar{\rho}}{dz}\frac{\partial^2\psi'}{\partial x^2}. \tag{13.85}$$

Assuming a sinusoidal solution propagating in the x-direction as

$$\psi' = \Psi(z)\exp\left[\tilde{\imath}k(x-ct)\right],$$

and substituting the above equation into (13.85), we get

$$(\bar{u}-c)\left(\frac{d^2}{dz^2}-k^2\right)\Psi + \left(\frac{N^2}{\bar{u}-c}-\frac{d^2\bar{u}}{dz^2}\right)\Psi = 0, \tag{13.86}$$

where $N^2 = -g/\rho_0\, d\bar{\rho}/dz$ and N is called the *Brunt–Väisälä frequency*. Equation (13.86) is known as the *Taylor–Goldstein equation*. We assume that the upper and the lower boundaries are rigid, namely

$$\Psi = 0 \quad \text{at} \quad z = 0,\ H. \tag{13.87}$$

Defining the new complex function Φ as

$$\Psi = \sqrt{\bar{u}-c}\,\Phi, \tag{13.88}$$

and substituting (13.88) into (13.86), we obtain

$$\frac{d}{dz}\left[(\bar{u}-c)\frac{d\Phi}{dz}\right]$$
$$-\left[k^2(\bar{u}-c)+\frac{1}{2}\frac{d^2\bar{u}}{dz^2}+\frac{1}{\bar{u}-c}\left\{\frac{1}{4}\left(\frac{d\bar{u}}{dz}\right)^2-N^2\right\}\right]\Phi = 0. \tag{13.89}$$

The boundary conditions are

$$\Phi = 0, \quad \text{at} \quad z = 0,\ H. \tag{13.90}$$

Multiplying (13.89) by $\Phi^\dagger$ (the complex conjugate of Φ) and integrating the resultant equation with respect to z from 0 to H, we get

$$\int_0^H \left[N^2-\frac{1}{4}\left(\frac{d\bar{u}}{dz}\right)^2\right]\frac{\Phi\Phi^\dagger}{\bar{u}-c}dz$$
$$= \int_0^H (\bar{u}-c)\left[\left|\frac{d\Phi}{dz}\right|^2+k^2|\Phi|^2\right]dz + \frac{1}{2}\int_0^H \frac{d^2\bar{u}}{dz^2}|\Phi|^2 dz. \tag{13.91}$$

Taking the imaginary part of (13.91), we find

$$c_i \int_0^H \left[N^2 - \frac{1}{4} \left(\frac{d\bar{u}}{dz} \right)^2 \right] \frac{|\Phi|^2}{|\bar{u} - c|^2} dz = -c_i \int_0^H \left[\left| \frac{d\Phi}{dz} \right|^2 + k^2 |\Phi|^2 \right] dz. \quad (13.92)$$

If $N^2 > 1/4 \, (d\bar{u}/dz)^2$ throughout the fluid domain ($z = 0, H$), it is necessary $c_i = 0$ for (13.92) to be valid. The *Richardson number* in the continuously stratified fluid is

$$R_i = \frac{N^2}{(d\bar{u}/dz)^2}. \quad (13.93)$$

The stably stratified shear flow is stable for the small disturbances, if $R_i > 1/4$ everywhere in the fluid domain. Therefore, the necessary condition for the shear flow to be unstable is that $R_i < 1/4$ anywhere in the fluid domain. This instability is called the Kelvin–Helmholtz instability, which is often observed at water surfaces under wind stress, the interface of atmospheric fronts, the top of stratus clouds, the interface of oceanic fronts, and around Jupiter's Great Red Spot.

13.4 Rayleigh–Bénard Convection

In 1900, Henri Bénard performed experiments on the convection of a thin liquid layer heated from below. He observed the honeycomb structure of convective cells, upward motion at the center of the hexagonal cell and downward motion at the rim. Bénard used whale oil or paraffin as working fluids and found that the beginning of the instability depended essentially on the temperature difference between the top and the bottom boundaries, and the viscosity of working fluids. In 1916, Lord Rayleigh solved the Bénard's problem using the linear stability theory.

As the basic state, we will suppose the motionless fluid layer of depth d with vertical linear temperature gradient γ.

$$\gamma = -\frac{d\bar{T}}{dz} = \frac{T_1 - T_2}{d}, \quad (13.94)$$

where T_1 and T_2 ($T_2 < T_1$) are the temperature at the bottom and top boundary. Letting u, w, T be the horizontal perturbation velocity, the vertical perturbation velocity and the perturbation temperature, respectively, the linearized momentum equations are

$$\frac{\partial u}{\partial t} = -\frac{1}{\rho_0} \frac{\partial p}{\partial x} + \nu \nabla^2 u, \quad (13.95)$$

$$\frac{\partial w}{\partial t} = -\frac{1}{\rho_0} \frac{\partial p}{\partial z} + \alpha g T + \nu \nabla^2 w. \quad (13.96)$$

And the thermodynamic energy equation is

$$\frac{\partial T}{\partial t} - \gamma w = \kappa \nabla^2 T, \tag{13.97}$$

where κ is the *coefficient of thermal diffusivity*. We will consider that the top and bottom boundaries are rigid and frictionless. The boundary conditions are

$$\frac{\partial u}{\partial z} = w = 0, \quad \text{at} \quad z = 0, \, d. \tag{13.98}$$

Differentiating (13.96) with respect to x and (13.95) with respect to z, and subtracting the latter from the former, we get

$$\frac{\partial}{\partial t}\left(\frac{\partial u}{\partial z} - \frac{\partial w}{\partial x}\right) = -\alpha g \frac{\partial T}{\partial x} + \nu \nabla^2 \left(\frac{\partial u}{\partial z} - \frac{\partial w}{\partial x}\right). \tag{13.99}$$

We will define the stream function ψ as

$$u = \frac{\partial \psi}{\partial z}, \quad w = -\frac{\partial \psi}{\partial x}. \tag{13.100}$$

Substituting from (13.100) into (13.99), we get

$$\alpha g \frac{\partial T}{\partial x} = -\left(\frac{\partial}{\partial t} - \nu \nabla^2\right)\nabla^2 \psi. \tag{13.101}$$

If $T = 0$, $\partial T/\partial x = 0$ at $z = 0, d$, we find

$$\nabla^2 \nabla^2 \psi = 0, \quad \text{at} \quad z = 0, \, d, \tag{13.102}$$

which are valid for any kind of boundaries. Operating $\alpha g\, \partial/\partial x$ on (13.97), we find

$$\left(\frac{\partial}{\partial t} - \kappa \nabla^2\right)\alpha g \frac{\partial T}{\partial x} = -\alpha \gamma g \frac{\partial^2 \psi}{\partial x^2}. \tag{13.103}$$

Substituting from (13.101) into (13.103), we obtain

$$\left(\frac{\partial}{\partial t} - \kappa \nabla^2\right)\left(\frac{\partial}{\partial t} - \nu \nabla^2\right)\nabla^2 \psi = \alpha \gamma g \frac{\partial^2 \psi}{\partial x^2}. \tag{13.104}$$

Here, we nondimensionalize dependent and independent variables as follows,

$$t = \frac{d^2}{\nu}t^*, \quad x = dx^*, \quad z = dz^*, \quad \psi = \nu \psi^*. \tag{13.105}$$

Substituting from (13.105) into (13.104) and omitting $*$, we get

$$\left(\frac{\partial}{\partial t} - \frac{1}{P_r}\nabla^2\right)\left(\frac{\partial}{\partial t} - \nabla^2\right)\nabla^2 \psi = G_r \frac{\partial^2 \psi}{\partial x^2}, \tag{13.106}$$

which contains two nondimensional parameters, the *Prandtl number*

$$P_{\mathrm{r}} = \frac{\nu}{\kappa},$$ (13.107)

and the *Grashof number*

$$G_{\mathrm{r}} = \frac{\alpha \gamma g d^4}{\nu^2}.$$ (13.108)

Assuming the stationary state, (13.106) becomes

$$\nabla^2 \nabla^2 \nabla^2 \psi = R_{\mathrm{a}} \frac{\partial^2 \psi}{\partial x^2},$$ (13.109)

where

$$R_{\mathrm{a}} = G_{\mathrm{r}} P_{\mathrm{r}} = \frac{\alpha \gamma g d^4}{\kappa \nu}$$ (13.110)

is known as the *Rayleigh number*. Supposing a wave structure in the x-direction and the uniform structure in the y-direction, i.e.,

$$\psi = \Psi(z) \sin kx$$

and substituting it into (13.109), we find

$$\left(D^2 - k^2\right)^3 \Psi = -k^2 R_{\mathrm{a}} \Psi,$$ (13.111)

where $D \equiv d/dz$.

Problem 4. Supposing the stationary state, derive the following relation between the temperature perturbation T and the perturbation of the stream function Ψ.

$$T = \frac{\nu^2}{k \alpha \gamma g d^4} \left(D^2 - k^2\right)^2 \Psi = \frac{1}{k G_{\mathrm{r}}} \left(D^2 - k^2\right)^2 \Psi.$$ (13.112)

We will distinguish two kinds of boundary surfaces: one is free surfaces on which no tangential stress acts, and the other is rigid surfaces which are slipless. We will consider solutions for three combinations of two kinds of boundaries: 1. both free boundaries, 2. both rigid boundaries, and 3. one rigid and one free boundary. We assume that the temperature perturbation is zero at all kinds of boundaries, which yields from (13.103)

$$\left(D^2 - k^2\right)^2 \Psi = 0.$$ (13.113)

1. Both free boundaries
Suppose that the fluid is confined between free boundaries at $z = 0$ and $z = 1$. The boundary conditions are

$$w = 0, \quad Du = 0, \quad \text{at} \quad z = 0, 1,$$

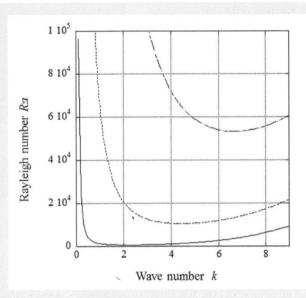

Figure 13.7: Stability diagram for the Rayleigh–Bénard convection with both free boundaries. Dependence of the Rayleigh number R_a on the horizontal wavenumber k for the vertical wavenumber $n = 1$, 2, 3. The *solid line* is for $n = 1$, the *dotted line* is for $n = 2$, and the *dash-dotted line* is for $n = 3$.

which become

$$\Psi = 0, \quad D^2\Psi = 0, \quad \text{at} \quad z = 0, 1, \tag{13.114}$$

Supposing the sinusoidal structure in the vertical direction,

$$\Psi(z) = \Psi_0 \sin n\pi z, \quad (n = 1, 2, 3, \cdots), \tag{13.115}$$

which satisfy the boundary conditions (13.113) and (13.114). Substituting (13.115) into (13.111), we get

$$R_a = \frac{\{k^2 + (n\pi)^2\}^3}{k^2}. \tag{13.116}$$

Equation (13.116) gives the relation between the wavenumber k and the Rayleigh number R_a, at which the instability occurs, and is shown in Fig. 13.7 for $n = 1, 2, 3$. The minimum Rayleigh number and the corresponding wavenumber, which are referred to as the marginal Rayleigh number R_{ac} and the marginal wavenumber k_c, are

$$R_{ac} = \frac{\{(\pi/\sqrt{2})^2 + \pi^2\}^3}{(\pi/\sqrt{2})^2} = 657.5, \tag{13.117}$$

$$k_c = \frac{\pi}{\sqrt{2}} = 2.221. \tag{13.118}$$

2. Both rigid boundaries

Suppose that the fluid is confined between rigid boundaries at $z = -1/2$ and $z = 1/2$. The boundary conditions are

$$w = u = 0, \quad \text{at} \quad z = -1/2,\ 1/2,$$

which yield

$$\Psi = 0, \quad D\Psi = 0, \quad \text{at} \quad z = -1/2,\ 1/2. \tag{13.119}$$

It is obvious that the lowest state is the even mode with no nodes and the first exited state is the odd mode with no node at $z = 0$. The general solution of (13.111) is expressed as a linear combination of the following functions.

$$\Psi = \exp(\pm qz), \tag{13.120}$$

where q^2 is a root of the equation

$$(q^2 - k^2)^3 = -R_a k^2. \tag{13.121}$$

Letting

$$R_a k^2 = r^3 k^6. \tag{13.122}$$

We can find three roots of (13.122) as follows

$$\begin{cases} q^2 = k^2(1 - r), \\ q^2 = k^2 \left[1 + \dfrac{1}{2} r \left(1 \pm i\sqrt{3} \right) \right]. \end{cases} \tag{13.123}$$

Taking the square roots of (13.123), we have six roots as

$$\pm i q_0, \quad \pm q. \quad \pm q^\dagger, \tag{13.124}$$

$$\begin{cases} q_0 = k\sqrt{r - 1}, \\ \Re\{q\} = q_1 = \dfrac{k}{\sqrt{2}} \left\{ \sqrt{1 + r + r^2} + \left(1 + \dfrac{1}{2} r \right) \right\}^{1/2}, \\ \Im\{q\} = q_2 = \dfrac{k}{\sqrt{2}} \left\{ \sqrt{1 + r + r^2} - \left(1 + \dfrac{1}{2} r \right) \right\}^{1/2}, \end{cases} \tag{13.125}$$

where $q^\dagger$ is the complex conjugate of q, q_1, and q_2 are the real part and the imaginary part of q, respectively. From (13.123), we obtain the following convenient relations

$$(q_0{}^2 + k^2)^2 = k^4 r^2. \tag{13.126}$$

$$(q^2 - k^2)^2 = \dfrac{1}{2} k^4 r^2 (-1 \pm i\sqrt{3}). \tag{13.127}$$

(1) The even solutions

The even solution can be written as

$$\Psi = A_0 \cos q_0 z + A \cosh qz + A^\dagger \cosh q^\dagger z, \tag{13.128}$$

where A_0 is a real constant, A is a complex constant, and $A^\dagger$ is a complex conjugate of A. Before applying (13.128) to the boundary conditions, we will prepare two equations.

$$D\Psi = -A_0 q_0 \sin q_0 z + Aq \sinh qz + A^\dagger q^\dagger \sinh q^\dagger z, \tag{13.129}$$

$$\begin{aligned}
\left(D^2 - k^2\right)^2 \Psi &= A_0 (q_0{}^2 + k^2)^2 s \cos q_0 z + A(q^2 - k^2)^2 \cosh qz \\
&\quad + A^\dagger (q^{\dagger 2} - k^2)^2 \cosh q^\dagger z \\
&= A_0 k^4 r^4 \cos q_0 z + A \frac{1}{2} k^4 r^2 (-1 + \tilde{\imath}\sqrt{3}) \cosh qz \\
&\quad + A^\dagger \frac{1}{2} k^4 r^2 (-1 - \tilde{\imath}\sqrt{3}) \cosh q^\dagger z \\
&= \frac{1}{2} k^4 r^2 \{ 2A_0 \cos q_0 z + (\tilde{\imath}\sqrt{3} - 1) A \cosh qz \\
&\quad - (\tilde{\imath}\sqrt{3} + 1) A^\dagger \cosh q^\dagger z \}. \tag{13.130}
\end{aligned}$$

In the derivation of the above equations, use is made of (13.126) and (13.127). Applying (13.128) to the boundary conditions (13.113) and (13.119), we obtain

$$\begin{cases}
A_0 \cos \frac{1}{2} q_0 + A \cosh \frac{1}{2} q + A^\dagger \cosh \frac{1}{2} q^\dagger = 0, \\
-A_0 q_0 \sin \frac{1}{2} q_0 + Aq \sinh \frac{1}{2} q + A^\dagger q^\dagger \sinh \frac{1}{2} q^\dagger = 0, \\
A_0 \cos \frac{1}{2} q_0 + \frac{1}{2} \left(\tilde{\imath}\sqrt{3} - 1\right) A \cosh \frac{1}{2} q - \frac{1}{2} \left(\tilde{\imath}\sqrt{3} + 1\right) A^\dagger \cosh \frac{1}{2} q^\dagger = 0.
\end{cases}$$

The above equations are expressed in the matrix form

$$\begin{pmatrix}
\cos \frac{1}{2} q_0 & \cosh \frac{1}{2} q & \cosh \frac{1}{2} q^\dagger \\
-q_0 \sin \frac{1}{2} q_0 & q \sinh \frac{1}{2} q & q^\dagger \sinh \frac{1}{2} q^\dagger \\
\cos \frac{1}{2} q_0 & \frac{1}{2}(\tilde{\imath}\sqrt{3} - 1) \cosh \frac{1}{2} q & -\frac{1}{2}(\tilde{\imath}\sqrt{3} + 1) \cosh \frac{1}{2} q^\dagger
\end{pmatrix}$$

$$\times \begin{pmatrix} A_0 \\ A \\ A^\dagger \end{pmatrix} = 0. \tag{13.131}$$

The necessary condition for A_0, A, and $A^\dagger$ to have nontrivial solutions is that the determinant of the matrix of (13.131) is zero, i.e.,

$$\begin{vmatrix}
\cos \frac{1}{2} q_0 & \cosh \frac{1}{2} q & \cosh \frac{1}{2} q^\dagger \\
-q_0 \sin \frac{1}{2} q_0 & q \sinh \frac{1}{2} q & q^\dagger \sinh \frac{1}{2} q^\dagger \\
\cos \frac{1}{2} q_0 & \frac{1}{2}(\tilde{\imath}\sqrt{3} - 1) \cosh \frac{1}{2} q & -\frac{1}{2}(\tilde{\imath}\sqrt{3} + 1) \cosh \frac{1}{2} q^\dagger
\end{vmatrix} = 0.$$

Deforming the above determinant continuously, we find

$$
0 = \begin{vmatrix} 1 & 1 & 1 \\ -q_0 \tan \frac{1}{2} q_0 & q \tanh \frac{1}{2} q & q^\dagger \tanh \frac{1}{2} q^\dagger \\ 1 & \frac{1}{2}(\tilde{\imath}\sqrt{3} - 1) & -\frac{1}{2}(\tilde{\imath}\sqrt{3} + 1) \end{vmatrix}
$$

$$
= \begin{vmatrix} 1 & 1 & 1 \\ -q_0 \tan \frac{1}{2} q_0 & q \tanh \frac{1}{2} q & q^\dagger \tanh \frac{1}{2} q^\dagger \\ 0 & \frac{\sqrt{3}}{2}(\tilde{\imath} - \sqrt{3}) & -\frac{\sqrt{3}}{2}(\tilde{\imath} + \sqrt{3}) \end{vmatrix}
$$

$$
= \begin{vmatrix} 1 & 1 & 1 \\ -q_0 \tan \frac{1}{2} q_0 & q \tanh \frac{1}{2} q & q^\dagger \tanh \frac{1}{2} q^\dagger \\ 0 & \sqrt{3} - \tilde{\imath} & \sqrt{3} + \tilde{\imath} \end{vmatrix} \tag{13.132}
$$

Expanding (13.132), we obtain

$$
(\sqrt{3} + \tilde{\imath}) q \tanh \frac{1}{2} q - (\sqrt{3} - \tilde{\imath}) q^\dagger \tanh \frac{1}{2} q^\dagger
$$
$$
+ (\sqrt{3} + \tilde{\imath}) q_0 \tan \frac{1}{2} q_0 - (\sqrt{3} - \tilde{\imath}) q_0 \tan \frac{1}{2} q_0 = 0,
$$
$$
\Im \left\{ (\sqrt{3} + \tilde{\imath}) q \tanh \frac{1}{2} q \right\} + q_0 \tan \frac{1}{2} q_0 = 0. \tag{13.133}
$$

Deforming (13.133), we find

$$
-q_0 \tan \frac{1}{2} q_0 = \Im \left\{ (\sqrt{3} + \tilde{\imath})(q_1 + \tilde{\imath} q_2) \frac{\sinh \frac{1}{2}(q_1 + \tilde{\imath} q_2)}{\cosh \frac{1}{2}(q_1 + \tilde{\imath} q_2)} \right\}
$$
$$
= \frac{(q_1 + \sqrt{3} q_2) \sinh q_1 + (\sqrt{3} q_1 - q_2) \sin q_2}{\cosh q_1 + \cos q_2}. \tag{13.134}
$$

Equation (13.134) gives the relation between k and r (i.e., R_a) implicitly, which can be solved numerically. The result is shown in Fig. 13.8. The marginal Rayleigh number and wavenumber are

$$
R_{ac} = 1707.762, \quad k_c = 3.117. \tag{13.135}
$$

The stream function and temperature perturbation at the marginal stability state are obtained from (13.128) and (13.112). Putting $A_0 = 1.0$, A_1 and A_2 are calculated using (13.131).

$$
\Psi = 1.0 \cos q_0 z - 6.1540 \times 10^{-2} \cos q_2 z \cosh q_1 z
$$
$$
+ 1.0393 \times 10^{-1} \sin q_2 z \sinh q_1 z, \tag{13.136}
$$
$$
(k G_r) T = 1.0 \cos q_0 z + 1.2078 \times 10^{-1} \cos q_2 z \cosh q_1 z
$$
$$
+ 1.3293 \times 10^{-3} \sin q_2 z \sinh q_1 z. \tag{13.137}
$$

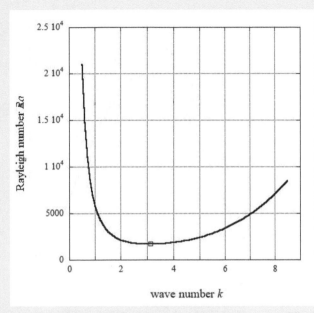

Figure 13.8: Stability diagram for the Rayleigh–Bénard convection with both rigid boundaries and for an even mode. Dependence of the Rayleigh number R_a on the horizontal wavenumber k.

The behavior of these functions is shown in Fig. 13.9.

(2) *The odd solutions*
The odd solution can be written as

$$\Psi = A_0 \sin q_0 z + A \sinh qz + A^\dagger \sinh q^\dagger z. \tag{13.138}$$

Applying (13.138) to the boundary conditions (13.113) and (13.119), we obtain

$$
\begin{cases}
A_0 \sin \frac{1}{2} q_0 + A \sinh \frac{1}{2} q + A^\dagger \sinh \frac{1}{2} q^\dagger = 0 , \\
A_0 q_0 \cos \frac{1}{2} q_0 + A q \cosh \frac{1}{2} q + A^\dagger q^\dagger \cosh \frac{1}{2} q^\dagger = 0 , \\
A_0 \sin \frac{1}{2} q_0 + \frac{1}{2} \left(i\sqrt{3} - 1 \right) A \sinh \frac{1}{2} q - \frac{1}{2} \left(i\sqrt{3} + 1 \right) A^\dagger \sinh \frac{1}{2} q^\dagger = 0 .
\end{cases}
$$

The above equations are expressed in the matrix form as

$$
\begin{pmatrix}
\sin \frac{1}{2} q_0 & \sinh \frac{1}{2} q & \sinh \frac{1}{2} q^\dagger \\
q_0 \cos \frac{1}{2} q_0 & q \cosh \frac{1}{2} q & q^\dagger \cosh \frac{1}{2} q^\dagger \\
\sin \frac{1}{2} q_0 & \frac{1}{2}(i\sqrt{3} - 1) \sinh \frac{1}{2} q & -\frac{1}{2}(i\sqrt{3} + 1) \sinh \frac{1}{2} q^\dagger
\end{pmatrix}
$$

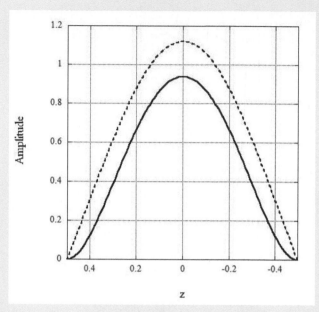

Figure 13.9: The perturbed stream function Ψ (*a solid line*) and temperature $(kG_r)\,T$ (*a broken line*) at the marginal stability state. For both rigid boundaries and the even mode.

$$\times \begin{pmatrix} A_0 \\ A \\ A^\dagger \end{pmatrix} = 0.$$ (13.139)

The necessary condition for A_0, A, and $A^\dagger$ to have nontrivial solutions is that the determinant of the matrix of (13.139) is zero.

$$\begin{vmatrix} \sin\tfrac{1}{2}q_0 & \sinh\tfrac{1}{2}q & \sinh\tfrac{1}{2}q^\dagger \\ q_0\cos\tfrac{1}{2}q_0 & q\cosh\tfrac{1}{2}q & q^\dagger\cosh\tfrac{1}{2}q^\dagger \\ \sin\tfrac{1}{2}q_0 & \tfrac{1}{2}(\tilde{\imath}\sqrt{3}-1)\sinh\tfrac{1}{2}q & -\tfrac{1}{2}(\tilde{\imath}\sqrt{3}+1)\sinh\tfrac{1}{2}q^\dagger \end{vmatrix} = 0.$$

Deforming the above determinant continuously,

$$0 = \begin{vmatrix} 1 & 1 & 1 \\ q_0\cot\tfrac{1}{2}q_0 & q\coth\tfrac{1}{2}q & q^\dagger\coth\tfrac{1}{2}q^\dagger \\ 1 & \tfrac{1}{2}(\tilde{\imath}\sqrt{3}-1) & -\tfrac{1}{2}(\tilde{\imath}\sqrt{3}+1) \end{vmatrix}$$

$$= \begin{vmatrix} 1 & 1 & 1 \\ q_0\cot\tfrac{1}{2}q_0 & q\coth\tfrac{1}{2}q & q^\dagger\coth\tfrac{1}{2}q^\dagger \\ 0 & \tfrac{\sqrt{3}}{2}(\tilde{\imath}-\sqrt{3}) & -\tfrac{\sqrt{3}}{2}(\tilde{\imath}+\sqrt{3}) \end{vmatrix}$$

$$= \begin{vmatrix} 1 & 1 & 1 \\ q_0 \cot \frac{1}{2} q_0 & q \coth \frac{1}{2} q & q^\dagger \coth \frac{1}{2} q^\dagger \\ 0 & \sqrt{3} - \tilde{\imath} & \sqrt{3} + \tilde{\imath} \end{vmatrix} \tag{13.140}$$

Expanding (13.140), we get

$$(\sqrt{3} + \tilde{\imath}) q \coth \frac{1}{2} q - (\sqrt{3} - \tilde{\imath}) q^\dagger \coth \frac{1}{2} q^\dagger$$
$$-(\sqrt{3} + \tilde{\imath}) q_0 \cot \frac{1}{2} q_0 + (\sqrt{3} - \tilde{\imath}) q_0 \cot \frac{1}{2} q_0 = 0,$$
$$\Im \left\{ (\sqrt{3} + \tilde{\imath}) q \coth \frac{1}{2} q \right\} - q_0 \cot \frac{1}{2} q_0 = 0. \tag{13.141}$$

Deforming (13.141), we find

$$q_0 \cot \frac{1}{2} q_0 = \Im \left\{ (\sqrt{3} + \tilde{\imath})(q_1 + \tilde{\imath} q_2) \frac{\cosh \frac{1}{2}(q_1 + \tilde{\imath} q_2)}{\sinh \frac{1}{2}(q_1 + \tilde{\imath} q_2)} \right\}$$
$$= \frac{(q_1 + \sqrt{3} q_2) \sinh q_1 - (\sqrt{3} q_1 - q_2) \sin q_2}{\cosh q_1 - \cos q_2}. \tag{13.142}$$

Equation (13.142) gives the relation between k and r (i.e., R_a) implicitly, which can be solved numerically. The result is shown in Fig. 13.10. The marginal Rayleigh number and wavenumber are

$$R_{ac} = 17610.393, \quad k_c = 5.365. \tag{13.143}$$

The stream function and temperature perturbation at the marginal stability state are obtained from (13.138) and (13.112). Putting $A_0 = 1.0$, A_1 and A_2 are calculated using (13.139).

$$\Psi = 1.0 \sin q_0 z - 1.7076 \times 10^{-2} \cos q_2 z \sinh q_1 z$$
$$+3.4566 \times 10^{-3} \sin q_2 z \cosh q_1 z, \tag{13.144}$$
$$(kG_r) T = 1.0 \sin q_0 z + 1.1531 \times 10^{-2} \cos q_2 z \sinh q_1 z$$
$$+1.3060 \times 10^{-2} \sin q_2 z \cosh q_1 z. \tag{13.145}$$

The behavior of these functions is shown in Fig. 13.11.

3. One rigid and one free boundary
Suppose that the fluid is confined between a rigid boundary at $z = -1$ and a free boundary at $z = 0$. This situation is the same as that of the Bénard's experiments. As the lower half region ($-1/2 \le z \le 0$) of the function (13.138) satisfies the upper and lower boundary conditions, we will choose this function as the solution, which is written again.

$$\Psi = A_0 \sin q_0 z + A \sinh q z + A^\dagger \sinh q^\dagger z. \tag{13.146}$$

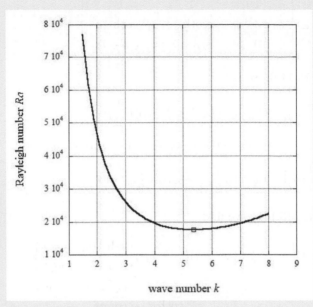

Figure 13.10: Stability diagram for the Rayleigh–Bénard convection with both rigid boundaries and for the odd mode. Dependence of the Rayleigh number R_a on the horizontal wavenumber k.

The boundary conditions are

$$\Psi = 0, \quad D\Psi = 0, \quad D^2\Psi = 0, \quad \text{at} \quad z = -1, \tag{13.147}$$

$$\Psi = 0, \quad D^2\Psi = 0, \quad \text{at} \quad z = 0. \tag{13.148}$$

Applying (13.146) to the lower boundary conditions (13.147), we find

$$\begin{cases} A_0 \sin q_0 + A \sinh q + A^\dagger \sinh q^\dagger = 0, \\ A_0 q_0 \cos q_0 + A q \cosh q + A^\dagger q^\dagger \cosh q^\dagger = 0, \\ 2A_0 \sin q_0 + \left(i\sqrt{3} - 1\right) A \sinh q - \left(i\sqrt{3} + 1\right) A^\dagger \sinh q^\dagger = 0. \end{cases}$$

The above equations are expressed in the matrix form as

$$\begin{pmatrix} \sin q_0 & \sinh q & \sinh q^\dagger \\ q_0 \cos q_0 & q \cosh q & q^\dagger \cosh q^\dagger \\ \sin q_0 & \frac{1}{2}(i\sqrt{3} - 1) \sinh q & -\frac{1}{2}(i\sqrt{3} + 1) \sinh q^\dagger \end{pmatrix}$$

$$\times \begin{pmatrix} A_0 \\ A \\ A^\dagger \end{pmatrix} = 0.$$

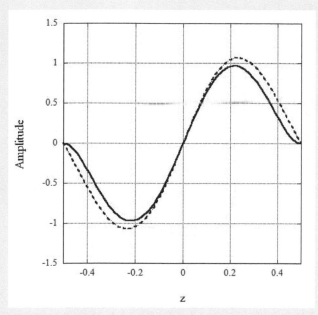

Figure 13.11: The perturbed stream function Ψ (*a solid line*) and temperature $(kG_r)\, T$ (*a broken line*) at the marginal stability state. For both rigid boundaries and the odd mode.

The necessary condition for A_0, A, and $A^\dagger$ to have nontrivial solutions is that the determinant of the matrix of (13.145) is zero.

$$
\begin{vmatrix}
\sin q_0 & \sinh q & \sinh q^\dagger \\
q_0 \cos q_0 & q \cosh q & q^\dagger \cosh q^\dagger \\
\sin q_0 & \tfrac{1}{2}(i\sqrt{3}-1)\sinh q & -\tfrac{1}{2}(i\sqrt{3}+1)\sinh q^\dagger
\end{vmatrix} = 0.
$$

Deforming the above determinant continuously,

$$
0 =
\begin{vmatrix}
1 & 1 & 1 \\
q_0 \cot q_0 & q \coth q & q^\dagger \coth q^\dagger \\
1 & \tfrac{1}{2}(i\sqrt{3}-1) & -\tfrac{1}{2}(i\sqrt{3}+1)
\end{vmatrix}
$$

$$
=
\begin{vmatrix}
1 & 1 & 1 \\
q_0 \cot q_0 & q \coth q & q^\dagger \coth q^\dagger \\
0 & \sqrt{3}-i & \sqrt{3}+i
\end{vmatrix}
\tag{13.149}
$$

Expanding (13.149) and deforming the resultant equation, we get

$$
q_0 \cot q_0 = \frac{(q_1 + \sqrt{3}q_2)\sinh 2q_1 - (\sqrt{3}q_1 - q_2)\sin 2q_2}{\cosh 2q_1 - \cos 2q_2}.
\tag{13.150}
$$

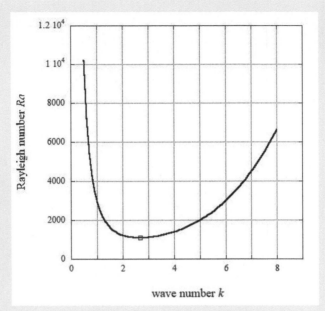

Figure 13.12: Stability diagram for the Rayleigh–Bénard convection with one rigid boundary and one free boundary. Dependence of the Rayleigh number R_a on the horizontal wavenumber k.

Equation (13.150) gives the relation between k and r (i.e., R_a) implicitly, which can be solved numerically. The result is shown in Fig. 13.12. The marginal Rayleigh number and wavenumber are

$$R_{ac} = 1100.650, \quad k_c = 2.682. \tag{13.151}$$

Problem 5. The marginal Rayleigh number and wavenumber for one rigid and one free boundary are obtained more smart way from the result of those for the odd mode (13.145) of both rigid boundaries. How can you do this?

The stream function and temperature perturbation at the marginal stability state are obtained from (13.147) and (13.112). Putting $A_0 = 1.0$, A_1 and A_2 are calculated using (13.148).

$$\Psi = 1.0 \sin q_0 z - 1.7084 \times 10^{-2} \cos q_2 z \sinh q_1 z$$
$$+3.4595 \times 10^{-3} \sin q_2 z \cosh q_1 z, \tag{13.152}$$
$$(kG_r)\, T = 1.0 \sin q_0 z + 1.1538 \times 10^{-2} \cos q_2 z \sinh q_1 z$$
$$+1.3065 \times 10^{-2} \sin q_2 z \cosh q_1 z. \tag{13.153}$$

The behavior of these functions is shown in Fig. 13.13.

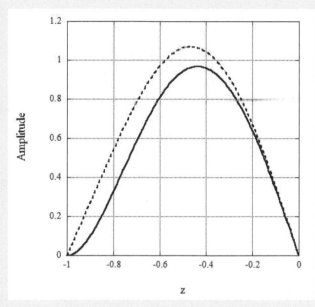

Figure 13.13: The perturbed stream function Ψ (*a solid line*) and temperature $(kG_r)\,T$ (*a broken line*) at the marginal stability state. For one rigid and one free boundary.

13.5 Taylor Vortices

Suppose an incompressible viscous fluid of annulus confined between two coaxial cylinders rotating with different angular velocities. A flow confined between two rigid parallel plates moving with different velocities is called the *Couette flow*, and the laminar flow confined between two coaxial cylinders is referred to as the *circular Couette flow* or the *Taylor–Couette flow*.

In 1923, G.I. Taylor performed experiments concerning the instability of a Taylor–Couette flow and found the criterion of the transition from the laminar flow (the non-periodic flow) to the axisymmetric toroidal flow (the singly periodic flow) (Fig. 13.14), and from the wavy to the non-axisymmetric wavy flow (the doubly periodic flow). In 1975, Gollub and Swinney investigated the phase transition from the doubly periodic flow to the turbulent flow. The reason why many researchers are fond of the Taylor–Couette flow for phase transition experiments is that it has the typical four bifurcations from laminar to turbulent flow and it is easy to control the external parameters.

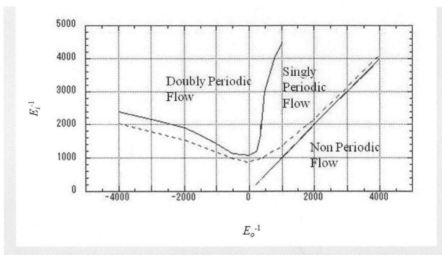

Figure 13.14: Stability diagram of Taylor–Couette flow. The *dash-dotted line* is the Rayleigh's criterion. The *dotted line* is the Taylor boundary (transition from non-periodic to singly periodic flow) and the *straight line* is the second boundary (transition from singly periodic to doubly periodic flow) obtained by Coles (1965). As the turbulent flow regime begins at $E_i^{-1} = 15,000$ at $E_o^{-1} = 0$, the turbulent flow regime is out of range in this figure.

13.5.1   Rayleigh's Criterion

We will discuss the stationary circular flow of an incompressible inviscid fluid between two coaxial cylinders rotating with different angular velocities (Fig. 13.15). Suppose that a circular fluid chain of unit mass and radius r_1 is displaced axisymmetrically outward to radius r_2 by the central force without disturbing the ambient flow field. Let the azimuthal velocity at r_1 be v_1, the azimuthal velocity at r_2 be v_2, and the azimuthal velocity of the fluid chain displaced from r_1 to r_2 be v_1'. The forces exerting on the fluid chain are the central force, so that the angular momentum of the fluid chain is conserved.

$$r_1 v_1 = r_2 v_1',$$
$$v_1' = \frac{r_1}{r_2} v_1. \tag{13.154}$$

If the centrifugal force exerting on the displaced fluid chain is smaller than the ambient pressure gradient force, the net force acts until the fluid chain returns to its original radius r_1. Namely, the flow field is stable for the infinitesimal perturbation. The pressure gradient force at $r = r_2$ is equal and opposite to

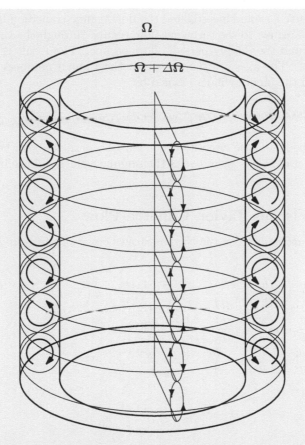

Figure 13.15: An illustration of Taylor vortices in the singly periodic regime. The inner cylinder rotates at larger angular velocity $\Omega + \Delta\Omega$ than that of the inner cylinder Ω. Arrows show the direction of the azimuth velocity component.

the centrifugal force, so that the condition for stability is written as

$$\frac{v_2{}^2}{r_2} > \frac{v_1'{}^2}{r_2} = \frac{r_1{}^2}{r_2{}^3}v_1{}^2,$$

$$(r_2 v_2)^2 > (r_1 v_1)^2 . \tag{13.155}$$

Therefore, the necessary and sufficient condition for stability is

$$\frac{\partial (rv)^2}{\partial r} > 0 \tag{13.156}$$

everywhere in the fluid domain. Equation (13.156) is referred to as the *Rayleigh's criterion*. Coles (1965) performed precise experiments of the

Taylor–Couette flow and investigated the flow regime transition from the non-periodic flow regime to the turbulent flow regime through the singly periodic flow regime and the doubly periodic flow regime. The result is summarized in Fig. 13.14. The axes are the inverse of the Ekman numbers of the outer cylinder and the inner cylinder defined by

$$E_{\text{o}} = \frac{\nu}{\Omega_o r_o{}^2}, \quad E_{\text{i}} = \frac{\nu}{\Omega_i r_i{}^2},$$

where ν is the kinematic viscosity coefficient, Ω_o is the angular velocity of the outer cylinder of radius r_o and Ω_i is the angular velocity of the inner cylinder of radius r_i.

13.5.2 Viscous Taylor–Couette Flow

For axisymmetric motions the Navier–Stokes equations in cylindrical coordinates are,

$$\frac{\partial u}{\partial t} + u\frac{\partial u}{\partial r} + w\frac{\partial u}{\partial z} - \frac{v^2}{r} = -\frac{1}{\rho_0}\frac{\partial p}{\partial r} + \nu\left(\nabla^2 u - \frac{u}{r^2}\right), \tag{13.157}$$

$$\frac{\partial v}{\partial t} + u\frac{\partial v}{\partial r} + w\frac{\partial v}{\partial z} + \frac{uv}{r} = \nu\left(\nabla^2 v - \frac{v}{r^2}\right), \tag{13.158}$$

$$\frac{\partial w}{\partial t} + u\frac{\partial w}{\partial r} + w\frac{\partial w}{\partial z} = -\frac{1}{\rho_0}\frac{\partial p}{\partial z} - g + \nu\nabla^2 w, \tag{13.159}$$

where

$$\nabla^2 = \frac{\partial^2}{\partial r^2} + \frac{\partial}{r\partial r} + \frac{\partial^2}{\partial z^2}, \tag{13.160}$$

and (u, v, w) are velocity components of r-, θ-, z-directions. The continuity equation is

$$\frac{\partial}{r\partial r}(ru) + \frac{\partial w}{\partial z} = 0, \tag{13.161}$$

Dividing field variables into basic portions and perturbation portions,

$$u = u', \quad v = \bar{v}(r) + v', \quad w = w', \quad p = \bar{p}(r,z) + p'. \tag{13.162}$$

Substituting from (13.160) into (13.154), we get the basic state equation for $\bar{v}$,

$$\frac{d^2\bar{v}}{dr^2} + \frac{1}{r}\frac{d\bar{v}}{dr} - \frac{\bar{v}}{r^2} = 0. \tag{13.163}$$

The boundary conditions are

$$\begin{cases} \bar{v} = r_i\Omega_i, & \text{at} \quad r = r_i, \\ \bar{v} = r_o\Omega_o, & \text{at} \quad r = r_o, \end{cases} \tag{13.164}$$

where Ω_i and Ω_o are angular velocities at r_i and r_o, respectively.
The particular solution of (13.163) satisfying the boundary conditions (13.164)
is

$$\bar{v} = Ar + \frac{B}{r}, \tag{13.165}$$

where

$$A = \frac{\Omega_o r_o{}^2 - \Omega_i r_i{}^2}{r_o{}^2 - r_i{}^2}, \tag{13.166}$$

$$B = \frac{(\Omega_i - \Omega_o)r_i{}^2 r_o{}^2}{r_o{}^2 - r_i{}^2}. \tag{13.167}$$

Linearized perturbation equations become as follows,

$$\frac{\partial u'}{\partial t} - \frac{2\bar{v}v'}{r} = -\frac{1}{\rho_0}\frac{\partial p'}{\partial r} + \nu\left(\nabla^2 u' - \frac{u'}{r^2}\right),$$

$$\frac{\partial v'}{\partial t} + \frac{\bar{v}u'}{r} = \nu\left(\nabla^2 v' - \frac{v'}{r^2}\right),$$

$$\frac{\partial w'}{\partial t} = -\frac{1}{\rho_0}\frac{\partial p}{\partial z} + \nu\nabla^2 w',$$

$$\frac{\partial}{r\partial r}(ru') + \frac{\partial w'}{\partial z} = 0.$$

Taking the limit $r \to \infty$, we obtain $\bar{v} = Ar$. Using this result and omitting $'$,
the above perturbation equations become,

$$\frac{\partial u}{\partial t} - 2Av = -\frac{1}{\rho_0}\frac{\partial p}{\partial r} + \nu\nabla^2 u, \tag{13.168}$$

$$\frac{\partial v}{\partial t} + Au = \nu\nabla^2 v, \tag{13.169}$$

$$\frac{\partial w}{\partial t} = -\frac{1}{\rho_0}\frac{\partial p}{\partial z} + \nu\nabla^2 w, \tag{13.170}$$

$$\frac{\partial u}{\partial r} + \frac{u}{r} + \frac{\partial w}{\partial z} = 0, \tag{13.171}$$

where

$$\nabla^2 = \frac{\partial^2}{\partial r^2} + \frac{1}{r}\frac{\partial}{\partial r} + \frac{\partial^2}{\partial z^2}. \tag{13.172}$$

Differentiating (13.168) with respect to z and (13.170) with respect to r, and
subtracting the latter from the former, we get

$$\frac{\partial}{\partial t}\left(\frac{\partial u}{\partial z} - \frac{\partial w}{\partial r}\right) - 2A\frac{\partial v}{\partial z} = \nu\nabla^2\left(\frac{\partial u}{\partial z} - \frac{\partial w}{\partial r}\right). \tag{13.173}$$

We will define the stream function ψ as

$$u = \frac{\partial \psi}{\partial z}, \quad w = -\frac{\partial \psi}{\partial r}. \tag{13.174}$$

Substituting from (13.174) into (13.173), we get

$$\nabla^2 \left(\frac{\partial}{\partial t} - \nu \nabla^2 \right) \psi = 2A \frac{\partial v}{\partial z}. \tag{13.175}$$

From (13.169), we obtain

$$\left(\frac{\partial}{\partial t} - \nu \nabla^2 \right) v = -A \frac{\partial \psi}{\partial z}. \tag{13.176}$$

Eliminating v from (13.175) and (13.176), we find

$$\nabla^2 \left(\frac{\partial}{\partial t} - \nu \nabla^2 \right)^2 \psi = -2A^2 \frac{\partial^2 \psi}{\partial z^2}. \tag{13.177}$$

We will nondimensionalize dependent and independent variables as follows,

$$t = \frac{d^2}{\nu} t^*, \quad r = dr^*, \quad z = dz^*, \quad \psi = \nu \psi^*, \tag{13.178}$$

where $d = r_o - r_i$. Substituting from (13.178) into (13.177) and omitting $*$ yields

$$\nabla^2 \left(\frac{\partial}{\partial t} - \nabla^2 \right)^2 \psi = -T_{\mathrm{a}} \frac{\partial^2 \psi}{\partial z^2}, \tag{13.179}$$

where

$$T_{\mathrm{a}} = \frac{2d^4 A^2}{\nu^2} \tag{13.180}$$

is called the *Taylor number*. When $d \ll r_o$, T_{a} is approximated

$$T_{\mathrm{a}} = \frac{(\Omega_o - \Omega_i)^2 d^2 r_o^{\,2}}{2\nu^2}. \tag{13.181}$$

Assuming the stationary state, (13.179) becomes

$$\nabla^2 \nabla^2 \nabla^2 \psi = -T_{\mathrm{a}} \frac{\partial^2 \psi}{\partial z^2}. \tag{13.182}$$

Equation (13.182) is the same as (13.109) for the Rayleigh–Bénard convection. Therefore, the Taylor–Couette flow becomes unstable when the Taylor number exceeds 1707.8, because the boundaries are both rigid and slipless. The Rayleigh–Bénard convection occurs due to the unstable stratification, while the Taylor vortices takes place owing to the unstable distribution of the angular momentum. The Taylor vortices are called the symmetric instability because the structure of the vortices is axisymmetric, or the inertial instability because the instability is caused by the imbalance between the pressure gradient force and the inertial force.

13.6 Problems

1. Suppose that a fluid in a cylindrical vessel is in the rigid body rotation of angular velocity ω. Derive the depth of the fluid H at radius r, letting the density of the fluid be ρ, the depth of the fluid at the center be H_0 and the magnitude of the acceleration due to gravity be g.

2. Derive the atmospheric Ekman flow corresponding to (13.19) and (13.20) for the Southern Hemisphere. Let the Coriolis parameter be $-|f|$.

3. Draw a hodograph of the Ekman flow for the Southern Hemisphere.

4. Obtain the oceanic Ekman flow corresponding to (13.31) and (13.32) for the Southern Hemisphere.

5. When you stir tea in a teacup by a spoon, you will observe that tea leaves gather to the center of the bottom. Explain this phenomenon supposing that tea is in the rigid body rotation except in the viscous boundary layer near the bottom.

6. What is the depth of the Oceanic Ekman layer at 45° latitude, supposing the eddy viscosity of the ocean to be $\nu = 1.0 \times 10^{-3}\,\mathrm{m^2\,s^{-1}}$?

7. Obtain the convergent Ekman flow corresponding to (13.44) and (13.45) for the Southern Hemisphere. Using the result, calculate the vertical motion at the top of the Ekman layer.

8. What is the spin-down time of the midlatitude depressions, supposing that the eddy viscosity of the atmosphere is $1.0 \times 10^1\,\mathrm{m^2\,s^{-1}}$, the depth of the troposphere is $1.0 \times 10^4\,\mathrm{m}$ and the Coriolis parameter is $f = 1.0 \times 10^{-4}\,\mathrm{s^{-1}}$?

9. Prove that (13.165) is the general solution of the second-order ordinary differential equation (13.163).

13.7 References

1. Chandrasekhar S.: *Hydrodynamic and Hydromagnetic Stability*, Dover Publications Inc., New York (1961).

2. Coles D.: *Transition in Circular Couette Flow*, J. Fluid Mech., **21**, 385–425 (1965).

3. Gollub J. P. and H. L. Swinny: *Onset of Turbulence in a Rotating Fluid*, Phys. Rev. Lett., **35**, 927–930 (1975).

4. von Helmholtz H.: *Über atmosphärische Bewegungen I, Sitzungsberichte Akad. Wissenshaften Berlin*, **3**, 647–663 (1888).

5. Hough S. S.: *The Oscillations of a Rotating Ellipsoidal Shell Containing Fluid, Philos. Trans. R. Soc. London*, **A 186**, 469–506 (1895).

Chapter 14

Phenomena in Geophysical Fluids: Part II

At the first section of this chapter, the phase velocity and the group velocity are explained precisely. Following this section, we will discuss various kinds of gravity waves; shallow water gravity waves, internal gravity waves in two-layer fluids, internal gravity waves in continuously stratified fluids, inertio-gravity waves, and mountain waves. The name of gravity waves comes from the fact that the restoring force is the Earth's gravity. The term of buoyancy waves is often used for internal gravity wave because the restoring force is cooperation of the Earth's gravity and buoyancy force.

14.1   Phase Velocity and Group Velocity

We will preliminarily consider the phase velocity and group velocity before discussing gravity waves.

14.1.1   Phase Velocity

Suppose a plane wave propagating in the x-y plane, whose wave length and frequency are L and σ, respectively. In Fig. 14.1, we show constant phase lines $\phi(t)$ at time t and $\phi(t+T)$ at $t+T$ ($T = 2\pi/\sigma$). The phase velocity $\mathbf{c}$ is given by

$$\mathbf{c} = \frac{L}{T}\frac{\mathbf{k}}{|\mathbf{k}|} = \frac{2\pi/|\mathbf{k}|}{2\pi/\sigma}\frac{\mathbf{k}}{|\mathbf{k}|} = \frac{\sigma}{k^2 + l^2}\mathbf{k}, \tag{14.1}$$

where $\mathbf{k} = (k, l)$ is the wave vector.

The phase speed of the x- and y-directions are given by

$$c_x = \frac{L_x}{T} = \frac{2\pi/k}{2\pi/\sigma} = \frac{\sigma}{k}. \tag{14.2}$$

DOI: 10.1201/9781003310068-14

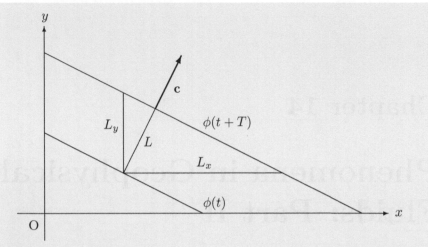

Figure 14.1: Phase velocity in the x-y plane. L is the wave length, L_x and L_y are the wave length in the x- and y-directions, respectively. The constant phase lines $\phi(t)$ at t and $\phi(t+T)$ at $t+T$ are propagating in the positive x- and y-directions.

$$c_y = \frac{L_y}{T} = \frac{2\pi/l}{2\pi/\sigma} = \frac{\sigma}{l}. \tag{14.3}$$

It is obvious from the above discussion that c_x and c_y are not the components of the phase velocity $\mathbf{c}$. The relation between $|\mathbf{c}|$, c_x, and c_y is given by

$$\frac{1}{|\mathbf{c}|^2} = \frac{k^2 + l^2}{\sigma^2} = \frac{1}{c_x{}^2} + \frac{1}{c_y{}^2}. \tag{14.4}$$

14.1.2 Wave Dispersion and Group Velocity

The frequency of waves depends on the wavenumber and the physical property of the medium in which they propagate. Thus, the phase speed also depends on them, and the formula relating the frequency σ and wavenumber k, referred to as the *dispersion relation*, becomes very important.

In general, a wave is a superposition of various waves of different amplitude, wavenumbers, and frequencies. Some special waves, having the relation $\sigma \propto k$, do not change their shape during the propagation. They are called nondispersive waves. Acoustic waves is one of such waves, so that they are used as the tool of transferring information.

On the other hand, there is another type of waves called dispersive waves. Their component waves have different phase speed and change their shape while they are propagating. During propagation, they make wave packets which have longer wave length and smaller frequency than those of component waves. Most wave energy is concentrated in the wave packet and the velocity of the wave packet is called the group velocity.

We will derive the group velocity considering the superposition of two plane waves ψ_1 and ψ_2 propagating in the x-direction. Suppose that they have same amplitude but slightly different wavenumbers and frequencies. Namely,

$$\psi_1(t, x) = \hat{\psi} \exp[\tilde{\imath}\{(k + \delta k)x - (\sigma + \delta\sigma)t\}], \tag{14.5}$$

$$\psi_2(t, x) = \hat{\psi} \exp[\tilde{\imath}\{(k - \delta k)x - (\sigma - \delta\sigma)t\}]. \tag{14.6}$$

Adding (14.5) and (14.6), we obtain the superposition of two waves.

$$
\begin{aligned}
\psi(t, x) &= \psi_1(t, x) + \psi_2(t, x) \\
&= \hat{\psi}[\exp \tilde{\imath}(\delta k x - \delta\sigma t) + \exp\{-\tilde{\imath}(\delta k x - \delta\sigma t)\}] \exp \tilde{\imath}(kx - \sigma t)] \\
&= 2\hat{\psi} \cos(\delta k x - \delta\sigma t) \exp \tilde{\imath}(kx - \sigma t). \tag{14.7}
\end{aligned}
$$

The wave is the product of a carrier wave with high frequency σ and large wavenumber k and an envelope with low frequency $\delta\sigma$ and small wavenumber δk. The speed of the wave packet is called the group velocity, which is given by

$$c_{gx} = \frac{\delta\sigma}{\delta k}.$$

Taking the limit $\delta k \to 0$, we get

$$c_{gx} = \frac{\partial\sigma}{\partial k}. \tag{14.8}$$

The above discussion is valid also for y- and z-directions,

$$c_{gy} = \frac{\partial\sigma}{\partial l}, \tag{14.9}$$

$$c_{gz} = \frac{\partial\sigma}{\partial m}. \tag{14.10}$$

The phase speed and the group velocity have different value in general, but they are the same for the special waves having the dispersion relation $\sigma \propto k$.

14.2 Shallow Water Gravity Waves

Suppose a fluid layer whose horizontal scale is far larger than the depth scale. For simplicity, we will assume that the bottom is flat, the upper boundary is free surface, the average fluid depth is constant, density of the fluid is constant, and the effect of the Earth rotation is negligible (Fig. 14.2). Neglecting viscous

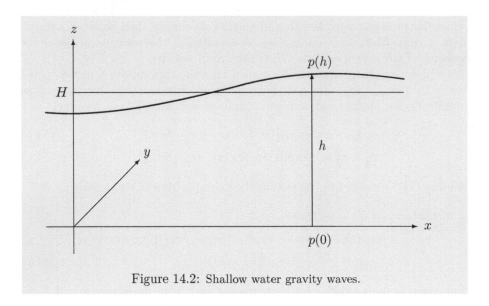

Figure 14.2: Shallow water gravity waves.

diffusion terms, Navier–Stokes equations are

$$\frac{\partial u}{\partial t} + u\frac{\partial u}{\partial x} + v\frac{\partial u}{\partial y} = -\frac{1}{\rho}\frac{\partial p}{\partial x}, \tag{14.11}$$

$$\frac{\partial v}{\partial t} + u\frac{\partial v}{\partial x} + v\frac{\partial v}{\partial y} = -\frac{1}{\rho}\frac{\partial p}{\partial y}. \tag{14.12}$$

The ratio of the fluid depth and the horizontal scale is far small, so that the vertical velocity is negligible and the hydrostatic equation is valid.

$$\frac{\partial p}{\partial z} = -\rho g. \tag{14.13}$$

The continuity equation is

$$\frac{\partial u}{\partial x} + \frac{\partial v}{\partial y} + \frac{\partial w}{\partial z} = 0. \tag{14.14}$$

Integrating (14.13) in the z-direction from 0 to h, we find

$$p(0) = \rho g h.$$

 The pressure gradient at the bottom is proportional to the gradient of h, so that it has the same value at all depth.

$$-\frac{1}{\rho}\frac{\partial p}{\partial x} = -g\frac{\partial h}{\partial x}, \qquad -\frac{1}{\rho}\frac{\partial p}{\partial y} = -g\frac{\partial h}{\partial y}. \tag{14.15}$$

Substituting from (14.15) into (14.11) and (14.12), we obtain

$$\frac{\partial u}{\partial t} + u\frac{\partial u}{\partial x} + v\frac{\partial u}{\partial y} = -g\frac{\partial h}{\partial x}, \tag{14.16}$$

$$\frac{\partial v}{\partial t} + u\frac{\partial v}{\partial x} + v\frac{\partial v}{\partial y} = -g\frac{\partial h}{\partial y}. \tag{14.17}$$

Supposing that u and v are independent of z initially, u and v are independent of z. Integrating (14.14) in the z-direction from 0 to h, we find following equation considering that $w = 0$ at $z = 0$.

$$h\left(\frac{\partial u}{\partial x} + \frac{\partial v}{\partial y}\right) = -\{w(h) - w(0)\} = -\frac{dh}{dt}$$

$$= -\left(\frac{\partial h}{\partial t} + u\frac{\partial h}{\partial x} + v\frac{\partial h}{\partial y}\right).$$

$$\frac{\partial h}{\partial t} + \frac{\partial(hu)}{\partial x} + \frac{\partial(hv)}{\partial y} = 0. \tag{14.18}$$

We separate dependent variables into the basic portions and the perturbed portions.

$$u = \bar{u} + u', \quad v = v', \quad h = H + h'. \tag{14.19}$$

Substituting from (14.19) into (14.16)–(14.18), and linearizing the resultant equations, we get

$$\left(\frac{\partial}{\partial t} + \bar{u}\frac{\partial}{\partial x}\right)u' = -g\frac{\partial h'}{\partial x}, \tag{14.20}$$

$$\left(\frac{\partial}{\partial t} + \bar{u}\frac{\partial}{\partial x}\right)v' = -g\frac{\partial h'}{\partial y}, \tag{14.21}$$

$$\left(\frac{\partial}{\partial t} + \bar{u}\frac{\partial}{\partial x}\right)h' + H\left(\frac{\partial u'}{\partial x} + \frac{\partial v'}{\partial y}\right) = 0. \tag{14.22}$$

Eliminating u' and v' from (14.20)–(14.22)

$$\left(\frac{\partial}{\partial t} + \bar{u}\frac{\partial}{\partial x}\right)^2 h' - gH\left(\frac{\partial^2}{\partial x^2} + \frac{\partial^2}{\partial y^2}\right)h' = 0. \tag{14.23}$$

Assuming a plane wave propagating to the x- and y-directions as,

$$h' = \hat{h}\exp\tilde{\imath}(kx + ly - \sigma t). \tag{14.24}$$

Substituting (14.24) into (14.23), we get the dispersion relation

$$\hat{\sigma} = \sigma - k\bar{u} = \pm\sqrt{gH(k^2 + l^2)}. \tag{14.25}$$

We call $\hat{\sigma}$ the *intrinsic frequency*, which is the frequency relative to the basic flow. The phase speeds in the x- and y-directions are

$$
\begin{cases}
c_x = \dfrac{\hat{\sigma}}{k} = \pm\sqrt{gH}\,\dfrac{\sqrt{k^2+l^2}}{k} \, , \\[3mm]
c_y = \dfrac{\hat{\sigma}}{l} = \pm\sqrt{gH}\,\dfrac{\sqrt{k^2+l^2}}{l} \, .
\end{cases}
\tag{14.26}
$$

Using (14.4), the magnitude of the phase velocity is

$$
|\mathbf{c}| = \sqrt{gH}.
$$

Thus, the phase velocity is given by

$$
\mathbf{c} = \pm\sqrt{gH}\,\frac{\mathbf{k}}{|\mathbf{k}|}.
\tag{14.27}
$$

The x- and y-components of the group velocity are

$$
\begin{cases}
c_{gx} = \dfrac{\partial \hat{\sigma}}{\partial k} = \pm\sqrt{gH}\,\dfrac{k}{\sqrt{k^2+l^2}} \, , \\[3mm]
c_{gy} = \dfrac{\partial \hat{\sigma}}{\partial l} = \pm\sqrt{gH}\,\dfrac{l}{\sqrt{k^2+l^2}} \, .
\end{cases}
\tag{14.28}
$$

The group velocity is

$$
\mathbf{c}_g = \pm\sqrt{gH}\,\frac{\mathbf{k}}{|\mathbf{k}|}.
\tag{14.29}
$$

Shallow water gravity waves have the phase velocity and group velocity of the same direction and magnitude.

14.3 Internal Gravity Waves of Two Layer Fluids

We will discuss gravity waves caused at the interface of two fluid layers. We assume that the density of the upper layer fluid ρ_2 is smaller than that of the lower layer fluid ρ_1, the bottom is flat and there is no disturbance at top surface.

We will find the pressure gradient force in the x-direction making use of Fig. 14.3. Let point A, B, C, and D are vertices of a rectangle in the x-z plane, A and D are on the interface, B is in the lower layer, and C is in the upper layer. The distance of AB and BD are δx and δz, respectively. Letting

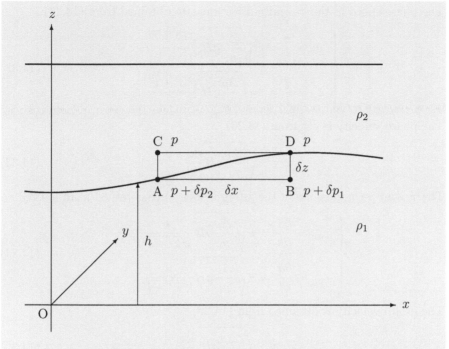

Figure 14.3: Internal gravity waves at the interface of two layers.

the pressure at A, B, C, and D be $p + \delta p_2$, $p + \delta p_1$, p, and p. Then, we find

$$
\begin{cases}
p_A = p + \delta p_2 = p + \rho_2 g \dfrac{\partial h}{\partial x} \delta x \, , \\[2mm]
p_B = p + \delta p_1 = p + \rho_1 g \dfrac{\partial h}{\partial x} \delta x \, , \\[2mm]
p_C = p \, , \\[2mm]
p_D = p \, .
\end{cases}
\tag{14.30}
$$

Taking the limit $\delta x \to 0$, we get the pressure gradient force in the lower layer.

$$
-\frac{1}{\rho_1} \frac{p_B - p_A}{\delta x} = -\frac{\Delta \rho}{\rho_1} g \frac{\partial h}{\partial x},
\tag{14.31}
$$

where $\Delta \rho = \rho_1 - \rho_2$.
Comparing (14.31) and the right hand-side of (14.20), it is obvious that all discussions in the previous section are valid replacing the acceleration due to gravity g by reduced gravity $(\Delta \rho / \rho_1) g$.

The phase speed in the x- and y-directions are obtained from (14.26).

$$\begin{cases} c_x = \dfrac{\hat{\sigma}}{k} = \pm\sqrt{\dfrac{\Delta\rho}{\rho_1}gH}\dfrac{\sqrt{k^2+l^2}}{k} \; , \\[4mm] c_y = \dfrac{\hat{\sigma}}{l} = \pm\sqrt{\dfrac{\Delta\rho}{\rho_1}gH}\dfrac{\sqrt{k^2+l^2}}{l} \; . \end{cases} \tag{14.32}$$

The phase velocity is get from (14.26).

$$\mathbf{c} = \pm\sqrt{\dfrac{\Delta\rho}{\rho_1}gH}\dfrac{\mathbf{k}}{|\mathbf{k}|}. \tag{14.33}$$

The x- and y-components of the group velocity are obtained from (14.28).

$$\begin{cases} c_{gx} = \dfrac{\partial\hat{\sigma}}{\partial k} = \pm\sqrt{\dfrac{\Delta\rho}{\rho_1}gH}\dfrac{k}{\sqrt{k^2+l^2}} \; , \\[4mm] c_{gy} = \dfrac{\partial\hat{\sigma}}{\partial l} = \pm\sqrt{\dfrac{\Delta\rho}{\rho_1}gH}\dfrac{l}{\sqrt{k^2+l^2}} \; . \end{cases} \tag{14.34}$$

The group velocity is obtained from (14.29)

$$\mathbf{c}_g = \pm\sqrt{\dfrac{\Delta\rho}{\rho_1}gH}\dfrac{\mathbf{k}}{|\mathbf{k}|}. \tag{14.35}$$

Dead water
From the old days, it was well known that a ship, navigating at a river mouth or a fjord, was suddenly reduced its speed by the unknown force. The phenomenon was called the *dead water*. This phenomenon was first described by Fridtjof Nansen in his navigation at the sea of the Nordenskiöld Archipelago in August 1893. His ship Fram could not travel at speed more than 1.5 knots in spite of her maximum speed of 6 to 7 knots in the normal condition. He asked Vilhelm Bjerknes to make clear this phenomenon scientifically. Bjerknes made Vagn Ekman to solve this problem. Ekman obtained the conclusion that the cause of the dead water is the internal gravity waves caused by a ship. When a depth of the ship's draft is almost equal to that of the interface of fresh and salty water, the ship produces internal gravity waves to waste its energy for propelling forward.

14.4 Internal Gravity Waves in the Continuously Stratified Fluid

In this section, we will consider internal gravity waves in the continuously stably stratified fluid. We will assume that the horizontal scale is small for the Coriolis force to be negligible.

14.4.1   Buoyancy Oscillation

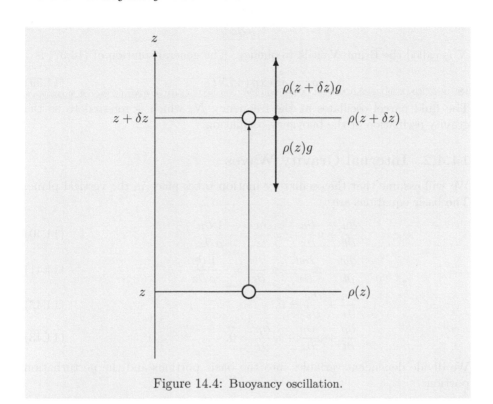

Figure 14.4: Buoyancy oscillation.

We will discuss the vertical oscillation of a stably stratified fluid. The restoring force is due to the buoyancy force, so that the oscillatory motion is called the buoyancy oscillation.

Suppose that a fluid parcel of unit volume is displaced upward from z to $z + \delta z$ (Fig. 14.4). The buoyancy force due to pressure is equal to the gravity exerting on a unit volume with the density of the ambient fluid $\rho(z + \delta z)g$ (the Archimedes' principle). While the downward force is the gravity exerting on the displaced parcel $\rho(z)g$. As the mass of the parcel is $\rho(z)$, the net force exerting on the parcel of a unit mass is

$$F_z = \frac{\rho(z + \delta z) - \rho(z)}{\rho(z)}g = \frac{1}{\rho}\frac{\partial \rho}{\partial z}g\delta z. \tag{14.36}$$

The equation of motion for the fluid parcel is

$$\frac{d^2\delta z}{dt^2} = \frac{1}{\rho}\frac{\partial \rho}{\partial z}g\delta z = -N^2\delta z, \tag{14.37}$$

where

$$N^2 = -\frac{1}{\rho}\frac{\partial\rho}{\partial z}g. \tag{14.38}$$

N is called the Brunt–Väisälä frequency. The general solution of (14.37) is

$$\delta z = \hat{\delta z}\exp\left(\pm\hat{i}Nt\right). \tag{14.39}$$

The fluid parcel oscillates at the frequency N, which is referred to as the gravity oscillation or the buoyancy oscillation.

14.4.2 Internal Gravity Waves

We will assume that the oscillatory motion takes place in the vertical plane. The basic equations are

$$\frac{\partial u}{\partial t} + u\frac{\partial u}{\partial x} + w\frac{\partial u}{\partial z} = -\frac{1}{\rho}\frac{\partial p}{\partial x}, \tag{14.40}$$

$$\frac{\partial w}{\partial t} + u\frac{\partial w}{\partial x} + w\frac{\partial w}{\partial z} = -\frac{1}{\rho}\frac{\partial p}{\partial z} - g, \tag{14.41}$$

$$\frac{\partial u}{\partial x} + \frac{\partial w}{\partial z} = 0, \tag{14.42}$$

$$\frac{\partial\rho}{\partial t} + u\frac{\partial\rho}{\partial x} + w\frac{\partial\rho}{\partial z} = 0. \tag{14.43}$$

We divide dependent variables into the basic portions and the perturbation portions.

$$\begin{cases} u = \bar{u} + u', \quad w = w', \\ p = \bar{p} + p', \quad \rho = \bar{\rho} + \rho'. \end{cases} \tag{14.44}$$

Substituting from (14.44) into (14.40)–(14.43) and linearizing the resultant equations, we obtain

$$\left(\frac{\partial}{\partial t} + \bar{u}\frac{\partial}{\partial x}\right)u' = -\frac{1}{\bar{\rho}}\frac{\partial p'}{\partial x}, \tag{14.45}$$

$$\left(\frac{\partial}{\partial t} + \bar{u}\frac{\partial}{\partial x}\right)w' = -\frac{1}{\bar{\rho}}\frac{\partial p'}{\partial z} - \frac{\rho'}{\bar{\rho}}g, \tag{14.46}$$

$$\frac{\partial u'}{\partial x} + \frac{\partial w'}{\partial z} = 0, \tag{14.47}$$

$$\left(\frac{\partial}{\partial t} + \bar{u}\frac{\partial}{\partial x}\right)\rho' + w'\frac{d\bar{\rho}}{dz} = 0. \tag{14.48}$$

We will show how (14.46) is derived from (14.41). The right-hand side of (14.41) is

$$-\frac{1}{\rho}\frac{\partial p}{\partial z} - g = -\frac{1}{\bar{\rho}}\left(1 + \frac{\rho'}{\bar{\rho}}\right)^{-1}\frac{\partial}{\partial z}(\bar{p} + p') - g$$

$$\simeq -\frac{1}{\bar{\rho}}\left(1 - \frac{\rho'}{\bar{\rho}}\right)\left(\frac{\partial \bar{p}}{\partial z} + \frac{\partial p'}{\partial z}\right) - g$$

$$= -\frac{1}{\bar{\rho}}\frac{\partial \bar{p}}{\partial z} - g + \frac{\rho'}{\bar{\rho}^2}\frac{\partial \bar{p}}{\partial z} - \frac{1}{\bar{\rho}}\frac{\partial p'}{\partial z}.$$

The basic equation is the hydrostatic balance.

$$\frac{1}{\bar{\rho}}\frac{\partial \bar{p}}{\partial z} + g = 0.$$

Making use of the above relation, the right-hand side of (14.46) is obtained.

$$\frac{\rho'}{\bar{\rho}^2}\frac{\partial \bar{p}}{\partial z} - \frac{1}{\bar{\rho}}\frac{\partial p'}{\partial z} = -\frac{1}{\bar{\rho}}\frac{\partial p'}{\partial z} - \frac{\rho'}{\bar{\rho}}g.$$

Subtracting $\partial(14.45)/\partial z$ from $\partial(14.46)/\partial x$, we can eliminate p' to get

$$\left(\frac{\partial}{\partial t} + \bar{u}\frac{\partial}{\partial x}\right)\left(\frac{\partial w'}{\partial x} - \frac{\partial u'}{\partial z}\right) + \frac{g}{\bar{\rho}}\frac{\partial \rho'}{\partial x} = 0. \tag{14.49}$$

Subtracting $g/\bar{\rho}\partial(14.48)/\partial x$ from $(\partial/\partial t + \bar{u}\partial/\partial x)(14.49)$, we can eliminate ρ' to obtain

$$\left(\frac{\partial}{\partial t} + \bar{u}\frac{\partial}{\partial x}\right)^2\left(\frac{\partial w'}{\partial x} - \frac{\partial u'}{\partial z}\right) - \frac{g}{\bar{\rho}}\frac{d\bar{\rho}}{dz}\frac{\partial w'}{\partial x} = 0. \tag{14.50}$$

From (14.47), we can define a perturbation stream function ψ' as

$$u' = -\frac{\partial \psi'}{\partial z}, \quad w' = \frac{\partial \psi'}{\partial x}. \tag{14.51}$$

Substituting from (14.51) into (14.50), we find

$$\left(\frac{\partial}{\partial t} + \bar{u}\frac{\partial}{\partial x}\right)^2\left(\frac{\partial^2}{\partial x^2} + \frac{\partial^2}{\partial z^2}\right)\psi' + N^2\frac{\partial^2 \psi'}{\partial x^2} = 0, \tag{14.52}$$

where $N^2 \equiv -gd\bar{\rho}/dz$.
We assume a plane wave propagating in the x- and z-directions.

$$\psi' = \hat{\psi}\exp\tilde{\imath}(kx + mz - \sigma t), \tag{14.53}$$

where σ is the frequency, k and m are x- and z-components of wave vector **k**. Substituting from (14.53) into (14.52), we obtain the dispersion relation

$$(\sigma - \bar{u}k)^2(k^2 + m^2) - N^2k^2 = 0,$$

$$\hat{\sigma} \equiv \sigma - \bar{u}k = \pm\frac{Nk}{\sqrt{k^2 + m^2}}, \tag{14.54}$$

where $\hat{\sigma}$ is the intrinsic frequency.

The phase speeds in the x- and z-directions are

$$\begin{cases} c_x = \dfrac{\hat{\sigma}}{k} = \pm\dfrac{N}{\sqrt{k^2 + m^2}} \,, \\[3mm] c_z = \dfrac{\hat{\sigma}}{m} = \pm\dfrac{Nk}{m\sqrt{k^2 + m^2}} \,. \end{cases} \qquad (14.55)$$

Using (14.4), phase velocityis obtained as

$$\mathbf{c} = \pm\frac{Nk}{(k^2 + m^2)^{3/2}}\mathbf{k}. \qquad (14.56)$$

The x- and z-components of the group velocity are

$$\begin{cases} c_{gx} = \dfrac{\partial \hat{\sigma}}{\partial k} = \pm\dfrac{Nm^2}{(k^2 + m^2)^{3/2}} \,, \\[3mm] c_{gz} = \dfrac{\partial \hat{\sigma}}{\partial m} = \pm\dfrac{-Nkm}{(k^2 + m^2)^{3/2}} \,. \end{cases} \qquad (14.57)$$

Thus, the group velocity is written as,

$$\mathbf{c}_g = \pm\frac{Nm}{(k^2 + m^2)^{3/2}}(m, -k). \qquad (14.58)$$

From (14.56) and (14.58), it is obvious that the phase velocity and the group velocity are orthogonal to each other.

Problem 1. Confirm that the scalar product of the phase velocity and the group velocity is zero.

We will assume a plane internal gravity wave whose phase propagation is in the positive x- and positive z-directions (Fig. 14.5). This means that we choose $k > 0$, $m > 0$, and the positive sign in (14.56) and (14.58). Then, we can see that the group velocity propagates downward. In the Earth's atmosphere, when internal gravity waves are caused by the surface topography the wave energy is transported upward while the phase propagates downward.

Sidewinder

The sidewinder, having the common names the horned rattlesnake, the sidewinder rattlesnake and sidewinder rattler, is a venomous pit viper living in the desert region of southwest United States and northwest Mexico. It moves in the very unique way, namely it waves longitudinally but moves its body laterally. Its movement is the same as the relation of the phase velocity and the group velocity of internal gravity waves, which is considered biologically the most efficient way moving on the sand.

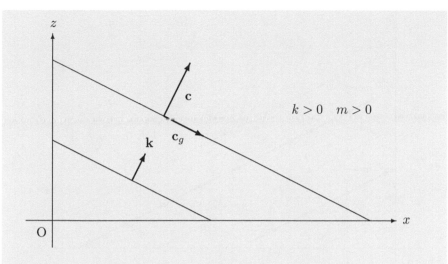

Figure 14.5: The phase velocity and group velocity of an internal gravity wave propagating in the positive x- and positive z-directions.

14.4.3 Structure of Internal Gravity Waves

We will consider the structure of an internal gravity wave propagating positive x- and z-directions, i.e., $k > 0$ and $m > 0$. Substituting from (14.53) into (14.51), we find

$$u' = -\tilde{\imath}m\psi',\tag{14.59}$$

$$w' = \tilde{\imath}k\psi'\ .\tag{14.60}$$

Then, the perturbation velocity in the x-z plane is

$$\mathbf{v}' = \tilde{\imath}\psi'(-m, k).\tag{14.61}$$

Substituting (14.59) and (14.60) into (14.49), we get

$$\rho' = -\frac{\bar{\rho}\hat{\sigma}}{gk}(k^2 + m^2)\psi'.\tag{14.62}$$

Substituting from (14.59) into (14.45) yields

$$p' = -\tilde{\imath}\frac{\bar{\rho}\hat{\sigma}m}{k}\psi'.\tag{14.63}$$

Thus, we can illustrate the structure of the internal gravity wave in Fig. 14.6

14.4.4 Mountain Waves

In the stably stratified Earth's atmosphere, when the constant basic flow $\bar{u}$ blows across the mountain range, internal gravity waves which is stationary

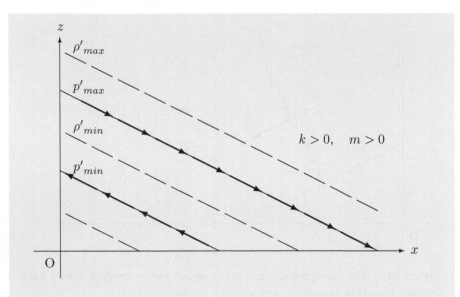

Figure 14.6: Structure of an internal gravity wave for $k > 0$ and $m > 0$. Solid lines indicate the maximum and minimum perturbation pressure, and dashed lines indicate the maximum and minimum perturbation density. Arrows indicate perturbation velocity.

relative to the ground. i.e., $\sigma = 0$ in (14.54) are induced. Then, we find

$$-\bar{u}k = \pm \frac{Nk}{\sqrt{k^2 + m^2}},\tag{14.64}$$

$$m^2 = \frac{N^2}{\bar{u}^2} - k^2.$$

In the case $|\bar{u}|k < N$, $m^2 > 0$, i.e., m is real and the solution (14.52) becomes

$$\psi' = \hat{\psi}\exp\tilde{\imath}(kx + mz).\tag{14.65}$$

The mountain waves can propagate vertically. If $k > 0$ and $\bar{u} > 0$, we must choose the minus sign in (14.59) and also in (14.54), (14.55), (14.56), (14.57), and (14.58). Then, m must be positive due to (14.57), as the energy source is at the lower boundary and the group velocity should have the upward component. As a consequence, the phase velocity has the downward component. If $k > 0$ and $\bar{u} < 0$, we should choose the plus sign in (14.59), and also in (14.54), (14.55), (14.56), (14.57), and (14.58). So, m must be negative for the group velocity to propagate upward.

Clear air turbulence (CAT)

In the Earth's atmosphere, the air density decreases exponentially with height, so that the amplitude of mountain waves increases exponentially with height. When the water vapor is sufficient to make clouds at the updraft region, mountain waves make clouds or cloud trains over the topography. However, when the upper air is dry, there are invisible large amplitude waves, which sometimes break to make turbulence. They are called clear air turbulence (CAT) and occasionally cause serious aircraft accidents.

In the case $|\bar{u}|k > N$, $m^2 < 0$, i.e., m is imaginary and the solution (14.52) becomes

$$\psi' = \hat{\psi} \exp\left(\pm|m|z\right) \exp \tilde{i}kx,$$

As the energy source is at the lower boundary, the minus sign in the above equation should be chosen.

$$\psi' = \hat{\psi} \exp\left(-|m|z\right) \exp\left(\tilde{i}kx\right), \tag{14.66}$$

The disturbance caused by the surface topography decays exponentially and the phase line is vertical. The stationary waves are called trapped waves or evanescent.

14.4.5 Physical Derivation of the Intrinsic Frequency of Internal Gravity Waves

We assume an internal gravity wave propagating in the negative x- and positive z-directions. We will take ξ-axis upward along the constant phase line tilted at an angle α to the vertical (Fig. 14.7). Supposing that the fluid parcel of the unit mass is displaced $\delta\xi = \delta z / \cos \alpha$ upward along ξ-axis, the ξ-component force exerting on the parcel is

$$F_\xi = -N^2 \delta z \cos \alpha = -N^2 \cos^2 \alpha \delta\xi,$$

Then, the equation of motion for the parcel is

$$\frac{d^2 \delta\xi}{dt^2} = -N^2 \cos^2 \alpha \delta\xi. \tag{14.67}$$

The general solution of (14.67) is

$$\delta\xi = \hat{\delta\xi} \exp\left(\pm\tilde{i}N \cos \alpha t\right). \tag{14.68}$$

Equation (14.68) shows the oscillatory motion of the frequency

$$\hat{\sigma} = \pm N \cos \alpha = \pm N \frac{k}{|\mathbf{k}|}, \tag{14.69}$$

where $\mathbf{k}$ is the wave vector.

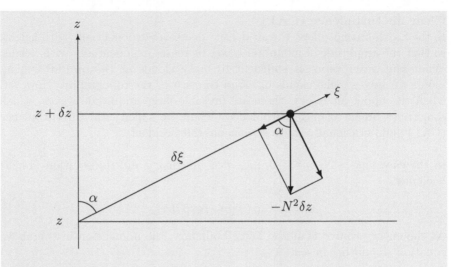

Figure 14.7: Oscillatory motion on the constant phase line inclined at an angle α from the vertical.

14.5 Inertio-Gravity Waves

Gravity waves having large horizontal scale (the Rossby number is smaller than 0.1) are influenced by the Earth's rotation.

14.5.1 Derivation from the Governing Equations

The perturbation equations are readily derived from (14.45) to (14.48) extending from two–dimensions to three–dimensions, adding Coriolis force, and taking the x-axis to the direction of the basic flow.

$$\left(\frac{\partial}{\partial t} + \bar{u}\frac{\partial}{\partial x}\right) u' - fv' = -\frac{1}{\bar{\rho}}\frac{\partial p'}{\partial x}, \tag{14.70}$$

$$\left(\frac{\partial}{\partial t} + \bar{u}\frac{\partial}{\partial x}\right) v' + fu' = -\frac{1}{\bar{\rho}}\frac{\partial p'}{\partial y}, \tag{14.71}$$

$$\left(\frac{\partial}{\partial t} + \bar{u}\frac{\partial}{\partial x}\right) w' = -\frac{1}{\bar{\rho}}\frac{\partial p'}{\partial z} - \frac{\rho'}{\bar{\rho}}g, \tag{14.72}$$

$$\frac{\partial u'}{\partial x} + \frac{\partial v'}{\partial y} + \frac{\partial w'}{\partial z} = 0, \tag{14.73}$$

$$\left(\frac{\partial}{\partial t} + \bar{u}\frac{\partial}{\partial x}\right) \rho' + w'\frac{d\bar{\rho}}{dz} = 0. \tag{14.74}$$

Subtracting $g/\bar{\rho}(14.74)$ from $\left(\frac{\partial}{\partial t} + \bar{u}\frac{\partial}{\partial x}\right)(14.72)$

$$\left(\frac{\partial}{\partial t} + \bar{u}\frac{\partial}{\partial x}\right)^2 w' + N^2 w' = -\frac{1}{\bar{\rho}}\left(\frac{\partial}{\partial t} + \bar{u}\frac{\partial}{\partial x}\right)\frac{\partial p'}{\partial z}. \tag{14.75}$$

Subtracting $\partial(14.70)/\partial y$ from $\partial(14.71)/\partial x$

$$\left(\frac{\partial}{\partial t} + \bar{u}\frac{\partial}{\partial x}\right)\left(\frac{\partial v'}{\partial x} - \frac{\partial u'}{\partial y}\right) + f\left(-\frac{\partial w'}{\partial z}\right) = 0, \tag{14.76}$$

where use is made of (14.73).
Adding $\partial(14.70)/\partial x$ and $\partial(14.71)/\partial y$, we get

$$\left(\frac{\partial}{\partial t} + \bar{u}\frac{\partial}{\partial x}\right)\left(-\frac{\partial w'}{\partial z}\right) - f\left(\frac{\partial v'}{\partial x} - \frac{\partial u'}{\partial y}\right) = -\frac{1}{\bar{\rho}}\left(\frac{\partial^2}{\partial x^2} + \frac{\partial^2}{\partial y^2}\right)p'. \tag{14.77}$$

Adding $\left(\frac{\partial}{\partial t} + \bar{u}\frac{\partial}{\partial x}\right)$ (14.77) and $f\times$(14.76),

$$-\left(\frac{\partial}{\partial t} + \bar{u}\frac{\partial}{\partial x}\right)^2\frac{\partial w'}{\partial z} - f^2\frac{\partial w'}{\partial z} = -\frac{1}{\bar{\rho}}\left(\frac{\partial^2}{\partial x^2} + \frac{\partial^2}{\partial y^2}\right)\left(\frac{\partial}{\partial t} + \bar{u}\frac{\partial}{\partial x}\right)p'. \tag{14.78}$$

Subtracting $\partial(14.78)/\partial z$ from $\left(\frac{\partial^2}{\partial x^2} + \frac{\partial^2}{\partial y^2}\right)(14.75)$,

$$\left(\frac{\partial}{\partial t} + \bar{u}\frac{\partial}{\partial x}\right)^2\left(\frac{\partial^2}{\partial x^2} + \frac{\partial^2}{\partial y^2} + \frac{\partial^2}{\partial z^2}\right)w'$$
$$+ \left\{N^2\left(\frac{\partial^2}{\partial x^2} + \frac{\partial^2}{\partial y^2}\right) + f^2\frac{\partial^2}{\partial z^2}\right\}w' = 0. \tag{14.79}$$

We assume a plane wave propagating x-, y-, and z-directions,

$$w' = \hat{w}\exp\tilde{\imath}(kx + ly + mz - \sigma t). \tag{14.80}$$

Substituting (14.80) into (14.79), we obtain dispersion relations,

$$\hat{\sigma} \equiv \sigma - \bar{u}k = \pm\frac{\left\{N^2(k^2 + l^2) + f^2 m^2\right\}^{1/2}}{(k^2 + l^2 + m^2)^{1/2}}, \tag{14.81}$$

where $\hat{\sigma}$ is the intrinsic frequency of inertial-gravity waves.
The phase speed in the x-, y-, and z-directions are

$$c_x = \frac{\hat{\sigma}}{k} = \pm\frac{\left\{N^2(k^2 + l^2) + f^2 m^2\right\}^{1/2}}{k(k^2 + l^2 + m^2)^{1/2}}, \tag{14.82}$$

$$c_y = \frac{\hat{\sigma}}{l} = \pm\frac{\left\{N^2(k^2 + l^2) + f^2 m^2\right\}^{1/2}}{l(k^2 + l^2 + m^2)^{1/2}}, \tag{14.83}$$

$$c_z = \frac{\hat{\sigma}}{m} = \pm\frac{\left\{N^2(k^2 + l^2) + f^2 m^2\right\}^{1/2}}{m(k^2 + l^2 + m^2)^{1/2}}. \tag{14.84}$$

The magnitude of the phase velocity $|\mathbf{c}|$ is obtained extending (14.4) to three dimensions.

$$\frac{1}{|\mathbf{c}|^2} = \frac{1}{c_x{}^2} + \frac{1}{c_y{}^2} + \frac{1}{c_z{}^2} = \frac{(k^2 + l^2 + m^2)^2}{N^2(k^2 + l^2) + f^2 m^2},$$

$$|\mathbf{c}| = \pm \frac{\{N^2(k^2 + l^2) + f^2 m^2\}^{1/2}}{k^2 + l^2 + m^2}.$$

Therefore,

$$\mathbf{c} = \pm \frac{\{N^2(k^2 + l^2) + f^2 m^2\}^{1/2}}{(k^2 + l^2 + m^2)^{3/2}} \mathbf{k}. \tag{14.85}$$

The x, y, and z-components of the group velocity are

$$c_{gx} = \frac{\partial \hat{\sigma}}{\partial k} = \pm \frac{(N^2 - f^2)\, km^2}{(k^2 + l^2 + m^2)^{3/2}\, \{N^2(k^2 + l^2) + f^2 m^2\}^{1/2}}, \tag{14.86}$$

$$c_{gy} = \frac{\partial \hat{\sigma}}{\partial l} = \pm \frac{(N^2 - f^2)\, lm^2}{(k^2 + l^2 + m^2)^{3/2}\, \{N^2(k^2 + l^2) + f^2 m^2\}^{1/2}}, \tag{14.87}$$

$$c_{gz} = \frac{\partial \hat{\sigma}}{\partial m} = \pm \frac{-(N^2 - f^2)\, (k^2 + l^2)m}{(k^2 + l^2 + m^2)^{3/2}\, \{N^2(k^2 + l^2) + f^2 m^2\}^{1/2}}. \tag{14.88}$$

The group velocity is

$$\mathbf{c}_g = \pm \frac{(N^2 - f^2)\, m}{(k^2 + l^2 + m^2)^{3/2}\, \{N^2(k^2 + l^2) + f^2 m^2\}^{1/2}} \\ \times (km, lm, -(k^2 + l^2))\ . \tag{14.89}$$

Taking the scalar product of $\mathbf{c}$ and $\mathbf{c}_g$,

$$\mathbf{c} \cdot \mathbf{c}_g = \frac{\{N^2(k^2 + l^2) + f^2 m^2\}^{1/2}\, (N^2 - f^2)}{(k^2 + l^2 + m^2)^3} \\ \times \{k^2 m + l^2 m - (k^2 + l^2)m\} = 0.$$

We find that the phase velocity and the group velocity of inertio-gravity waves are orthogonal to each other. The horizontal components of two vectors are in the same direction, while the vertical component of them is in opposite direction. Namely, when the energy source exists on the lower boundary and inertio-gravity waves transport energy upward, their phase propagates downward.

14.5.2 Physical Derivation of the Intrinsic Frequency of Inertio-Gravity Waves

Suppose a stably stratified fluid, the Coriolis parameter is increasing in the y-direction, and the basic flow in the x-direction is constant. We assume that

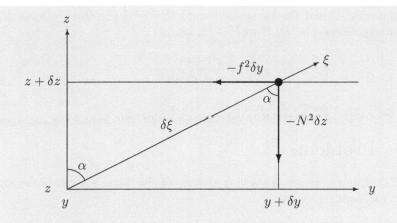

Figure 14.8: Physical derivation of the frequency of inertio-gravity waves.

an inertio-gravity wave is propagating in the y-z plane. As is shown in Fig. 14.8, we will consider the motion of fluid parcel of unit mass along the ξ-axis (a constant phase line, which is inclined α from the vertical). At some instance, the fluid parcel is displaced $\delta\xi$, which corresponds to the displacement δz in the z-direction and δy in the y-direction. The relations between them are,

$$\delta z = \delta\xi \cos\alpha, \quad \delta y = \delta\xi \sin\alpha. \tag{14.90}$$

The restoring force due to buoyancy in the z-direction is $F_z = -N^2\delta z$ and the restoring force due to Coriolis force in the y-direction is $F_y = -f^2\delta y$ (refer subsection 15.3.3). Therefore, the total restoring force in the ξ-direction is obtained using (14.90)

$$F_\xi = F_z \cos\alpha + F_y \sin\alpha = -\left(N^2\cos^2\alpha + f^2\sin^2\alpha\right)\delta\xi$$

$$= -\frac{1}{l^2 + m^2}\left(m^2 N^2 + l^2 f^2\right)\delta\xi, \tag{14.91}$$

where use is made of the relation $\tan\alpha = L_z/L_y = l/m$.
The equation of motion of the fluid parcel is

$$\frac{d^2\delta\xi}{dt^2} - -\frac{1}{l^2 + m^2}\left(m^2 N^2 + l^2 f^2\right)\delta\xi, \tag{14.92}$$

The general solution of (14.92) is

$$\delta\xi = \hat{\delta\xi} \exp\left(\tilde{i}\hat{\sigma}\right),$$

$$\hat{\sigma} = \pm\frac{(l^2 N^2 + m^2 f^2)^{1/2}}{(l^2 + m^2)^{1/2}}. \tag{14.93}$$

We can readily extend this two-dimensional illustration to three–dimensional wave propagation replacing l^2 to $k^2 + l^2$, i.e.,

$$\hat{\sigma} = \pm \frac{\{N^2(k^2 + l^2) + m^2 f^2\}^{1/2}}{(k^2 + l^2 + m^2)^{1/2}},$$

which is the same dispersion relation as (14.81).

14.6 Problems

1. Calculate the Brunt–Väisälä frequency of the Earth's atmosphere solving following problems.

   (1) Suppose that the air is an ideal gas composed of 80% nitrogen and 20% oxygen. Calculate the density of the air at the standard condition ($0\,°C$, $1\,atm.{=}1.01325{\times}10^3\,hPa$), using the fact that an ideal gas of 1 mol has the volume of $22.4{\times}10^{-3}\,m^3$ at the standard condition. .

   (2) Calculate the density of the atmosphere ρ_0 at the Earth's surface of some point of midlatitude ($45°$). Suppose that the temperature is $288\,K$, and the pressure is $1.00{\times}10^3\,hPa$.

   (3) At the altitude of $1.5\,km$ above the point, the pressure is $8.50{\times} 10^2\,hPa$ and the temperature is $279\,K$. What is the density of the air?

   (4) Calculate the Brunt–Väisälä frequency N between the surface and the altitude of $1.5\,km$, supposing that the air density decreases linearly with height.

2. The 1960 Valdivia earthquake, happened on 22 May 1960 (19:11GMT) and at $28.29°S$, $73.05°W$, had the largest magnitude ($M\,9.5$) ever recorded. It caused big tsunamis, which attacked Chile, Hawaii, Philippines, Japan, Aleutian Islands, New Zealand, and southeast Australia. The distance between Chile and Japan is around $1.64 \times 10^7\,m$. Obtain the time in which tsunamis traveled from Chile to Japan, supposing the depth of the Pacific Ocean is $4.00 \times 10^3\,m$. Let the magnitude of the acceleration due to gravity be $9.80\,ms^{-2}$.

3. Suppose that the surface height perturbation h' in a one-dimensional shallow water gravity wave is given by
   $h' = \hat{h} \exp\{\tilde{\imath}(kx - \sigma t)\}$
   Assuming $k > 0$, find the velocity perturbation u', and the horizontal convergence $(-\partial u'/\partial x)$ for a wave propagating in the positive x-direction ($\hat{\sigma} > 0$) and a wave propagating in the negative x-direction ($\hat{\sigma} < 0$). Discuss the structure of the wave.

4. At the river mouth of the Amazon, the fresh water of density $1.00 \times 10^3 \, \mathrm{kgm^{-3}}$ lies over the salty water of density $1.03 \times 10^3 \, \mathrm{kgm^{-3}}$ without mixing. What is the phase speed of an internal gravity wave at the interface of two layers? Let the distance from the interface to the bottom be 10.0 m, and the magnitude of the acceleration due to gravity be $9.80 \, \mathrm{ms^{-2}}$.

5. Derive (14.25).

6. Derive (14.86)–(14.88).

7. The Brunt–Väisälä frequency of the midlatitude troposphere is $1.20 \times 10^{-2} \, \mathrm{s^{-1}}$. Calculate the angle θ between the constant phase line of the inertio-gravity waves and a horizontal plane for the restoring forces due to the buoyancy and the Coriolis force to be equal.

Chapter 15

Phenomena in Geophysical Fluids: Part III

In this chapter, we will present several dynamic phenomena in the Earth's atmosphere, oceans, and laboratory fluid systems following the previous two chapters. They are inertial oscillations, Rossby waves, barotropic instability, baroclinic instability, and geostrophic turbulence. When we discuss barotropic instability, two important integral theorems, Rayleigh's inflection point theorem and Howard's semi-circle theorem, are introduced. In the section of the geostrophic turbulence, we will show that there is two inertial subranges in the two-dimensional turbulence, one is the downward energy cascading rage obeying a $-5/3$ power law and the other is the upward enstrophy cascading range obeying -3 power law.

15.1 Inertial Oscillations

In the Earth's oceans and large lakes, there is an oscillatory motion analogous to the Foucault pendulum. Suppose that the time scale of the motion is comparable to the period of the Earth's rotation, the horizontal scale is much smaller than the Earth's radius and the horizontal pressure gradient is negligible. We may use the f-plane approximation and the governing equations are obtained omitting small terms in (12.28) and (12.29),

$$\frac{\partial u}{\partial t} = fv, \tag{15.1}$$

$$\frac{\partial v}{\partial t} = -fu. \tag{15.2}$$

Multiplying (15.2) by the imaginary unit $\tilde{\imath}$ and adding the result to (15.1), we obtain

$$\frac{\partial w}{\partial t} = -\tilde{\imath} f w, \tag{15.3}$$

where $w = u + \tilde{\imath} v$. The general solution of (15.3) is

$$w = \hat{w} \exp(-\tilde{\imath} f t), \tag{15.4}$$

where $\hat{w} = w_r + \tilde{\imath} w_i$ is the complex amplitude. Separating (15.4) into the real and imaginary parts, we find

$$u = V \cos(\phi - ft), \tag{15.5}$$
$$v = V \sin(\phi - ft), \tag{15.6}$$

where $V = \sqrt{w_r{}^2 + w_i{}^2}$ is the tangential velocity of the uniform circular motion and $\phi = \tan^{-1}(w_i/w_r)$ is the initial phase. The solutions (15.5) and (15.6) represent clockwise rotation in the Northern Hemisphere ($f > 0$) and counterclockwise rotation in the Southern Hemisphere ($f < 0$). The period is

$$T = \frac{2\pi}{|f|} = \frac{\pi}{\Omega |\sin\theta|}, \tag{15.7}$$

which is equal to the time for the Foucault pendulum to turn through a half circle, hence it is called *one-half pendulum day*.
The radius of the circle is

$$R = \frac{V}{2\pi/T} = \frac{V}{|f|}. \tag{15.8}$$

This phenomenon is known as the *inertial oscillation*, which is often observed in the Earth's oceans and large lakes but is seldom observed in the atmosphere.

15.2   Rossby Waves

Rossby waves are the wave motion due to the conservation of the absolute vorticity, so the latitudinal variation of the Coriolis parameter f is crucial. Therefore, it is adequate to use the quasi-geostrophic vorticity equation (12.66) on the midlatitude β-plane.

15.2.1   Non-Divergent Rossby Waves

If the upper and lower boundaries are rigid surfaces, there is no horizontal divergence and (12.66) becomes

$$\frac{d\zeta}{dt} + \beta v = 0, \tag{15.9}$$

where $\beta = df/dy$.

Now we assume that the zonal velocity $\bar{u}$ is constant. Field variables are divided into a basic portion and perturbation portions as

$$u = \bar{u} + u', \quad v = v'. \tag{15.10}$$

We will define a perturbation stream function ψ' as

$$u' = -\frac{\partial \psi'}{\partial y}, \quad v' = \frac{\partial \psi'}{\partial x}. \tag{15.11}$$

A perturbation vorticity is expressed as

$$\zeta' = \frac{\partial v'}{\partial x} - \frac{\partial u'}{\partial y} = \nabla_{\mathrm{H}}^2 \psi'.$$

Substituting the above relation into (15.9), we obtain

$$\frac{d}{dt} \nabla_{\mathrm{H}}^2 \psi' + \beta \frac{\partial \psi'}{\partial x} = 0, \tag{15.12}$$

Assuming a plane wave propagating in the x- and y-directions,

$$\psi' = \hat{\psi} \exp \tilde{\imath}(kx + ly - \sigma t), \tag{15.13}$$

where k and l are wavenumbers in the x- and y-directions and σ is the frequency. Substituting (15.11) and (15.13) into (15.12), we get

$$-\tilde{\imath}(k\bar{u} - \sigma)(k^2 + l^2)\hat{\psi} + \tilde{\imath}k\beta\hat{\psi} = 0,$$

from which we obtain the intrinsic frequency

$$\hat{\sigma} = \sigma - \bar{u}k = -\frac{\beta k}{k^2 + l^2}. \tag{15.14}$$

The phase speed of the x- and y-directions are given by

$$\begin{cases} c_x = \dfrac{\hat{\sigma}}{k} = -\dfrac{\beta}{k^2 + l^2} \, , \\[2mm] c_y = \dfrac{\sigma}{l} = -\dfrac{\beta k}{l(k^2 + l^2)} \, . \end{cases} \tag{15.15}$$

The magnitude of the phase velocity is obtained using (14.4)

$$|\mathbf{c}| = \frac{\beta k}{(k^2 + l^2)^{3/2}}.$$

Then, the phase velocity is

$$\mathbf{c} = \frac{\beta k}{(k^2 + l^2)^2} \mathbf{k}. \tag{15.16}$$

The x- and y-components of the group velocity are given by

$$\begin{cases} c_{gx} = \dfrac{\partial \hat{\sigma}}{\partial k} = \dfrac{\beta(k^2 - l^2)}{(k^2 + l^2)^2} \,, \\[3mm] c_{gy} = \dfrac{\partial \hat{\sigma}}{\partial l} = \dfrac{2\beta kl}{(k^2 + l^2)^2} \,. \end{cases} \tag{15.17}$$

The group velocity is given by

$$\mathbf{c}_g = \frac{\beta}{(k^2 + l^2)^2}(k^2 - l^2, 2kl). \tag{15.18}$$

We will investigate the propagation of Rossby waves on the reference frame moving at the velocity $\bar{u}$. Equation (15.14) becomes

$$\left(k + \frac{\beta}{2\hat{\sigma}}\right)^2 + l^2 = \left(\frac{\beta}{2\hat{\sigma}}\right)^2. \tag{15.19}$$

Letting $k < 0$, $l > 0$ and the angle between the wave vector $\mathbf{k} = (k, l)$ and the k-axis be θ, wavenumbers should be on a circle of radius $\beta/2\sigma$ centering at $(-\beta/2\sigma, 0)$ in the wavenumber space as shown in Fig. 15.1. The direction of the group velocity $\mathbf{c}_g$ is 2θ from the k-axis.

The energy of waves is transferred by the group velocity, so that long wavelength Rossby waves ($k < l$) transfer energy to the negative k-direction, while short wavelength Rossby waves ($k > l$) transfer energy to the positive k-direction.

Example 1. What is the direction of the group velocity of the non-divergent Rossby waves?

Answer
Using θ and κ, components of the wave vector are expressed by

$$k = \kappa \cos \theta, \quad l = \kappa \sin \theta.$$

Components of the group velocity are given by

$$c_{gx} = \frac{\beta(\cos^2\theta - \sin^2\theta)}{\kappa^2} = |c_x| \cos 2\theta,$$

$$c_{gy} = \frac{2\beta \cos \theta \sin \theta}{\kappa^2} = |c_x| \sin 2\theta.$$

Therefore, the angle between the group velocity and the x-axis is 2θ. In Fig. 15.2, the phase velocity and the group velocity of a Rossby wave are shown in the x-y plane. L_x and L_y are wavelengths in the x- and y-directions.

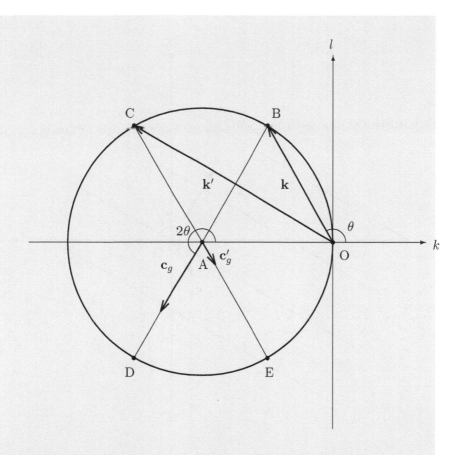

Figure 15.1: Phase velocity and group velocity in wavenumber space. A is the center of the circle defined by (15.19). $\overline{OB}$ represents wave vector $\mathbf{k}$, C is the symmetric point of B with respect to the line perpendicular to the k-axis and passing through A, D is the symmetric point of B with respect to A, and E is the symmetric point of C with respect to A.

15.2.2 The Reflection of Rossby Waves

Let us consider the reflection of Rossby waves at the rigid boundaries. There are no boundaries in the Earth's atmosphere but there are in oceans, large lakes, and laboratory systems. Let the stream function of the incidental wave be

$$\psi = \Psi \exp \tilde{\imath}(kx + ly - \sigma t), \tag{15.20}$$

and the stream function of the reflected wave be

$$\psi' = \Psi' \exp \tilde{\imath}(k'x + l'y - \sigma't). \tag{15.21}$$

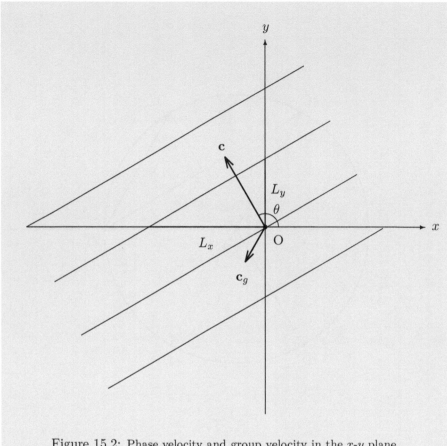

Figure 15.2: Phase velocity and group velocity in the x-y plane.

Suppose that the boundary is perpendicular to the x-axis taking the origin at the boundary. The boundary condition is that the normal velocity is zero, namely

$$u + u' = 0, \tag{15.22}$$

where u, u' are the x-component of the velocity of the incidental and reflected waves. Using (15.20) and (15.21), (15.22) becomes

$$[u + u']_{x=0} = \left[-\frac{\partial \psi}{\partial y} - \frac{\partial \psi'}{\partial y} \right]_{x=0}$$
$$= -\tilde{\imath}[l\Psi \exp \tilde{\imath}(ly - \sigma t) + l'\Psi \exp \tilde{\imath}(l'y - \sigma' t)]. \tag{15.23}$$

Therefore,

$$l = l', \quad \sigma = \sigma', \quad \Psi = -\Psi'. \tag{15.24}$$

If the wave vector is given by Fig. 15.1, the wave vector of the reflected wave should be on the circle. Then, the wave vector and the group velocity of the reflected wave are given by $\mathbf{k}'$ and $\mathbf{c}'_g$ in Fig. 15.1. The incidental Rossby wave transfers energy westward, while the reflected Rossby wave transfers energy eastward shrinking its wavelength. Further, we can find

$$\angle \text{BAO} = \angle \text{EAO}.$$

Namely, the *law of reflection* holds for the reflection of the group velocity of Rossby waves.

15.2.3 Rossby Waves with Free Surface

In this subsection, we will discuss Rossby waves on the β-plane with free surface and constant density. Then, we may start from the conservation of the potential vorticity equation (12.68), dividing field variables into basic portions and perturbation portions as

$$u = \bar{u} + u', \quad v = v', \quad \zeta = \zeta', \quad H = \bar{H} + h'. \tag{15.25}$$

Substituting from (15.25) into (12.68) and linearizing the resultant equation,

$$\left(\frac{\partial}{\partial t} + \bar{u} \frac{\partial}{\partial x} \right) \zeta' + \beta v' - \frac{f}{H} \left(\frac{\partial}{\partial t} + \bar{u} \frac{\partial}{\partial x} \right) h' = 0. \tag{15.26}$$

Integrating the hydrostatic equation (12.47) from z to H with respect to z, we obtain

$$p = \rho g (\bar{H} + h' - z),$$
$$p' = \rho g h'. \tag{15.27}$$

The geostrophic equations for perturbation variables are

$$u' = -\frac{1}{f\rho} \frac{\partial p'}{\partial y} = -\frac{g}{f} \frac{\partial h'}{\partial y}, \tag{15.28}$$

$$v' = \frac{1}{f\rho} \frac{\partial p'}{\partial x} = \frac{g}{f} \frac{\partial h'}{\partial x}. \tag{15.29}$$

Then, the perturbation vorticity is

$$\zeta' = \frac{\partial v'}{\partial x} - \frac{\partial u'}{\partial y} = \frac{g}{f} \nabla_{\text{H}}{}^2 h'. \tag{15.30}$$

Substituting from (15.29) and (15.30) into (15.26), we obtain

$$\left(\frac{\partial}{\partial t} + \bar{u} \frac{\partial}{\partial x} \right) \left(\nabla_{\text{H}}{}^2 - \frac{f^2}{g\bar{H}} \right) h' + \beta \frac{\partial h'}{\partial x} = 0. \tag{15.31}$$

Assuming a plane wave propagating in the x- and y-directions as,

$$h' = \hat{h} \exp \tilde{\imath}(kx + ly - \sigma t). \tag{15.32}$$

Substituting (15.32) into (15.31), we get the intrinsic frequency

$$\hat{\sigma} = \sigma - \bar{u}k = \frac{\beta k}{k^2 + l^2 + 1/\lambda^2}, \tag{15.33}$$

where

$$\lambda = \frac{\sqrt{g\bar{H}}}{f} \tag{15.34}$$

is called the *Rossby radius of deformation*. The x- and y-components of the group velocity are

$$c_{gx} = \frac{\partial \hat{\sigma}}{\partial k} = \frac{k^2 - l^2 - 1/\lambda^2}{(k^2 + l^2 + 1/\lambda^2)^2}, \tag{15.35}$$

$$c_{gy} = \frac{\partial \hat{\sigma}}{\partial l} = \frac{2\beta kl}{(k^2 + l^2 + 1/\lambda^2)^2}. \tag{15.36}$$

The dispersion relation is shown in Fig. 15.3 for the case of $l = 0$. When the phase line of Rossby waves is parallel to the y-axis ($l = 0$), the

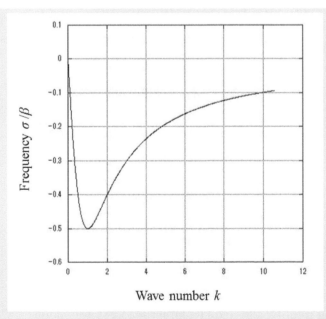

Figure 15.3: The intrinsic frequency of the divergent Rossby waves in the case of $l = 0$ and $\lambda = 1$.

energy flux is westward for long waves ($k < 1/\lambda$) and eastward for short waves ($k > 1/\lambda$).

15.2.4 Rossby Waves in the Laboratory System

Rossby waves can be simulated in experimental vessels by making fluid depth vary monotonically in the y-direction in spite that the Coriolis parameter f is constant everywhere. We will expand (12.68) to obtain

$$\frac{d\zeta}{dt} + \beta^* v = 0, \tag{15.37}$$

where

$$\beta^* = -\frac{f}{H_0}\frac{dH}{dy}. \tag{15.38}$$

The most simple example is a vessel known as a sliced cylinder (Fig. 15.4) which is the rotating cylinder with a sliced flat bottom. Rossby waves are induced by this simple equipment at the spin-up and spin-down. In this case, the bottom inclines linearly to the positive y-direction, i.e., the fluid depth becomes

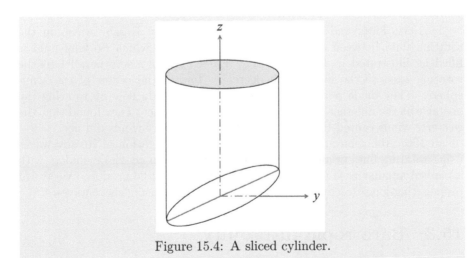

Figure 15.4: A sliced cylinder.

$$H = H_0 - \alpha y.$$

Substituting the above equation into (15.38), we get $\beta^* = \alpha f/H_0$.

Comparing (15.37) with (15.9), we find that β^* has the same dynamic effect as β.

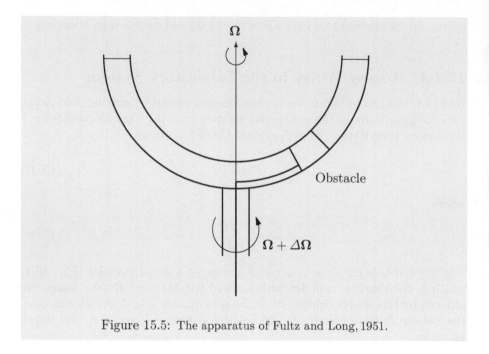

Figure 15.5: The apparatus of Fultz and Long, 1951.

In 1951, Fultz and Long studied the behavior of Rossby waves in the rotating fluid between coaxial semi-spheres between which working fluid is filled as illustrated in Fig. 15.5. The Earth's gravity acts parallel to the rotating axis, so the fluid depth decreases toward the center of the semi-sphere. They made relative westerly flow and easterly flow by rotating the obstacle at the different angular velocity with the vessel. They found that the westerly winds caused Rossby waves while the easterly winds did not.

In 1967, Ibbetson and Phillips studied the propagation of Rossby waves using rotating fluid of annulus contained between two coaxial cylinders with a banked annular base (Fig. 15.6). An oscillating paddle produced westward propagating long Rossby waves and eastward propagating short waves.

15.3   Barotropic Instability

We examine the stability of a barotropic fluid on the midlatitude β-plane. Suppose that the fluid depth is constant, the density is uniform and the basic zonal flow is a function of y. If the perturbation can grow in the barotropic fluid, its kinetic energy is supplied only from the basic zonal flow, so that instability is referred to as *barotropic instability*.

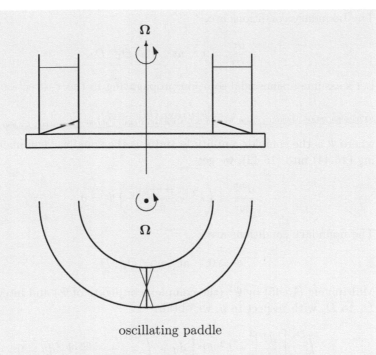

oscillating paddle

Figure 15.6: The apparatus of Ibbetson and Phillips, 1967.

15.3.1 Rayleigh's Inflection Point Theorem

We will seek the condition for barotropic instability to occur. Linearizing the quasi-geostrophic equations (12.50) and (12.51), we obtain

$$\frac{\partial u}{\partial t} + \bar{u}\frac{\partial u}{\partial x} + v\bar{u}_y - fv = -fv_g, \tag{15.39}$$

$$\frac{\partial v}{\partial t} + \bar{u}\frac{\partial v}{\partial x} + fu = fu_g, \tag{15.40}$$

where $\bar{u}_y = d\bar{u}/dy$. The boundary conditions are

$$v = 0, \quad \text{at} \quad y = D_1, D_2. \tag{15.41}$$

Differentiating (15.40) with respect to x and (15.39) with respect to y and subtracting the latter from the former yields the vorticity equation

$$\left(\frac{\partial}{\partial t} + \bar{u}\frac{\partial}{\partial x}\right)\nabla_{\mathrm{H}}^2\psi + (\beta - \bar{u}_{yy})\frac{\partial\psi}{\partial x} = 0, \tag{15.42}$$

where $\bar{u}_{yy} = d^2\bar{u}/dy^2$ and ψ is the stream function defined by

$$v = \frac{\partial\psi}{\partial x}, \quad u = -\frac{\partial\psi}{\partial y}.$$

The boundary conditions are,

$$\frac{\partial \psi}{\partial x} = 0, \quad \text{at} \quad y = D_1 , D_2. \tag{15.43}$$

Let's assume a sinusoidal solution propagating in the x-direction

$$\psi = \Psi(y) \exp \tilde{\imath}k(x - ct), \tag{15.44}$$

where Ψ is the complex amplitude and k is the zonal wavenumber. Substituting (15.44) into (15.42), we get

$$\frac{d^2\Psi}{dy^2} - \left(k^2 - \frac{\beta - \bar{u}_{yy}}{\bar{u} - c}\right)\Psi = 0. \tag{15.45}$$

The boundary conditions are

$$\Psi = 0, \quad \text{at} \quad y = D_1 , D_2. \tag{15.46}$$

Multiplying (15.45) by $\Psi^\dagger$ (the complex conjugate of Ψ) and integrating from D_1 to D_2 with respect to y, we obtain

$$\int_{D_1}^{D_2} \left[\left|\frac{d\Psi}{dy}\right|^2 + k^2|\Psi|^2\right] dy - \int_{D_1}^{D_2} \frac{\beta - \bar{u}_{yy}}{\bar{u} - c}|\Psi|^2 dy = 0. \tag{15.47}$$

Taking the imaginary part of (15.47), we find

$$c_i \int_{D_1}^{D_2} \frac{\beta - \bar{u}_{yy}}{|\bar{u} - c|^2}|\Psi|^2 dy = 0. \tag{15.48}$$

The necessary condition for instability to occur is that $\beta - \bar{u}_{yy}$ must change its sign somewhere in the fluid domain. When $\beta = 0$, (15.48) is called *Rayleigh's inflection point theorem*, which states that the necessary condition for instability to occur is that the flow must have an inflection point somewhere in the fluid domain. Equation (15.48) is the geophysical version of Rayleigh's inflection point theorem.

15.3.2   Howard's Semi-Circle Theorem

We will find another integration theorem concerning barotropic instability. Multiplying (15.45) by $(\bar{u} - c)$, we obtain

$$(\bar{u} - c)\frac{d^2\Psi}{dy^2} - \left\{k^2(\bar{u} - c) - (\beta - \bar{u}_{yy})\right\}\Psi = 0. \tag{15.49}$$

Letting

$$\Psi = (\bar{u} - c)\Phi, \tag{15.50}$$

and substituting (15.50) into (15.49), we get

$$\frac{d}{dy}\left\{(\bar{u}-c)^2\frac{d\Phi}{dy}\right\} - k^2(\bar{u}-c)^2\Phi + \beta(\bar{u}-c)\Phi = 0. \tag{15.51}$$

And the boundary conditions are

$$\Phi = 0, \quad \text{at} \quad y = D_1, \ D_2. \tag{15.52}$$

Multiplying (15.51) by $\Phi^\dagger$ (the complex conjugate of Φ) and integrating from D_1 to D_2 with respect to y, we find

$$\int_{D_1}^{D_2} (\bar{u}-c)^2\left[\left|\frac{d\Phi}{dy}\right|^2 + k^2|\Phi|^2\right]dy - \beta\int_{D_1}^{D_2}(\bar{u}-c)|\Phi|^2dy = 0. \tag{15.53}$$

Taking the imaginary part of (15.53), we get

$$\int_{D_1}^{D_2} (\bar{u}-c_r)\left[\left|\frac{d\Phi}{dy}\right|^2 + k^2|\Phi|^2\right]dy = \frac{\beta}{2}\int_{D_1}^{D_2}|\Phi|^2dy. \tag{15.54}$$

Because $\beta > 0$, the necessary condition for (15.54) to hold is

$$c_r < \bar{u}_{max}, \tag{15.55}$$

where $\bar{u}_{max}$ is the maximum value of $\bar{u}$. When $\beta = 0$, $c_r = \bar{u}$ must hold somewhere in the channel. In general, the following inequality holds if $f(x)$ is a continuous and differentiable function satisfying the boundary conditions $f(x) = 0$ at $x = D_1, D_2$,

$$\int_{D_1}^{D_2}\left|\frac{df(x)}{dy}\right|^2 dy \geq \frac{\pi^2}{(D_2-D_1)^2}\int_{D_1}^{D_2}|f(x)|^2dy, \tag{15.56}$$

From (15.54), we find

$$\frac{\beta}{2}\int_{D_1}^{D_2}|\Phi|^2dy \geq (\bar{u}_{min}-c_r)\int_{D_1}^{D_2}\left[\left|\frac{d\Phi}{dy}\right|^2 + k^2|\Phi|^2\right]dy, \tag{15.57}$$

where $\bar{u}_{min}$ is the minimum value of $\bar{u}$. Applying (15.56) to the inequality (15.57), we obtain

$$\bar{u}_{min} - \frac{\beta/2}{k^2 + \pi^2/(D_2-D_1)^2} \leq c_r < \bar{u}_{max}, \tag{15.58}$$

where we take into account (15.55). Taking the real part of (15.53),

$$\int_{D_1}^{D_2}\{(\bar{u}-c_r)^2 - c_i^2\}Pdy - \beta\int_{D_1}^{D_2}(\bar{u}-c_r)Qdy = 0, \tag{15.59}$$

where

$$P = \left|\frac{d\Phi}{dy}\right|^2 + k^2|\Phi|^2, \quad Q = |\Phi|^2. \tag{15.60}$$

Substituting from (15.60) into (15.54), we get

$$\int_{D_1}^{D_2} \bar{u}P\,dy = \frac{\beta}{2}\int_{D_1}^{D_2} Q\,dy + c_r\int_{D_1}^{D_2} P\,dy, \tag{15.61}$$

Substituting (15.61) into (15.59) yields

$$\int_{D_1}^{D_2} \bar{u}^2 P\,dy = (c_r{}^2 + c_i{}^2)\int_{D_1}^{D_2} P\,dy + \beta\int_{D_1}^{D_2} \bar{u}Q\,dy. \tag{15.62}$$

It is obvious that the following inequality is valid,

$$0 \geq \int_{D_1}^{D_2} (\bar{u} - \bar{u}_{max})(\bar{u} - \bar{u}_{min})P\,dy$$

$$= \int_{D_1}^{D_2} \bar{u}^2 P\,dy - (\bar{u}_{max} + \bar{u}_{min})\int_{D_1}^{D_2} \bar{u}P\,dy + \bar{u}_{max}\bar{u}_{min}\int_{D_1}^{D_2} P\,dy. \tag{15.63}$$

Substituting from (15.61) and (15.62) into (15.63), we find

$$0 \geq \left(c_r - \frac{\bar{u}_{max} + \bar{u}_{min}}{2}\right)^2 \int_{D_1}^{D_2} P\,dy + c_i{}^2 \int_{D_1}^{D_2} P\,dy$$

$$+ \beta\int_{D_1}^{D_2} \left(\bar{u} - \frac{\bar{u}_{max} + \bar{u}_{min}}{2}\right) Q\,dy - \left(\frac{\bar{u}_{max} - \bar{u}_{min}}{2}\right)^2 \int_{D_1}^{D_2} P\,dy$$

$$\geq \left[\left(c_r - \frac{\bar{u}_{max} + \bar{u}_{min}}{2}\right)^2 + c_i{}^2\right]\int_{D_1}^{D_2} P\,dy$$

$$- \beta\frac{\bar{u}_{max} - \bar{u}_{min}}{2}\int_{D_1}^{D_2} Q\,dy - \left(\frac{\bar{u}_{max} - \bar{u}_{min}}{2}\right)^2 \int_{D_1}^{D_2} P\,dy. \tag{15.64}$$

Using (15.56) for the integral of P, we find

$$\int_{D_1}^{D_2} P\,dy \geq \left[k^2 + \left(\frac{\pi}{D_2 - D_1}\right)^2\right]\int_{D_1}^{D_2} Q\,dy. \tag{15.65}$$

Substituting from (15.65) into (15.64), we get

$$0 \geq \left[\left(c_r - \frac{\bar{u}_{max} + \bar{u}_{min}}{2}\right)^2 + c_i{}^2\right.$$

$$\left. - \frac{\beta}{k^2 + \pi^2/(D_2 - D_1)^2}\left(\frac{\bar{u}_{max} - \bar{u}_{min}}{2}\right) - \left(\frac{\bar{u}_{max} - \bar{u}_{min}}{2}\right)^2\right]\int_{D_1}^{D_2} P\,dy.$$

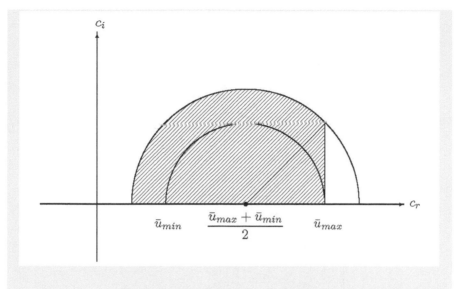

Figure 15.7: Howard's semi-circle theorem.

The integral term of the right-hand side is positive, so that the bracketed quantity must be negative, namely

$$
\left(\frac{\bar{u}_{max} - \bar{u}_{min}}{2} \right)^2 + \frac{\beta}{k^2 + \pi^2/(D_2 - D_1)^2} \left(\frac{\bar{u}_{max} - \bar{u}_{min}}{2} \right)
$$

$$
\geq \left(c_r - \frac{\bar{u}_{max} + \bar{u}_{min}}{2} \right) + c_i{}^2. \tag{15.66}
$$

The inequality (15.66) together with (15.55) implies the region where the complex phase velocity can exist in the phase space (c_r, c_i). Since we are concerned with unstable modes, c_i must be positive. Therefore, the complex phase velocity of the unstable wave must be in the hatched area in Fig. 15.7 when β is positive. This result was derived by Pedlosky (1964), and the case of $\beta = 0$ is called *Howard's semi-circle theorem* (Howard, 1961).

15.3.3 Physical Interpretation of Barotropic Instability

Let us start from the quasi-geostrophic momentum equations (12.50) and (12.51), supposing that the basic flow is zonal in the x-direction and depends on y.

$$
\frac{du}{dt} = fv, \tag{15.67}
$$

$$
\frac{dv}{dt} = f(\bar{u} - u), \tag{15.68}
$$

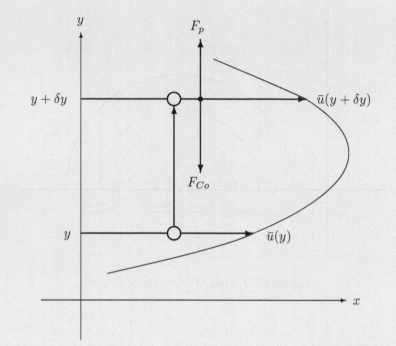

Figure 15.8: Consideration of the instability by a parcel method.

where u_g is replaced by $\bar{u}$. We consider that a fluid parcel is displaced from y to $y + \delta y$ crossing the zonal flow (Fig. 15.8), and the change of the zonal velocity of the parcel is obtained using the finite difference form of (15.67),

$$\frac{\delta u}{\delta t} = f \frac{\delta y}{\delta t},$$
$$\delta u = f \delta y.$$

The zonal velocity of the displaced parcel is

$$\bar{u} + \delta u = \bar{u} + f \delta y.$$

While the ambient zonal flow is obtained expanding $\bar{u}(y)$ in a Taylor series

$$\bar{u}(y + \delta y) = \bar{u}(y) + \frac{d\bar{u}}{dy} \delta y, \tag{15.69}$$

neglecting all terms of order δy^2 and higher. In the ambient flow field, the pressure gradient force and the Coriolis force are balanced, so that the pressure gradient force exerting on the displaced parcel is

$$F_p = f\bar{u}(y + \delta y) = f \left\{ \bar{u}(y) + \frac{d\bar{u}}{dy} \delta y \right\}. \tag{15.70}$$

While the Coriolis force acting on the parcel is

$$F_{\text{Co}} = -f\{\bar{u}(y) + f\delta y\}. \tag{15.71}$$

If $|F_{\text{Co}}| > |F_p|$, the restoring force exerting on the particle until it returns to the original position. The stability condition is written as

$$f\{\bar{u}(y) + f\delta y\} > f\left\{\bar{u}(y) + \frac{d\bar{u}}{dy}\delta y\right\},$$

$$f\left(f - \frac{d\bar{u}}{dy}\right) > 0. \tag{15.72}$$

Therefore, the condition is summarized as follows:

$$f\left(f - \bar{u}_y\right) \begin{cases} > 0 & \text{stable} \\ = 0 & \text{neutral} \\ < 0 & \text{unstable} \end{cases}. \tag{15.73}$$

The condition (15.73) is consistent with Rayleigh's inflection point theorem, because if the theorem holds there must be a region where $(f - \bar{u}_y) < 0$ somewhere in the channel. In the Earth's atmosphere of the Northern Hemisphere $(f > 0)$, the westerly jet is unstable in the southern streak and the easterly jet is in the northern streak. Readers may notice that barotropic instability is essentially the same as the overturning of the Taylor vortices, so that barotropic instability is referred to as symmetric instability or inertial instability.

15.4   Baroclinic Instability

When the perturbation vorticity grows due to the solenoid term (see Subsection 12.10.1), the instability is called *baroclinic instability*. In this section, we will discuss the simplest model of baroclinic instability and its laboratory experiments.

15.4.1   Eady's Model

In 1949, Eady revealed the mechanism and structure of the midlatitude synoptic-scale disturbances. His model is very simple and presented clearly the essence of baroclinic instability. The model assumptions are as follows:
(1) f-plane approximation is applied.
(2) The quasi-geostrophic approximation is used.
(3) The basic zonal flow has a linear vertical shear.
(4) The basic density depends linearly on y and z.
(5) The upper and lower boundaries are rigid.
We assume that the geostrophic balance and the hydrostatic balance hold for

the basic field

$$\bar{u} = -\frac{1}{f\rho_0}\frac{\partial \bar{p}}{\partial y}, \tag{15.74}$$

$$0 = -\frac{1}{\bar{\rho}}\frac{\partial \bar{p}}{\partial z} - g, \tag{15.75}$$

where ρ_0 is the mean density. The perturbation momentum equations are

$$\frac{\partial u}{\partial t} + \bar{u}\frac{\partial u}{\partial x} - fv = -\frac{1}{\rho_0}\frac{\partial p}{\partial x}, \tag{15.76}$$

$$\frac{\partial v}{\partial t} + \bar{u}\frac{\partial v}{\partial x} + fu = -\frac{1}{\rho_0}\frac{\partial p}{\partial y}. \tag{15.77}$$

The continuity equation is

$$\frac{\partial u}{\partial x} + \frac{\partial v}{\partial y} + \frac{\partial w}{\partial z} = 0. \tag{15.78}$$

The thermodynamic energy equation is

$$\frac{\partial \rho}{\partial t} + \bar{u}\frac{\partial \rho}{\partial x} + v\frac{\partial \bar{\rho}}{\partial y} + w\frac{\partial \bar{\rho}}{\partial z} = 0, \tag{15.79}$$

in which the temperature perturbation T is replaced by density perturbation ρ, because T and ρ are linearly related by the thermal expansion equation. We assume that the geostrophic balance and the hydrostatic equation hold for perturbation field variables,

$$u = -\frac{1}{f\rho_0}\frac{\partial p}{\partial y}. \tag{15.80}$$

$$v = \frac{1}{f\rho_0}\frac{\partial p}{\partial x}, \tag{15.81}$$

$$0 = -\frac{1}{\rho}\frac{\partial p}{\partial z} - g. \tag{15.82}$$

Differentiating (15.77) with respect to x and (15.76) with respect to y and subtracting the latter from the former, we obtain the vorticity equation

$$\left(\frac{\partial}{\partial t} + \bar{u}\frac{\partial}{\partial x}\right)\zeta - f\frac{\partial w}{\partial z} = 0, \tag{15.83}$$

where

$$\zeta = \frac{\partial v}{\partial x} - \frac{\partial u}{\partial y}.$$

Using (15.80) and (15.81), we can express ζ by pressure perturbation p

$$\zeta = \frac{\partial v}{\partial x} - \frac{\partial u}{\partial y} = \frac{1}{f\rho_0}\nabla_{\mathrm{H}}^2 p. \tag{15.84}$$

Differentiating (15.79) with respect to z, we get

$$\left(\frac{\partial}{\partial t} + \bar{u}\frac{\partial}{\partial x}\right)\frac{\partial \rho}{\partial z} + \frac{\partial \bar{u}}{\partial z}\frac{\partial \rho}{\partial x} + \frac{\partial v}{\partial z}\frac{\partial \bar{\rho}}{\partial y} + \frac{\partial w}{\partial z}\frac{\partial \bar{\rho}}{\partial z} = 0.$$

Using (15.74), (15.75), (15.81), and (15.82), the above equation becomes

$$\frac{\partial w}{\partial z} = -\frac{1}{\rho_0 N^2}\left(\frac{\partial}{\partial t} + \bar{u}\frac{\partial}{\partial x}\right)\frac{\partial^2 p}{\partial z^2}, \tag{15.85}$$

where

$$N^2 = -\frac{g}{\rho_0}\frac{\partial \bar{\rho}}{\partial z}, \tag{15.86}$$

and N is the Brunt–Väisälä frequency. Substituting from (15.84) and (15.85) into (15.83), we find

$$\left(\frac{\partial}{\partial t} + \bar{u}\frac{\partial}{\partial x}\right)\left(\nabla_H^2 + \frac{f^2}{N^2}\frac{\partial^2}{\partial z^2}\right)p = 0, \tag{15.87}$$

The boundary conditions are

$$w = 0, \quad \text{at} \quad z = 0, H.$$

Applying the above condition to (15.79) and using (15.81) and (15.82), we get

$$\left(\frac{\partial}{\partial t} + \bar{u}\frac{\partial}{\partial x}\right)\frac{\partial p}{\partial z} - \frac{g}{f\rho_0}\frac{\partial \bar{\rho}}{\partial y}\frac{\partial p}{\partial x} = 0, \quad \text{at} \quad z = 0, H. \tag{15.88}$$

Differentiating (15.74) with respect to z and using (15.75), we obtain

$$\frac{d\bar{u}}{dz} = \lambda = -\frac{1}{f\rho_0}\frac{\partial}{\partial y}\frac{\partial \bar{\rho}}{\partial z} = \frac{g}{f\rho_0}\frac{\partial \bar{\rho}}{\partial y}. \tag{15.89}$$

Substituting (15.89) into (15.88), we find

$$\left(\frac{\partial}{\partial t} + \bar{u}\frac{\partial}{\partial x}\right)\frac{\partial p}{\partial z} - \lambda\frac{\partial p}{\partial x} = 0, \quad \text{at} \quad z = 0, H. \tag{15.90}$$

Assuming a sinusoidal solution propagating in the x-direction as

$$p(x, y, z, t) = \hat{p}(z)\cos ly \exp \tilde{\imath}k(x - ct). \tag{15.91}$$

Substituting (15.91) into (15.87), we obtain

$$\frac{d^2\hat{p}}{dz^2} - \varepsilon^2\hat{p} = 0, \tag{15.92}$$

where

$$\varepsilon^2 = \frac{N^2}{f^2}(k^2 + l^2). \tag{15.93}$$

The general solution of (15.92) is

$$\hat{p} = \alpha \sinh \varepsilon z + \beta \cosh \varepsilon z. \tag{15.94}$$

Then, the solution (15.91) becomes

$$p = (\alpha \sinh \varepsilon z + \beta \cosh \varepsilon z) \cos ly \exp \tilde{i}k(x - ct). \tag{15.95}$$

Substituting (15.95) into (15.90), we find

$$- \varepsilon c \alpha - \lambda \beta = 0, \tag{15.96}$$

$$\varepsilon(\lambda H - c)\{\cosh (\varepsilon H)\alpha + \sinh (\varepsilon H)\beta\}$$
$$- \lambda\{\sinh (\varepsilon H)\alpha + \cosh (\varepsilon H)\beta\} = 0. \tag{15.97}$$

The necessary condition for α and β to have nontrivial solutions is that the determinant of the coefficients of (15.96) and (15.97) is zero, namely

$$\begin{vmatrix} \varepsilon c & \lambda \\ \varepsilon(\lambda H - c) \cosh \varepsilon H - \lambda \sinh \varepsilon H & \varepsilon(\lambda H - c) \sinh \varepsilon H - \lambda \cosh \varepsilon H \end{vmatrix} = 0.$$

The above equation leads to a quadratic equation of the phase speed c.

$$c^2 - \lambda H c + \frac{\lambda^2}{\varepsilon^2} [\varepsilon H \coth \varepsilon H - 1] = 0. \tag{15.98}$$

Solving (15.98), we obtain

$$c = \frac{\lambda H}{2} \pm \frac{\lambda H}{2} \left[1 - \frac{4}{\varepsilon H} \coth \varepsilon H + \frac{4}{\varepsilon^2 H^2} \right]^{1/2}. \tag{15.99}$$

The solution (15.91) is unstable when $c_i \neq 0$, namely

$$1 - \frac{4}{\varepsilon H} \coth \varepsilon H + \frac{4}{\varepsilon^2 H^2} < 0,$$

$$\left[\frac{\varepsilon H}{2} - \tanh \left(\frac{\varepsilon H}{2} \right) \right] \left[\frac{\varepsilon H}{2} - \coth \left(\frac{\varepsilon H}{2} \right) \right] < 0. \tag{15.100}$$

Since

$$\frac{\varepsilon H}{2} - \tanh \left(\frac{\varepsilon H}{2} \right) > 0 ,$$

therefore,

$$\frac{\varepsilon H}{2} < \coth \left(\frac{\varepsilon H}{2} \right) \tag{15.101}$$

is the condition for instability to occur. When $c_i = 0$,

$$\frac{\varepsilon_c H}{2} = \coth \left(\frac{\varepsilon_c H}{2} \right) \tag{15.102}$$

and the flow is said to be critically stable, which turns out

$$\varepsilon_c H = 2.40.\tag{15.103}$$

Example 1. What is the critical wavelength of baroclinic waves in the Earth's atmosphere?

Answer
For the troposphere $N = 1.2 \times 10^{-2}\,\mathrm{s}^{-1}$

$$\varepsilon H = \frac{N}{f}\kappa_c H = \frac{1.2 \times 10^{-2}}{5.0 \times 10^{-5}}\kappa_c H = 2.4,$$

$$L_c = \frac{2\pi}{\kappa_c} = \frac{2\pi \times 2.4 \times 10^2}{2.4} \times 1.0 \times 10^4 = 6.3 \times 10^6\,[\mathrm{m}].$$

The complex phase velocity is shown in Fig. 15.9. As nondimensional wavenumber εH decreases, the phase velocity of the perturbation due to the lower boundary c_r^- increases and approaches $\lambda H/2$ while the phase velocity of the perturbation due to the upper boundary c_r^+ decreases and approaches $\lambda H/2$. Finally, at $\varepsilon H = 2.4$, they become $\lambda H/2$ which is the same as the basic zonal velocity at $z = H/2$, and baroclinic instability takes place. This is the result

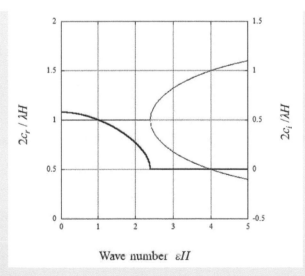

Figure 15.9: The dependence of the nondimensional complex phase velocity of the Eady type baroclinic wave on the nondimensional wavenumber εH. The *heavy solid line* indicates the imaginary part c_i (the right ordinate), and *thin solid lines* are the real part c_r (the left ordinate). The unit is $\lambda H/2$.

that the perturbation vorticity is produced only at two boundaries due to the pressure density solenoid.

Assuming $k = l$, the growth rate $\sigma = kc_i$ of the unstable wave is given by

$$\sigma = \frac{\lambda H}{2} k \left[-1 + \frac{4}{\varepsilon H} \coth(\varepsilon H) - \frac{4}{\varepsilon^2 H^2} \right]^{1/2}$$

$$= \frac{\lambda f}{2\sqrt{2}N} \varepsilon H \left[-1 + \frac{4}{\varepsilon H} \coth(\varepsilon H) - \frac{4}{\varepsilon^2 H^2} \right]^{1/2}. \tag{15.104}$$

The growth rate is shown in Fig. 15.10, which becomes maximum at $\epsilon H = 1.605$, which corresponds to the wavenumber k_m,

$$\varepsilon H = 1.605 = \sqrt{2} \frac{NH}{f} k_m,$$

$$k_m = \frac{1.605}{\sqrt{2}} \frac{f}{NH}. \tag{15.105}$$

The growth rate is proportional to the vertical shear λ, namely the horizontal gradient of the basic density (temperature). In the Earth's atmosphere, the horizontal temperature gradient is very large at the east coast of the Eurasian Continent and the North American Continent, where midlatitude cyclones often occur and develop. These regions are called the *storm tracks*.

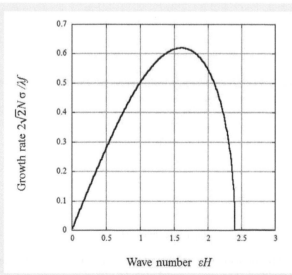

Figure 15.10: The dependence of the nondimensional growth rate $\frac{2\sqrt{2}N}{\lambda f}\sigma$ of the Eady type baroclinic wave on the nondimensional wavenumber εH.

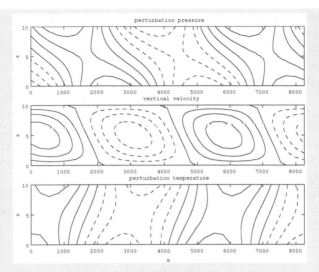

Figure 15.11: The structure of the fastest growing Eady wave. The upper panel is the perturbation pressure, the middle panel is the upward velocity, and the lower panel is the perturbation temperature. *Solid lines* indicate positive value and *dashed lines* are negative value.

We will show the structure of the fastest growing *Eady wave* in Fig. 15.11. A line connecting the minimum value of the perturbation pressure is called the trough, and a line connecting the maximum value of the perturbation pressure is called the ridge. Axes of the trough and ridge slope backward with height facing downstream of the basic flow, while the lines connecting the maximum and the minimum of the perturbation temperature slope forward. In front of the trough on the lower boundary, the y-component of the velocity and the ascending motion are maximum. Conversely, in front of the ridge on the lower boundary, the negative y-component of the velocity and the descending motion are maximum. Thus, the potential energy of the basic field is released and converted to the perturbation kinetic energy. In spite of the simplicity of the model, the Eady wave well documents the midlatitude synoptic-scale disturbances.

15.4.2 Laboratory Experiments of Baroclinic Waves

From the late 1940's, two kinds of experiments simulating atmospheric large-scale motions were begun. Fultz (1959) used a flat cylindrical vessel rotating on a turn table, heating its rim by candle flame and cooling its center by small ice cubes. His experiment was called *dishpan experiments*, in which it was not easy to control steadily the temperature contrast between the rim and the center of the vessel. Hide (1953) performed experiments using annular vessels, which consisted of three coaxial cylinders; the outer chamber was the warm heat bath, the middle chamber was the experimental chamber and the inner chamber was the cool heat bath. The experiments of Hide were called *annulus experiments*, in which it was easy to control the external parameters so that annulus experiments were widely conducted by a large number of researchers. Fowlis and Hide (1965) studied precisely the transition between flow regimes using three annuli with different inner radii, working fluids with three different kinematic viscosity, and various combinations of the temperature difference and the rotation rates. They found four flow regimes, the symmetric (laminar) flow regime, the steady wave flow regime, the oscillating wave (doubly periodic) flow regime and the irregular (turbulent) flow regime. As shown in Fig. 15.12, they are classified consistently on a log-log plot with the *thermal Rossby number* as the ordinate and the Taylor number as the abscissa. The thermal Rossby number is defined as

$$\Theta = \frac{gd\Delta\rho}{\rho_0 \Omega^2 (b-a)^2},\tag{15.106}$$

where $\Delta\rho$ is the difference of the density between the inner cylinder of radius a and the outer cylinder of radius b, ρ_0 is the mean density of the working fluid and Ω is the angular velocity of the rotating system. The Taylor number

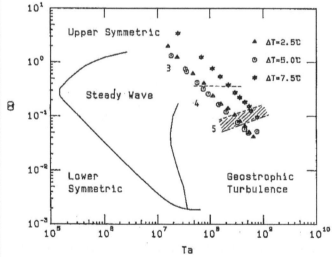

Figure 15.12: Stability diagram for baroclinic waves in a rotating fluid of annulus. Solid curves are after Fowlis and Hide, 1965. Numerals are zonal wavenumber which depends on the thermal Rossby number and the geometry of annular vessels (after Morita and Uryu, 1989). ©American Meteorological Society. Reproduced with permission of AMS.

is defined as

$$T_a = \frac{4\Omega^2(b-a)^4}{\nu^2}\frac{b-a}{d}, \tag{15.107}$$

which is the product multiplying the ordinary Taylor number by the aspect ratio. There are two symmetric flow regimes, one is the upper symmetric regime and the other is the lower symmetric regime. In the former, the stable stratification prevents the baroclinic overturning, while in the latter, the weak lateral temperature contrast cannot cause baroclinic instability, overcoming viscous resistive force. The critical thermal Rossby number Θ_c for the transition from the upper symmetric regime to the steady wave regime can be deduced from *Eady's criterion* (15.103). Letting the zonal wavenumber be k^*, the wavelength of baroclinic waves become

$$L = 2\pi\frac{a+b}{2}\frac{1}{k^*} = \frac{\pi(a+b)}{k^*}.$$

Then, the wavenumber is

$$k = \frac{2\pi}{L} = \frac{2k^*}{a+b}.$$

Replacing H by d in (15.103), taking the square of the resultant equation and using the above result, we find

$$5.76 = \varepsilon^2 d^2 = \frac{\Delta\rho g\pi}{4\Omega^2\rho_0 d} = \Theta_c 2k^*\left(\frac{b-a}{a+b}\right)^2,$$

$$\Theta_c = 2.88\left(\frac{a+b}{b-a}\right)^2\frac{1}{k^*}. \tag{15.108}$$

From (15.108), we can understand that the smaller the thermal Rossby number becomes, the larger the zonal wavenumber becomes. The zonal wavenumber also depends on the geometry of the annular vessel. Hide (1958) found the empirical formula concerning the geometrical dependence of the zonal wavenumber. The maximum zonal wavenumber is the nearest integer of

$$k^*{}_{max} = (0.67 \pm 0.02)\pi\frac{a+b}{b-a},$$

and the minimum wavenumber is given by

$$k^*{}_{min} \geq k^*{}_{max} - 5. \tag{15.109}$$

In Fig. 15.13, a streak photograph[1] of baroclinic waves in a rotating fluid of annulus of wavenumber 3 is shown. The surface flow pattern is visualized by aluminum powder coated with wax.

---

[1] A shutter speed is 1.0 s.

Figure 15.13: A streak photograph of a baroclinic wave (k=3) in a rotating fluid of annulus. Experimental parameters are, a=7.5cm, b=15.0 cm, d=10.0 cm, ΔT=9.0 K, and Ω=0.30 s^{-1}.

15.5   Geostrophic Turbulence

Turbulent motion which takes place in the fluid system under the geostrophic balance and hydrostatic balance is referred to as the geostrophic turbulence. That is to say, the fluid motion is highly two-dimensional so that it has two kinds of inertial range. The terminology, geostrophic turbulence, was first used by Charney (1971).

15.5.1   Three-Dimensional Turbulence

Suppose that the energy is injected at wavenumber k_f (spatial scale $l_f = 2\pi/k_f$) and is transferred to smaller scale eddies and is finally dissipated by viscosity at wavenumber k_d (spatial scale $l_d = 2\pi/k_d$). Let the energy dissipation rate per unit mass be ε and the kinematic viscosity be ν. The dimension of ε is

$$\varepsilon = \left(\frac{L}{T}\right)^2 \frac{1}{T} = L^2 T^{-3}, \tag{15.110}$$

where L is the length scale and T is the time scale. The viscous dissipation length scale or *Kolmogorov scale* depends only on ε and ν, so that we find by dimensional analysis,

$$l_d = \varepsilon^{-1/4}\nu^{3/4}, \tag{15.111}$$

Problem 1. Prove (15.111), letting the length scale be L and the time scale be T.

The wavenumber space of $k_f \ll k \ll k_d$ is called the inertial subrange, in which the turbulent motion is homogeneous, isotropic, inviscid, and self-similar. Kolmogorov (1941) assumed that the energy spectrum of unit mass $E(k)$ depends only on wavenumber k and ε in the inertial subrange. Dimension of $E(k)$ is

$$[E(k)] = \frac{L^2}{T^2}\frac{1}{L^{-1}} = L^3 T^{-2}. \qquad (15.112)$$

Letting $E(k) \propto \varepsilon^\alpha k^\beta$, we find

$$L^3 T^{-2} = (L^2 T^{-3})^\alpha (L^{-1})^\beta.$$

We get the following equations for α and β.

$$\begin{cases} 2\alpha - \beta = 3\,, \\ -3\alpha = -2\,. \end{cases} \qquad (15.113)$$

Solving (15.113), we get $\alpha = 2/3$ and $\beta = -5/3$. Thus,

$$E(k) \propto \varepsilon^{2/3} k^{-5/3}. \qquad (15.114)$$

Equation (15.114) is referred to as *Kolmogorov's spectrum*, which is confirmed by many studies of the turbulent motion of the Earth's atmosphere and laboratory systems.

15.5.2 Two-Dimensional Turbulence

In the two-dimensional fluid motion, the vertical component of the vorticity is conserved in the inviscid condition and without solenoidal term. Therefore, not only energy but also enstrophy[2] are constants of motion in the inertial subrange. As a consequence, there are two inertial subranges in the two-dimensional turbulence (Kraichnan, 1967; Leith, 1968; Batchelor, 1969). The rate of energy transfer per unit mass ε from larger eddies to smaller eddies (the downward cascade) in one inertial subrange with $k^{-5/3}$ power law, and the rate of enstrophy transfer per unit mass from smaller eddies to larger eddies (the upward cascade or inverse cascade) in another inertial subrange obeying k^{-3} law.

Letting $S(k)$ be the enstrophy spectrum and ϑ be the rate of enstrophy transfer per unit mass in the inertial subrange, the dimension of $S(k)$ and ϑ are

$$[S(k)] = \frac{T^{-2}}{L^{-1}} = LT^{-2}, \qquad (15.115)$$

$$[\vartheta] = \frac{T^{-2}}{T} = T^{-3}. \qquad (15.116)$$

---

[2] The enstrophy is defined by $\int_V \frac{1}{2} (\boldsymbol{\nabla} \times \mathbf{v} \cdot \boldsymbol{\nabla} \times \mathbf{v})\ dv$.

As $S(k)$ depends only on ϑ and k, we can put $S(k) \propto \vartheta^\alpha k^\beta$. We find a dimensional equation

$$LT^{-2} = T^{-3\alpha}L^{-\beta}. \tag{15.117}$$

We get the result $\alpha = 2/3$, $\beta = -1$. Thus, we obtain the enstrophy spectrum in the inertial subrange.

$$S(k) \propto \vartheta^{2/3}k^{-1}. \tag{15.118}$$

In the wavenumber space, the relation between $E(k)$ and $S(k)$ is

$$S(k) = k^2 E(k). \tag{15.119}$$

From (15.118) and (15.119), we obtain the energy spectrum in the enstrophy cascading subrange.

$$E(k) = \vartheta^{2/3}k^{-3}. \tag{15.120}$$

Equation (15.120) is called a -3 power law, and is confirmed in the large-scale[3] eddies in the Earth's atmosphere and laboratory systems.

15.5.3   Geostrophic Turbulence in Various Fluid Systems

There are many observational studies of the large scale motions in the Earth's atmosphere (Ogura 1958, Benton and Kahn 1958, Horn and Bryson 1963, Wiin–Nielsen 1967, Julian et al. 1970, Kao and Wendell 1970, Kao et al. 1970, Charney 1971, Baer 1972, Baer 1974, Boer and Shepherd 1983). They found that the energy spectrum in the inertial subrange of the large scale atmospheric motion obeys k^{-3} law. Charney (1971) called the large scale atmospheric turbulence the *geostrophic turbulence*.

As for the laboratory experiments, Pfeffer et al. (1980) made an annular vessel with 2016-thermistor network immersed in the fluid for temperature measurement, and obtained a k^{-4} and a k^{-5} law. Their experimental apparatus was the ultimate achievement of experiments of rotating annulus of fluids. Using the same experimental apparatus, Buzyna et al. (1984) found that the slope of the energy spectra in the wavenumber space depends significantly on the thermal Rossby number Θ ranging from -4.8 at $\Theta = 0.17$ to -2.4 at $\Theta = 0.02$. Further, they found that the energy spectrum in the frequency space obeys a -4 power law. While, Saffman (1971) predicted theoretically that many probes may develop the energy spectrum obeying k^{-4} law. Morita and Uryu (1989) performed experiments of rotating fluid of annulus measuring temperature fluctuation by four thermistor probes to minimize the probe origin disturbances. They obtained the result that the energy spectrum

---

[3]The Rossby number is smaller than 0.1, which is the guarantee of the two-dimensional motion.

Table 15.1: Preferred wavenumber k, thermal Rossby number Θ, and Taylor number T_a for each panel in Fig. 15.12.

number	k	Θ	T_a
a	3	1.989×10^0	1.577×10^7
b	3	6.722×10^{-1}	3.587×10^7
c	3	5.419×10^{-1}	1.560×10^8
d	5?	8.067×10^{-2}	3.634×10^8
e	5?	7.734×10^{-2}	3.280×10^8
f	5?	9.624×10^{-2}	7.757×10^8
g	?	5.019×10^{-2}	5.347×10^8
h	?	5.730×10^{-2}	4.427×10^8
i	?	5.266×10^{-2}	7.619×10^8

in the frequency space in the inertial subrange obeys a -3 power law in the wide range of the thermal Rossby number (Fig. 15.14). They confirmed that the *Taylor's frozen turbulence hypothesis* is valid due to high correlation coefficients between temperature fluctuations between two probes. Therefore, the frequency spectrum was able to be transformed to the wavenumber spectrum, so they proved that the energy spectrum in the inertial subrange of annulus experiments obeys the k^{-3} law.

A numerical simulation of two-dimensional turbulence was first performed by Lilly (1967), and he produced two energy spectra of a $k^{-5/3}$ range for $k > k_f$ and a k^{-3} range for $k < k_f$ (k_f is the wavenumber of forcing) predicted theoretically by Kraichnan (1967) and Leith (1968). Rhines (1975) simulated upward cascade of two-dimensional turbulence on the β–plane, and he found that the upward cascade nearly ceases at wavenumber $k_\beta = \sqrt{\beta/(2U)}$ (U is the root mean square of the initial speed of fluid parcel). This phenomenon was called later *Rhines effect*. Williams (1978) succeeded in simulating terrestrial and Jovian general circulation in which upward enstrophy transfer and the Rhines effect played the important role. So far, many studies of numerical simulations for two-dimensional turbulence are accomplished, e.g., Ohkitani (1991), Borune (1994), Schorghofer (2000), Kaneda and Ishihara (2001), Ishihara and Kaneda (2001), Iwayama et al. (2002), Smith et al. (2002), Scott (2007). Iwayama et al. (2019) proved Danilov inequality numerically using two-layer quasi-geostrophic equations.

Theoretical researches were started by Kraichnan (1967), Leith (1967), and Batchelor (1969), and they stated that there are two inertial subranges in the two-dimensional turbulence, one is energy cascading $k^{-5/3}$ range and the other is enstrophy cascading k^{-3} range in the energy spectrum. Iwayama and Watanabe (2016) derived analytically a k^{-1} spectrum in the enstrophy inertial range using the eddy damped quasinormal Marcovianized closure equation.

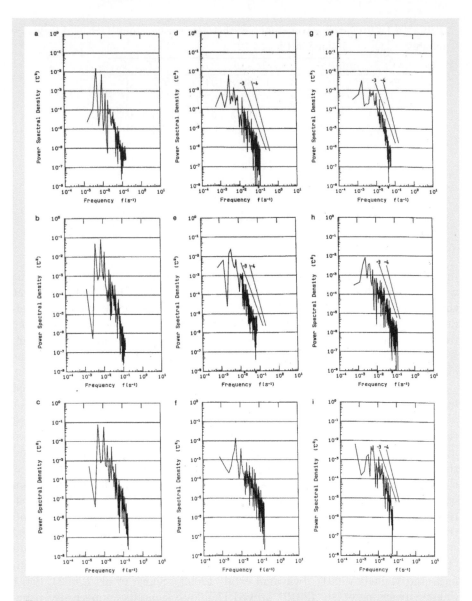

Figure 15.14: Energy spectrum in the inertial subrange of various flow regimes: (a), (b), (c) are in the steady wave regime, (d), (e), (f) are in the transition regime, (g), (h), (i) are in the turbulent regime. Thermal Rossby number Θ and Taylor number T_a for each figure are listed in Table 15.1 (after Morita and Uryu, 1989). ©American Meteorological Society. Reproduced with permission of AMS.

15.6 Problems

1. Consider the reason why the inertial oscillations are often observed in the oceans and large lakes, but are seldom observed in the Earth's atmosphere.

2. What is the magnitude of the group velocity of the non-divergent Rossby waves?

3. In the derivation of (15.85), prove that $\partial \bar{u}/\partial z \, \partial \rho/\partial x$ and $\partial v/\partial z \, \partial \bar{\rho}/\partial y$ cancel each other.

4. Derive (15.85) by differentiating (15.79) with respect to z.

5. Derive the thermal Rossby number Θ from the Rossby number R_o defined by (12.43), replacing U by the zonal velocity at the upper free boundary of the rotating fluid of annulus. In this calculation, make use of the thermal flow balance given by (12.53), supposing that the zonal velocity is a linear function of z and zero at the bottom.

15.7 References

1. Baer, F.: *An alternative scale representation of atmospheric energy spectra*, J. Atmos. Sci., **29**, 649–664 (1972).

2. Baer, F.: *Hemispheric spectral statistics of APE*, J. Atmos. Sci., **31**, 932–941 (1974).

3. Batchelor, G. K.: *Computation of the energy spectrum in homogeneous two-dimensional turbulence*,Phys. Fluids,**12**, 233–239 (1969).

4. Benton, G. T., and A. B. Kahn: *Spectra of large-scale atmospheric flow at 300 millibar*, J. Meteor., **15**, 404–410 (1958).

5. Boer, G. J. and T. G. Shepherd: *Largr-scale two-dimensional turbulence in the atmosphere*, J. Atmos. Sci., **40**, 164–184 (1983).

6. Borue, V.: *Inverse energy cascade in stationary two-dimensional homogeneous turbulence*, Phys. Rev. Lett., **72**, 1475– (1994).

7. Buzyna, G., R. L. Pfeffer and R. Kung: *Transition to geostrophic turbulence in a rotating differentially heated annulus of fluid*, J. Fluid Mech., **145**, 377–403 (1984).

8. Charney, J. G.: *Geostrophic turbulence*, J. Atmos. Sci., **28**, 1087–1095 (1971).

9. Fontane, D., G. Dritschel and R. K. Scott: *Vortical control of forced two-dimensional turbulence*, Phys. Fluids, **25**, 015101– (2013).

10. Fowlis, W. W. and R. Hide: *Thermal convection in a rotating annulus of liquid: effect of viscosity on the transition between axisymmetric and non-axisymmetric flow regimes*, J. Atmos. Sci., **22**, 541–558 (1965).

11. Frenkiel, F. N.: *The comparison between the longitudinal correlation in a turbulent flow*, Phys. Rev., **88**, 1380–1382 (1952).

12. Fultz, D. and R. R. Long: *Two-dimensional Flow around a Circular Barrier in a Rotating Shell*, Tellus, **3**, 61–68 (1951).

13. Hide R.: *Some experiments on thermal convection in a rotating liquid*, Q. J. R. Meteoro. Soc., **79**, 161, 294–297 (1953).

14. Hide R.: *The character of the equilibrium of a heavy, viscous, incompressible, rotating fluid of variable density. I and II*, Q. J. Mech. Appl. Math., **9**, 22–34, 35–50 (1956).

15. Hide R.: *An experimental study of thermal convection in a rotating liquid*, Philos. Trans. R. Soc. London, **A250**, 441–478 (1958).

16. Horn, L. H. and R. A. Bryson: *An analysis of the geostrophic kinetic energy spectrum of large scale atmospheric turbulence*, J. Geophys. Res., **68**, 1059–1064 (1963).

17. Howard, L. N.: *Note on a paper by John W. Miles*. J. Fluid Mech., **10**, 509-512 (1961).

18. Ibbetson A. and N. A. Phillips: *Some Laboratory Experiments on Rossby Waves with Application to the Ocean*, Tellus, **19**, 81–88 (1967).

19. Ishihara, T. and Y. Kaneda: *Energy spectrum in the enstrophy transfer range of two-dimensional forced turbulence*, Phys. Fluids, **13**, 544– (2001).

20. Iwayama, T. T. G. Shepherd and T. Watanabe: *An 'ideal' form of decaying two-dimensional turbulence*, J. Fluid Mech., **456**, 183 (2002).

21. Iwayama, T. and T. Watanabe: *Enstrophy inertial range dynamics in generalized two-dimensional turbulence*, Phys. Rev. Fluids, **1**, 034403 (2016).

22. Iwayama, T., S. Okazaki and T. Watanabe: *Numerical investigation of the Danilov inequality for two-layer quasi-geostrophic systems*, Fluid Dyn. Res., **51**, 055507 (2019).

23. Julian, P. R., W. M. Washington, L. Lambree and C. Ridley: *On the spectral distribution of large-scale atmospheric kinetic energy*, J. Atmos. Sci., **27**, 376–387 (1970).

24. Kaneda, Y. and T. Ishihara: *Nonuniversal k^{-3} energy spectrum in stationary two-dimensional homogeneous turbulence, Phys. Fluids*, **13**, 1431 (2001).

25. Kao, S.-K. and L. L. Wendell: *The kinetic energy of the large-scale atmospheric motion in wavenumber-frequency space. Part I: Northern Hemisphere, J. Atmos. Sci.*, **27**, 359–375 (1970).

26. Kao, S.-K., R. L. Jenne and J. F. Sagendorf: *The kinetic energy of large-scale atmospheric motion in wavenumber-frequency space: mid-troposphere of the Southern Hemisphere, J. Atmos. Sci.*, **27**, 1008–1020 (1970).

27. Kraichnan, R. H.: *Inertial subranges in two-dimensional turbulence, Phys. Fluids*, **10**, 1417–1423 (1967).

28. Leith, C. E.: *Diffusion approximation to inertial energy transfer in isotropic turbulence, Phys. Fluids*, **10**, 1409–1416, (1967).

29. Lilly, D. K.: *Numerical simulation of two-dimensional turbulence, Phys. Fluids*, **II** (Suppl.), 240–249, (1969).

30. Lilly, D. K.: *Numerical simulation of developing and decaying two-dimensional turbulence, J. Fluid Mech.*, **45**, 395–415, (1971).

31. Manabe, S., J. Smagorinsky, J. L. Holloway Jr. and H. M. Stone: *Simulated climatology by a general circulation model with a hydrologic cycle. Part III: Effects of increasing horizontal resolution, Mon. Wea. Rev.*, **98**, 175–212, (1970).

32. Morita, O. and M. Uryu: *Geostrophic turbulence in a rotating annulus of fluid, J. Atmos. Sci.*, **46**, 2349–2355 (1989).

33. Morita, O.: *Transition between Flow Regimes of Baroclinic Flow in a Rotating Annulus of Fluid, Phase Transitions*, **28**, 213–244 (1990).

34. Ogura, Y.: *The relation between the space- and time-correlation functions in turbulent flow, J. Meteor. Soc. Japan*, **31**, 2129–2149 (1953).

35. Ogura, Y.: *On the isotropy of the large scale disturbances in the upper troposphere functions in turbulent flow, J. Meteor.*, **15**, 375–382 (1958).

36. Ohkitani, K.: *Wave number space dynamics of enstrophy cascade in a forced two-dimensional turbulence, Phys. Fluids*, **3**, 1598–1611 (1991)

37. Pfeffer, R. L., G. Buzyna, and R. Kung: *Time-dependent modes of behavior of thermally driven rotating fluids, J. Atmos. Sci.*, **37**, 2129–2149 (1980).

38. Pedlosky, J.: *The Stability of Currents in the Atmosphere and the Ocean: Part I, J. Atmos. Sci.*, **21**, 201–219 (1964).

39. Rao, S. T. and C. B. Ketchum: *Spectral characteristics of the baroclinic annulus waves, J. Atmos. Sci.*, **32**, 698–711 (1975).

40. Rhines, P. B.: *Observations of the energy-containing oceanic eddies and theoretical models of waves and turbulence, Bound.-Layer Meteor.*, **4**, 345–360 (1973).

41. Rhines, P. B.: *Waves and turbulence on a beta-plane, J. Fluid Mech.*, **69**, 417–443 (1975).

42. Saffman, P. G.: *On the spectrum and decay of random two-dimensional vorticity distributions at large Reynolds number, Stud. Appl. Math.*, **L**, 377–383 (1971).

43. Scott, R. K. : *Nonrobustness of the two-dimensional turbulent inverse cascade, Phys. Rev.*, **E 75**, 046301 (2007).

44. Schorghofer, N.: *Energy spectra of steady two-dimensional turbulent flows, Phys. Rev.*, **E 61**, 6572 (2000).

45. Smith, K. S., G. Boccaletti, C. C. Henning, I. Marinov, C. Y. Tam, I. M. Held, and G. K. Vallis: *Turbulent diffusion in the geostrophic inverse cascade, J. Fluid Mech.*, **469**, 13 (2002).

46. Williams, G. P.: *Planetary circulations. 1. Barotropic representation of Jovian and terrestrial turbulence, J. Atmos. Sci.*, **35**, 1399–1426 (1978).

47. Wiin-Nielsen, A.: *On the annual variation and spectral distribution of atmosphere energy, Tellus*, **19**, 540–559 (1967).

Appendix A

Acceleration in Spherical Coordinates

We will expand the acceleration into the r-, ϕ-, θ-components in spherical coordinates. We set local Cartesian coordinates at point $\mathrm{P}(r, \phi, \theta)$, taking the x, y, z-axes as ϕ, θ, r-directions, respectively, where $z = r - a$ and a is the Earth's radius. Let basis vectors be $(\mathbf{e}_x, \mathbf{e}_y, \mathbf{e}_z)$ and velocity components be (u, v, w), then

$$\mathbf{v} = u\mathbf{e}_x + v\mathbf{e}_y + w\mathbf{e}_z.$$

The x, y, z-components of the velocity $\mathbf{v}$ is

$$u = \frac{dx}{dt}, \quad v = \frac{dy}{dt}, \quad w = \frac{dz}{dt}.$$

Using the relation $dx = r\cos\theta d\phi, dy = rd\theta, dz = dr$, we obtain

$$u = r\cos\theta\frac{d\phi}{dt}, \quad v = r\frac{d\theta}{dt}, \quad w = \frac{dr}{dt}. \tag{A.1}$$

Taking into account the position dependence of the basis vectors, the acceleration becomes

$$\frac{d\mathbf{v}}{dt} = \mathbf{e}_x\frac{du}{dt} + \mathbf{e}_y\frac{dv}{dt} + \mathbf{e}_z\frac{dw}{dt} + u\frac{d\mathbf{e}_x}{dt} + v\frac{d\mathbf{e}_y}{dt} + w\frac{d\mathbf{e}_z}{dt}. \tag{A.2}$$

The relationship between the total derivative and the local derivative is

$$\frac{d\Phi}{dt} = \frac{\partial\Phi}{\partial t} + u\frac{\partial\Phi}{\partial x} + v\frac{\partial\Phi}{\partial y} + w\frac{\partial\Phi}{\partial z} \tag{A.3}$$

for an arbitrary function Φ. We will first consider $d\mathbf{e}_x/dt$. Expanding the total derivative of $\mathbf{e}_x$ as in (A.3) and noting that $\mathbf{e}_x$ depends only on x, we obtain

$$\frac{d\mathbf{e}_x}{dt} = u\frac{\partial\mathbf{e}_x}{\partial x}.$$

DOI: 10.1201/9781003310068-A

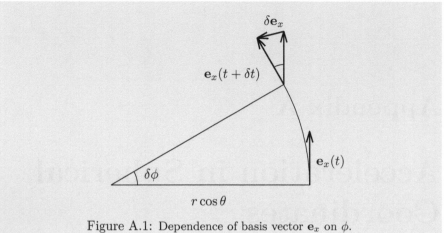

Figure A.1: Dependence of basis vector $\mathbf{e}_x$ on ϕ.

From Fig. A.1 we find that

$$|\delta \mathbf{e}_x| = \delta \phi$$

and $\delta \mathbf{e}_x$ is directed toward the z-axis.

From Fig. A.2, we see that $\delta \mathbf{e}_x$ is decomposed into the ϕ and r-directions as

$$\delta \mathbf{e}_x = \mathbf{e}_y \sin \theta \delta \phi - \mathbf{e}_z \cos \theta \delta \phi.$$

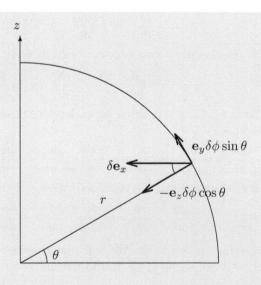

Figure A.2: Decomposition of $\delta \mathbf{e}_x$ into the θ- and r-directions.

Thus,

$$\frac{\partial \mathbf{e}_x}{\partial x} = \lim_{\delta x \to 0} \frac{\delta \mathbf{e}_x}{\delta x} = \lim_{\delta \phi \to 0} \frac{\mathbf{e}_y \delta \phi \sin \theta - \mathbf{e}_z \delta \phi \cos \theta}{r \cos \theta \delta \phi} = \frac{\mathbf{e}_y \sin \theta - \mathbf{e}_z \cos \theta}{r \cos \theta}.$$

Therefore,

$$\frac{d\mathbf{e}_x}{dt} = \frac{u}{r \cos \theta} (\mathbf{e}_y \sin \theta - \mathbf{e}_z \cos \theta). \tag{A.4}$$

Next we will consider $d\mathbf{e}_y/dt$. Expanding the total derivative of $\mathbf{e}_y$ and noting that $\mathbf{e}_y$ depends on x and y, we get

$$\frac{d\mathbf{e}_y}{dt} = u \frac{\partial \mathbf{e}_y}{\partial x} + v \frac{\partial \mathbf{e}_y}{\partial y}.$$

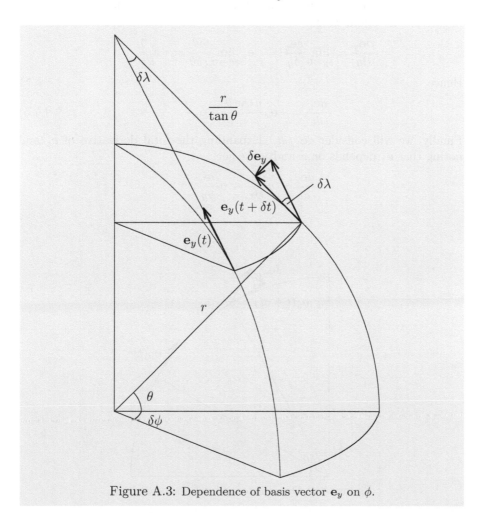

Figure A.3: Dependence of basis vector $\mathbf{e}_y$ on ϕ.

From Fig. A.3, we can find

$$\delta x = r \cos \theta \delta \phi = \frac{r}{\tan \theta} \delta \lambda, \quad \delta \lambda = \sin \theta \delta \phi,$$
$$|\delta \mathbf{e}_y| = \delta \lambda = \sin \theta \delta \phi,$$

for the motion of the ϕ-direction. Considering that $\delta \mathbf{e}_y$ directed in the negative ϕ-direction, we obtain

$$\frac{\partial \mathbf{e}_y}{\partial x} = \lim_{\delta x \to 0} \frac{\delta \mathbf{e}_y}{\delta x} = -\mathbf{e}_x \lim_{\delta \phi \to 0} \frac{\delta \phi \sin \theta}{r \cos \theta \delta \phi} = -\mathbf{e}_x \frac{\tan \theta}{r}.$$

From Fig. A.4, we find

$$|\delta \mathbf{e}_y| = \delta \theta$$

for the motion of the θ-direction. Taking into account that $\delta \mathbf{e}_y$ is in the negative r-direction, we get

$$\frac{\partial \mathbf{e}_y}{\partial y} = \lim_{\delta y \to 0} \frac{\delta \mathbf{e}_y}{\delta y} = -\mathbf{e}_z \lim_{\delta \theta \to 0} \frac{\delta \theta}{r \delta \theta} = -\mathbf{e}_z \frac{1}{r}.$$

Hence,

$$\frac{d \mathbf{e}_y}{dt} = -\mathbf{e}_x \frac{u \tan \theta}{r} - \mathbf{e}_z \frac{v}{r}. \tag{A.5}$$

Finally, we will consider $d\mathbf{e}_z/dt$. Expanding the total derivative of $\mathbf{e}_z$ and noting that $\mathbf{e}_z$ depends on x and y, we get

$$\frac{d \mathbf{e}_z}{dt} = u \frac{\partial \mathbf{e}_z}{\partial x} + v \frac{\partial \mathbf{e}_z}{\partial y}.$$

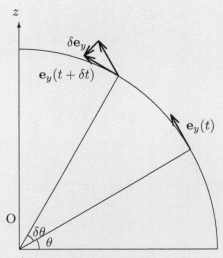

Figure A.4: Dependence of basis vector $\mathbf{e}_y$ on θ.

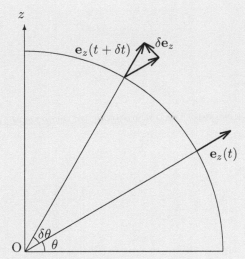

Figure A.5: Dependence of basis vector $\mathbf{e}_z$ on θ.

From Fig. A.5, for the motion of the θ-direction, we find

$$|\delta\mathbf{e}_z| = \delta\theta.$$

Considering that $\delta\mathbf{e}_z$ is in the positive θ-direction, we obtain

$$\frac{\partial\mathbf{e}_z}{\partial y} = \lim_{\delta y \to 0} \frac{\delta\mathbf{e}_z}{\delta y} = \mathbf{e}_y \lim_{\delta\theta \to 0} \frac{\delta\theta}{r\delta\theta} = \mathbf{e}_y\frac{1}{r}.$$

By similar arguments it is easily shown that

$$\frac{\partial\mathbf{e}_z}{\partial x} = \lim_{\delta x \to 0} \frac{\delta\mathbf{e}_z}{\delta x} = \mathbf{e}_x\frac{1}{r}.$$

Therefore,

$$\frac{d\mathbf{e}_z}{dt} = \mathbf{e}_x\frac{u}{r} + \mathbf{e}_y\frac{v}{r}. \tag{A.6}$$

Substituting (A.4), (A.5), and (A.6) into (A.2), we obtain the spherical coordinate expansion of the acceleration

$$\frac{d\mathbf{v}}{dt} = \mathbf{e}_x\left(\frac{du}{dt} - \frac{uv\tan\theta}{r} + \frac{uw}{r}\right) + \mathbf{e}_y\left(\frac{dv}{dt} + \frac{u^2\tan\theta}{r} + \frac{vw}{r}\right)$$
$$+ \mathbf{e}_z\left(\frac{dw}{dt} - \frac{u^2 + v^2}{r}\right). \tag{A.7}$$

Appendix B

Vector Analysis

B.1 Vector Identities

In this section, frequently used vector identities are shown.

1. $\mathbf{A} \times (\mathbf{B} \times \mathbf{C}) = (\mathbf{A} \cdot \mathbf{C})\mathbf{B} - (\mathbf{A} \cdot \mathbf{B})\mathbf{C}$

2. $\mathbf{A} \cdot (\mathbf{B} \times \mathbf{C}) = \mathbf{B} \cdot (\mathbf{C} \times \mathbf{A}) = \mathbf{C} \cdot (\mathbf{A} \times \mathbf{B})$

3. $\mathbf{\nabla} \times \mathbf{\nabla} \Phi = 0$

4. $\mathbf{\nabla} \cdot (\Phi \mathbf{A}) = \Phi \mathbf{\nabla} \cdot \mathbf{A} + \mathbf{A} \cdot \mathbf{\nabla} \Phi$

5. $\mathbf{\nabla} \times (\Phi \mathbf{A}) = \mathbf{\nabla} \Phi \times \mathbf{A} + \Phi \mathbf{\nabla} \times \mathbf{A}$

6. $\mathbf{\nabla} \cdot (\mathbf{\nabla} \times \mathbf{A}) = 0$

7. $(\mathbf{A} \cdot \mathbf{\nabla})\mathbf{A} = \dfrac{1}{2}\mathbf{\nabla}(\mathbf{A} \cdot \mathbf{A}) - \mathbf{A} \times (\mathbf{\nabla} \times \mathbf{A})$

8. $\mathbf{\nabla} \times (\mathbf{A} \times \mathbf{B}) = \mathbf{A}(\mathbf{\nabla} \cdot \mathbf{B}) - \mathbf{B}(\mathbf{\nabla} \cdot \mathbf{A}) - (\mathbf{A} \cdot \mathbf{\nabla})\mathbf{B} + (\mathbf{B} \cdot \mathbf{\nabla})\mathbf{A}$

Where Φ is an arbitrary scalar and $\mathbf{A}, \mathbf{B}, \mathbf{C}$ are arbitrary vectors.

B.2 Vector Operations in Various Coordinates

In this section, three vector operators are shown in Cartesian, cylindrical, and spherical coordinate systems.

B.2.1 Cartesian Coordinates

Vector operations in Cartesian coordinates are shown.

1. $\mathbf{\nabla}\Phi = \mathbf{e}_x \dfrac{\partial \Phi}{\partial x} + \mathbf{e}_y \dfrac{\partial \Phi}{\partial y} + \mathbf{e}_z \dfrac{\partial \Phi}{\partial z}$

DOI: 10.1201/9781003310068-B

2. $\nabla \cdot \mathbf{A} = \dfrac{\partial A_x}{\partial x} + \dfrac{\partial A_y}{\partial y} + \dfrac{\partial A_z}{\partial z}$

3. $\nabla \times \mathbf{A} = \mathbf{e}_x \left(\dfrac{\partial A_z}{\partial y} - \dfrac{\partial A_y}{\partial z} \right) + \mathbf{e}_y \left(\dfrac{\partial A_x}{\partial z} - \dfrac{\partial A_z}{\partial x} \right) + \mathbf{e}_z \left(\dfrac{\partial A_y}{\partial x} - \dfrac{\partial A_x}{\partial y} \right)$

Where $\varPhi$ is an arbitrary scalar, $\mathbf{e}_x, \mathbf{e}_y, \mathbf{e}_z$ are basis vectors in the x-, y-, z-directions and $\mathbf{A} = \mathbf{e}_x A_x + \mathbf{e}_y A_y + \mathbf{e}_z A_z$ is an arbitrary vector.

B.2.2   Cylindrical Coordinates

Vector operations in cylindrical coordinates are shown.

1. $\nabla \varPhi = \mathbf{e}_r \dfrac{\partial \varPhi}{\partial r} + \mathbf{e}_\theta \dfrac{\partial \varPhi}{r \partial \theta} + \mathbf{e}_z \dfrac{\partial \varPhi}{\partial z}$

2. $\nabla \cdot \mathbf{A} = \dfrac{1}{r} \dfrac{\partial (r A_r)}{\partial r} + \dfrac{1}{r} \dfrac{\partial A_\theta}{\partial \theta} + \dfrac{\partial A_z}{\partial z}$

3. $\nabla \times \mathbf{A} = \mathbf{e}_r \left(\dfrac{1}{r} \dfrac{\partial A_z}{\partial \theta} - \dfrac{\partial A_\theta}{\partial z} \right) + \mathbf{e}_\theta \left(\dfrac{\partial A_r}{\partial z} - \dfrac{\partial A_z}{\partial r} \right) + \mathbf{e}_z \dfrac{1}{r} \left\{ \dfrac{\partial (r A_\theta)}{r \partial r} - \dfrac{\partial A_r}{r \partial \theta} \right\}$

Where $\varPhi$ is an arbitrary scalar, $\mathbf{e}_r, \mathbf{e}_\theta, \mathbf{e}_z$ are basis vectors in the r-, θ-, z-directions and $\mathbf{A} = \mathbf{e}_r A_r + \mathbf{e}_\theta A_\theta + \mathbf{e}_z A_z$ is an arbitrary vector.

B.2.3   Spherical Coordinates

Vector operations in spherical coordinates are shown.

1. $\nabla \varPhi = \mathbf{e}_r \dfrac{\partial \varPhi}{\partial r} + \mathbf{e}_\phi \dfrac{1}{r \cos \theta} \dfrac{\partial \varPhi}{\partial \phi} + \mathbf{e}_\theta \dfrac{1}{r} \dfrac{\partial \varPhi}{\partial \theta}$

2. $\nabla \cdot \mathbf{A} = \dfrac{1}{r^2} \dfrac{\partial (r^2 A_r)}{\partial r} + \dfrac{1}{r \cos \theta} \dfrac{\partial A_\phi}{\partial \phi} + \dfrac{1}{r \cos \theta} \dfrac{\partial (\cos \theta A_\theta)}{\partial \theta}$

3. $\nabla \times \mathbf{A} = \mathbf{e}_r \dfrac{1}{r \cos \theta} \left\{ \dfrac{\partial (\cos \theta A_\phi)}{\partial \theta} - \dfrac{\partial A_\theta}{\partial \phi} \right\} + \mathbf{e}_\phi \dfrac{1}{r} \left\{ \dfrac{\partial (r A_\theta)}{\partial r} - \dfrac{\partial A_r}{\partial \theta} \right\}$
$$+ \mathbf{e}_\theta \dfrac{1}{r} \left\{ \dfrac{1}{\cos \theta} \dfrac{\partial A_r}{\partial \phi} - \dfrac{\partial (r A_\phi)}{\partial r} \right\}$$

Where $\varPhi$ is an arbitrary scalar, $\mathbf{e}_r, \mathbf{e}_\phi, \mathbf{e}_\theta$ are basis vectors in the r-, ϕ-, θ-directions and $\mathbf{A} = \mathbf{e}_r A_r + \mathbf{e}_\phi A_\phi + \mathbf{e}_\theta A_\theta$ is an arbitrary vector.

Appendix C

Useful Constants and Parameters

Gravitational constant: $G = (6.67259 \pm 0.00030) \times 10^{-11} \, \mathrm{m^3 \, kg^{-1} \, s^{-2}}$

Radius of the Earth at the equator: $a = 6.3781 \times 10^6 \, \mathrm{m}$

Radius of the Earth at the pole: $b = 6.3568 \times 10^6 \, \mathrm{m}$

Mean radius of the Earth: $a_0 = 6.3710 \times 10^6 \, \mathrm{m}$

Mass of the Earth: $M_E = 5.9724 \times 10^{24} \, \mathrm{kg}$

Rotational period of the Earth: $T_0 = 9.9727 \times 10^{-1} \, \mathrm{day} = 8.6164 \times 10^4 \, \mathrm{s}$

Angular velocity of the Earth rotation: $\Omega = 7.2921 \times 10^{-5} \, \mathrm{s^{-1}}$

Mean orbital radius of the Earth: $r_E = 1.4959787 \times 10^{11} \, \mathrm{m}$

Orbital period of the Earth: $T_E = 3.6526 \times 10^2 \, \mathrm{days}$

Mass of the Sun: $M_S = 1.9891 \times 10^{30} \, \mathrm{kg}$

Mass of the Moon: $M_M = 7.3483 \times 10^{22} \, \mathrm{kg}$

Radius of the Moon at the equator: $a_M = 1.7379 \times 10^6 \, \mathrm{m}$

Mean radius of the Moon: $a_{M0} = 1.7371 \times 10^6 \, \mathrm{m}$

Mean orbital radius of the Moon: $r_M = 3.8440 \times 10^8 \, \mathrm{m}$

Orbital period of the Moon: $T_M = 2.73217 \times 10 \, \mathrm{days} = 2.36059 \times 10^6 \, \mathrm{s}$

DOI: 10.1201/9781003310068-C

Appendix D

Answers to Problems

Chapter 1

Problem 1. $2.455 \times 10^5 \, \text{mm}^3$

Answers to Chapter End Problems

1.

$$S = \int_{-\pi/2}^{\pi/2} \int_0^{2\pi} a^2 \cos\theta \, d\phi \, d\theta = 2\pi a^2 \int_{-\pi/2}^{\pi/2} \cos\theta \, d\theta$$
$$= 2\pi a^2 \left[\sin\theta\right]_{-\pi/2}^{\pi/2} = 4\pi a^2$$

2.

$$V = \int_{-\pi/2}^{\pi/2} \int_0^{2\pi} \int_0^a r^2 \cos\theta \, d\phi \, d\theta \, dr = 2\pi \left[\frac{1}{3}r^3\right]_0^a \int_{-\pi/2}^{\pi/2} \cos\theta \, d\theta$$
$$= \frac{2\pi}{3} a^3 \left[\sin\theta\right]_{-\pi/2}^{\pi/2} = \frac{4}{3}\pi a^3$$

3. Expanding $\exp x$ in a Taylor series around $x = 0$, then
$$\exp x = 1 + \frac{1}{1!}x + \frac{1}{2!}x^2 + \frac{1}{3!}x^3 + \cdots = \sum_{m=0}^{\infty} \frac{1}{m!}x^m \,.$$
Putting $x \to \tilde{\imath}x$,
$$\exp(\tilde{\imath}x) = 1 + \tilde{\imath}\frac{1}{1!}x - \frac{1}{2!}x^2 - \tilde{\imath}\frac{1}{3!}x^3 + \cdots .$$
$$= 1 - \frac{1}{2!}x^2 + \frac{1}{4!}x^4 + \cdots + \tilde{\imath}\left(\frac{1}{1!}x - \frac{1}{3!}x^3 + \cdots\right) = \cos x + \tilde{\imath}\sin x \,.$$

4. $(\exp \tilde{\imath}\theta)^n = \exp(\tilde{\imath}n\theta)$.
Applying the Euler formula,
$(\cos\theta + \tilde{\imath}\sin\theta)^n = \cos n\theta + \tilde{\imath}\sin n\theta$.

DOI: 10.1201/9781003310068-D

5. -1.7×10^{-7}

Chapter 2

Problem 1.
$\mathbf{A} \cdot \mathbf{B} = (A_x \mathbf{e}_x + A_y \mathbf{e}_y + A_z \mathbf{e}_z) \cdot (B_x \mathbf{e}_x + B_y \mathbf{e}_y + B_z \mathbf{e}_z)$
$= A_x B_x + A_y B_y + A_z B_z$

Problem 2.
$\mathbf{A} \times \mathbf{B} = (A_x \mathbf{e}_x + A_y \mathbf{e}_y + A_z \mathbf{e}_z) \times (B_x \mathbf{e}_x + B_y \mathbf{e}_y + B_z \mathbf{e}_z)$
$= A_x B_y \mathbf{e}_z + A_x B_z(-\mathbf{e}_y) + A_y B_x(-\mathbf{e}_z) + A_y B_z \mathbf{e}_x + A_z B_x \mathbf{e}_y + A_z B_y(-\mathbf{e}_x)$
$= (A_y B_z - A_z B_y)\mathbf{e}_x + (A_z B_x - A_x B_z)\mathbf{e}_y + (A_x B_y - A_y B_x)\mathbf{e}_z$

Problem 3.
(the area of parallelogram OPQR) = (the area of parallelogram OPST)
= (the area of rectangle UPSX)
= (the area of rectangle UPVW) $-$ (the area of rectangle XSVW)
(the area of rectangleUPVW) $= A_x B_y$

(the area of rectangle XSVW) $= A_x \overline{SV} = A_x B_x \dfrac{A_y}{A_x} = B_x A_y$

(the area of parallelogram OPQR) $= A_x B_y - B_x A_y$

Problem 4.
$\mathbf{A} \times (\mathbf{B} \times \mathbf{C}) = \mathbf{A} \times \{(B_y C_z - B_z C_y)\mathbf{e}_x + (B_z C_x - B_x C_z)\mathbf{e}_y + (B_x C_y - B_y C_x)\mathbf{e}_z\}$
$= (B_x C_y A_y - C_x A_y B_y - C_x A_z B_z + B_x C_z A_z)\mathbf{e}_x$
$+ (B_y C_z A_z - C_y A_z B_z - C_y A_x B_x + C_x A_x B_y)\mathbf{e}_y$
$+ (B_z C_x A_x - C_z A_x B_x - C_z A_y B_y + B_z C_y A_y)\mathbf{e}_z$
$= B_x \mathbf{e}_x(\mathbf{C} \cdot \mathbf{A}) - C_x \mathbf{e}_x(\mathbf{A} \cdot \mathbf{B}) + B_y \mathbf{e}_y(\mathbf{C} \cdot \mathbf{A})$
$- C_y \mathbf{e}_y(\mathbf{A} \cdot \mathbf{B}) + B_z \mathbf{e}_z(\mathbf{C} \cdot \mathbf{A}) - C_z \mathbf{e}_z(\mathbf{A} \cdot \mathbf{B})$
$= \mathbf{B}(\mathbf{C} \cdot \mathbf{A}) - \mathbf{C}(\mathbf{A} \cdot \mathbf{B})$

Problem 5. 1. $34.3\,\text{m s}^{-1}$ 2. $44.1\,\text{m s}^{-1}$ 3. $9.8\,\text{m s}^{-2}$

Problem 6. $50.0\,\text{m s}^{-1}$

Problem 7.
From (2.26), $v = r\omega$
At point P, $v = v_y$, $r = x$, so that $v_y = \omega x$,
at point Q, $v = -v_x$, $r = y$, so that $v_x = -\omega y$.

Answers to Chapter End Problems

1. (1) $\mathbf{A} \cdot \mathbf{B} = -29$

   (2) $\mathbf{A} \times \mathbf{B} = 2\mathbf{e}_x - \mathbf{e}_y - 2\mathbf{e}_z$

   (3) $\dfrac{1}{3}(2\mathbf{e}_x - \mathbf{e}_y - 2\mathbf{e}_z)$

2. $\mathbf{A} \cdot (\mathbf{B} \times \mathbf{C}) = \mathbf{A} \cdot \begin{vmatrix} \mathbf{e}_x & \mathbf{e}_y & \mathbf{e}_z \\ B_x & B_y & B_z \\ C_x & C_y & C_z \end{vmatrix}$
$= A_x(B_y C_z - B_z C_y) + A_y(B_z C_x - B_x C_z) + A_z(B_x C_y - B_y C_x)$

$$= \begin{vmatrix} A_x & A_y & A_z \\ B_x & B_y & B_z \\ C_x & C_y & C_z \end{vmatrix},$$

$$\mathbf{B} \cdot (\mathbf{C} \times \mathbf{A}) = \begin{vmatrix} B_x & B_y & B_z \\ C_x & C_y & C_z \\ A_x & A_y & A_z \end{vmatrix} = - \begin{vmatrix} B_x & B_y & B_z \\ A_x & A_y & A_z \\ C_x & C_y & C_z \end{vmatrix} = \begin{vmatrix} A_x & A_y & A_z \\ B_x & B_y & B_z \\ C_x & C_y & C_z \end{vmatrix}$$

$$= \mathbf{A} \cdot (\mathbf{B} \times \mathbf{C}),$$

$$\mathbf{C} \cdot (\mathbf{A} \times \mathbf{B}) = \begin{vmatrix} C_x & C_y & C_z \\ A_x & A_y & A_z \\ B_x & B_y & B_z \end{vmatrix} = - \begin{vmatrix} A_x & A_y & A_z \\ C_x & C_y & C_z \\ B_x & B_y & B_z \end{vmatrix} = \begin{vmatrix} A_x & A_y & A_z \\ B_x & B_y & B_z \\ C_x & C_y & C_z \end{vmatrix}$$

$$= \mathbf{A} \cdot (\mathbf{B} \times \mathbf{C}).$$

3. $\mathbf{e}_x = \dfrac{\mathbf{A}}{|\mathbf{A}|}$, $\quad \mathbf{e}_y = \dfrac{\mathbf{B} - \mathbf{e}_x(\mathbf{e}_x \cdot \mathbf{B})}{|\mathbf{B} - \mathbf{e}_x(\mathbf{e}_x \cdot \mathbf{B})|}$

4. $-19.3 \, \mathrm{m\,s}^{-2}$

5. (1) $v_{0-10} = 5.75 \, \mathrm{m\,s}^{-1}, v_{10-20} = 10.1 \, \mathrm{m\,s}^{-1}, v_{20-30} = 11.1 \, \mathrm{m\,s}^{-1},$
   $v_{30-40} = 11.6 \, \mathrm{m\,s}^{-1}, v_{40-50} = 12.0 \, \mathrm{m\,s}^{-1}, v_{50-60} = 12.2 \, \mathrm{m\,s}^{-1},$
   $v_{60-70} = 12.3 \, \mathrm{m\,s}^{-1}, v_{70-80} = 12.2 \, \mathrm{m\,s}^{-1}, v_{80-90} = 12.0 \, \mathrm{m\,s}^{-1},$
   $v_{90-100} = 12.0 \, \mathrm{m\,s}^{-1}$

   (2) $\alpha_{5-15} = 3.45 \, \mathrm{m\,s}^{-2}, \alpha_{15-25} = 1.06 \, \mathrm{m\,s}^{-2}, \alpha_{25-35} = 0.568 \, \mathrm{m\,s}^{-2},$
   $\alpha_{35-45} = 0.472 \, \mathrm{m\,s}^{-2}, \alpha_{45-55} = 0.242 \, \mathrm{m\,s}^{-2}, \alpha_{55-65} = 0.123 \, \mathrm{m\,s}^{-2}$
   $\alpha_{65-75} = -0.123 \, \mathrm{m\,s}^{-2}, \alpha_{75-85} = -0.242 \, \mathrm{m\,s}^{-2}, \alpha_{85-95} = 0.000 \, \mathrm{m\,s}^{-2}$

   (3) the acceleration$=4.03 \, \mathrm{m\,s}^{-2}$, the maximum speed $= 12.7 \, \mathrm{m\,s}^{-1}$

6. $7.2921 \times 10^{-5} \, \mathrm{s}^{-1}, \quad 3.39 \times 10^{-2} \, \mathrm{m\,s}^{-2}$

7. $1.02 \times 10^{3} \, \mathrm{m\,s}^{-1}, \quad 2.72 \times 10^{-3} \, \mathrm{m\,s}^{-2}$

Chapter 3

Problem 1. $-1.04 \times 10^4 \, \mathrm{N}$
Problem 2. $-4.58 \times 10^3 \, \mathrm{m\,s}^{-2}, \quad -6.64 \times 10^2 \, \mathrm{N}$
Problem 3. $1.4 \times 10^2 \, \mathrm{m}$
Problem 4. When the falling speed becomes constant, the gravity balances with the resistive force. Therefore,
$0 = mg - kv_\infty, \qquad v_\infty = mg/k$
Problem 5. $\theta = \pi/4$
Problem 6. Because when the direction of motion is positive (negative), the θ-component force is negative (positive).
Problem 7. $2.01 \, \mathrm{s}$

Problem 8. Fixing one end of the connected springs to a rigid support, let's exert a force F on the other end of two springs. Suppose that the extension of spring 1 be Δl_1, the extension of spring 2 be Δl_2, the total extension of two springs be Δl and the equivalent spring constant of two springs be k, we find

$$\Delta l = \Delta l_1 + \Delta l_2, \qquad F = k_1 \Delta l_1 = k_2 \Delta l_2 = k \Delta l$$
$$\frac{F}{k} = \frac{F}{k_1} + \frac{F}{k_2}, \qquad \frac{1}{k} = \frac{1}{k_1} + \frac{1}{k_2}$$

Problem 9. $4.51°$

Answers to Chapter End Problems

1. $h = V_0{}^2/2g, \qquad t = 2V_0/g$

2. $y = (V_0 \sin \theta)^2/2g, \qquad x = V_0{}^2 \sin 2\theta/g$

3. (1) $27.8\,\mathrm{m\,s^{-1}}$
   (2) $2.44 \times 10^3\,\mathrm{N}$

4. The monkey cannot escape from the bullet because the falling distance of the monkey is equal to the falling distance of the bullet. This result is readily confirmed by simple calculations.

5. $S_1 = (m_1 + m_2)(g + a)$, $S_2 = m_2(g + a)$

6. $T = 2\pi\sqrt{\dfrac{l \cos \theta}{g}}$

7. (1) $2.00 \times 10^{20}\,\mathrm{N}$
   (2) $2.78 \times 10^{-4}\,\mathrm{times}$

8. $k = k_1 + k_2$

Chapter 4

Problem 1. The vehicle seems to travel to the northwest at $70.7\,\mathrm{kmh^{-1}}$.

Answers to Chapter End Problems

1. (1) 16.4 s   (2) 1.43×10^{-4}

2. 1.15×10^5 s $(1.33\,\mathrm{days})$

3. Letting the time that the stone falls to the ground be τ, the mass of the stone be m, the height of the deck be h and the magnitude of the acceleration due to gravity be g, we find

$$\tau = \sqrt{\frac{2h}{g}} = \sqrt{\frac{2 \times 4.50 \times 10^2}{9.80}} = 9.58\,[\mathrm{s}].$$

The falling speed of the stone at time t is $w = -gt$, so that the Coriolis force exerting on the stone is

$$\mathbf{F}_{Co} = -2m\Omega \cos\theta(-w)\mathbf{e}_x.$$

Thus, the equation of motion is

$$m\frac{d^2x}{dt^2} = 2m\Omega\cos\theta w = 2m\Omega\cos\theta gt.$$

Integrating the above equation with respect to t under the initial condition that $dx/dt = u = 0$ at $t = 0$, we get

$$\frac{dx}{dt} = u = \Omega\cos\theta gt^2.$$

Integrating the above equation with respect to t under the initial condition that $dx/dt = u = 0$ at $t = 0$, we get

$$\frac{dx}{dt} = u = \Omega\cos\theta gt^2.$$

Integrating the above equation with respect to t from $t = 0$ to $t = \tau$, we find

$$x = \frac{1}{3}\Omega\cos\theta g\tau^3$$
$$= \frac{1}{3} \times 7.27 \times 10^{-5}\cos 35.7° \times 9.80 \times 9.58^3 = 1.70 \times 10^{-1}\,[\text{m}].$$

The stone is displaced 1.70×10^{-1} m eastward from just below the released point.

4. Supposing that the eastward speed of the rocket is u and the time falling to the sea surface is τ, the Coriolis force exerting on the rocket is

$$\mathbf{F}_{Co} = -2m\Omega u\left(\sin\theta\mathbf{e}_y - \cos\theta\mathbf{e}_z\right),$$

and $\tau = 5.00 \times 10^3$ s. The equation of motion of the y-component is

$$m\frac{d^2y}{dt^2} = -2mu\Omega\sin\theta.$$

Integrating the above equation with respect to t under the initial condition that $dy/dt = v = 0$ at $t = 0$, we get

$$\frac{dy}{dt} = -2u\Omega\sin\theta t.$$

Integrating the above equation with respect to t from 0 to τ, we find

$$y = -u\Omega\sin\theta\tau^2 = -9.20 \times 10^5\,[\text{m}].$$

The rocket is deflected southward by 9.20×10^5 m from east.

Chapter 5

Problem 1. 0 (Because the Coriolis force is perpendicular to the direction of motion.)

Problem 2. $W = mgl \sin \theta$

Problem 3. $U = mgh$

Problem 4. 0.5 J

Answers to Chapter End Problems

1. $v = \sqrt{2gl(1 - \cos \theta_0)}$

2. $v = 42.0 \, \text{m s}^{-1}$

3. $y_{max} = (V_0 \sin \theta)^2 / 2g$

4. $F = 2.90 \times 10^4 \, \text{N}, \, \mu' = 1.97$

5. $V_0 \geq \sqrt{5ag}$

6. Taking T as the reference point of the potential energy and letting the tangential speed at point P be v, we find that
$$0 = \frac{1}{2}mv^2 - mga(1 - \cos \theta), \quad v^2 = 2ag(1 - \cos \theta) .$$
due to the law of mechanical energy conservation. The component of gravity vertical to the spherical surface is equal to the centrifugal force exerting on the particle at point P, so that we obtain
$$m\frac{v^2}{a} = mg \cos \theta, \quad \cos \theta = \frac{v^2}{a} = 2(1 - \cos \theta) ,$$
$$\cos \theta = \frac{2}{3}, \quad \theta = \cos^{-1}\frac{2}{3} = 48.2° .$$

Chapter 6

Answers to Chapter End Problems

1. Applying (6.4) to (6.5), we obtain
$$x_0 = \alpha + \beta. \tag{D.1}$$

Differentiating (6.4) with respect to t and applying the result to (6.5)
$$\frac{dx}{dt} = \exp(-\gamma t)\left\{ \left(\sqrt{\gamma^2 - \omega^2} - \gamma\right) \alpha \exp\left(\sqrt{\gamma^2 - \omega^2}\, t\right) \right.$$
$$\left. - \left(\sqrt{\gamma^2 - \omega^2} + \gamma\right) \beta \exp\left(-\sqrt{\gamma^2 - \omega^2}\, t\right) \right\},$$
$$0 = \left(\sqrt{\gamma^2 - \omega^2} - \gamma\right) \alpha - \left(\sqrt{\gamma^2 - \omega^2} + \gamma\right) \beta. \tag{D.2}$$

Eliminating β from (D.1) and (D.2), we find

$$\alpha = \frac{1}{2}\left(1 + \frac{\gamma}{\sqrt{\gamma^2 - \omega^2}}\right)x_0 . \qquad \beta = \frac{1}{2}\left(1 - \frac{\gamma}{\sqrt{\gamma^2 - \omega^2}}\right)x_0 .$$

Thus, (6.6) is proved.

2. Applying (6.8) to (6.9), we get

$$x_0 = A\cos\phi \qquad\qquad (D.3)$$

Differentiating (6.8) with respect to t and applying the result to (6.9) yields

$$\frac{d\Re[x]}{dt} = -A\gamma\exp(-\gamma t)\cos\left(\sqrt{\omega^2 - \gamma^2}\,t - \phi\right)$$
$$-A\sqrt{\omega^2 - \gamma^2}\exp(-\gamma t)\sin\left(\sqrt{\omega^2 - \gamma^2}\,t - \phi\right).$$
$$0 = -A\gamma\cos\phi + A\sqrt{\omega^2 - \gamma^2}\sin\phi. \qquad\qquad (D.4)$$

From (D.3) and (D.4), we find

$$A = \frac{\omega}{\sqrt{\omega^2 - \gamma^2}}x_0, \qquad \phi = \tan^{-1}\frac{\gamma}{\sqrt{\omega^2 - \gamma^2}}.$$

Thus, (6.10) is proved.

3. Applying (6.13) to (6.14), we get

$$x_0 = A. \qquad\qquad (D.5)$$

Differentiating (6.13) with respect to t and applying the result to (6.14) yields

$$\frac{dx}{dt} = \{-\gamma A + (1 - \gamma t)B\}\exp(-\gamma t),$$
$$0 = -\gamma A + B. \qquad\qquad (D.6)$$

From (D.5) and (D.6), we find
$$A = x_0, \qquad B = \gamma x_0.$$
Thus, (6.15) is proved.

4. As shown in Fig. D.1 let's take the x-axis perpendicular to the string. Letting the displacement of particle 1,2,3 be x_1, x_2, x_3, and the forces exerting on particle 1,2,3 be F_1, F_2, F_3, we find

$$F_1 = \frac{x_2 - x_1}{l}S - \frac{x_1}{l}S = \frac{x_2 - 2x_1}{l}S,$$
$$F_2 = -\frac{x_2 - x_1}{l}S - \frac{x_2 - x_3}{l}S = \frac{x_1 - 2x_2 + x_3}{l}S,$$
$$F_3 = \frac{x_2 - x_3}{l}S - \frac{x_3}{l}S = \frac{x_2 - 2x_3}{l}S.$$

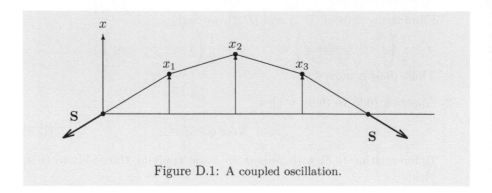

Figure D.1: A coupled oscillation.

The equations of motion of three particles are

$$m\frac{d^2x_1}{dt^2} = \frac{x_2 - 2x_1}{l}S,$$

$$m\frac{d^2x_2}{dt^2} = \frac{x_1 - 2x_2 + x_3}{l}S,$$

$$m\frac{d^2x_3}{dt^2} = \frac{x_2 - 2x_3}{l}S.$$

We will assume the exponential type solution of the angular frequency ω, namely $x_1 = \hat{x}_1 \exp \tilde{\imath}\omega t, x_2 = \hat{x}_2 \exp \tilde{\imath}\omega t, x_3 = \hat{x}_3 \exp \tilde{\imath}\omega t$. Substituting the assumed solution to the equation of motion, we find

$$-\omega^2\hat{x}_1 = \beta^2\left(\hat{x}_2 - 2\hat{x}_1\right),$$

$$-\omega^2\hat{x}_2 = \beta^2\left(\hat{x}_1 - 2\hat{x}_1 + \hat{x}_3\right),$$

$$-\omega^2\hat{x}_3 = \beta^2\left(\hat{x}_2 - 2\hat{x}_3\right),$$

where $\beta^2 = S/ml$. The necessary condition that $\hat{x}_1$, $\hat{x}_2$, $\hat{x}_3$ are nontrivial solutions is

$$\begin{vmatrix} 2\beta^2 - \omega^2 & -\beta^2 & 0 \\ -\beta^2 & 2\beta^2 - \omega^2 & -\beta^2 \\ 0 & -\beta^2 & 2\beta^2 - \omega^2 \end{vmatrix} = 0.$$

Three angular frequencies and the amplitudes corresponding to them are

$$\omega = \sqrt{2 - \sqrt{2}}\sqrt{\frac{S}{ml}}, \quad x_1 : x_2 : x_3 = 1 : \sqrt{2} : 1\,,$$

$$\omega = \sqrt{2}\sqrt{\frac{S}{ml}}, \quad x_1 : x_2 : x_3 = 1 : 0 : -1\,,$$

$$\omega = \sqrt{2 + \sqrt{2}}\sqrt{\frac{S}{ml}}, \quad x_1 : x_2 : x_3 = -1 : \sqrt{2} : -1\,.$$

5. Letting the displacement of particle 1 and particle 2 be x_1, x_2, the equations of motion of the particles are

$$m\frac{d^2x_1}{dt^2} = -mg\frac{x_1}{l} - k(x_1 - x_2),$$

$$m\frac{d^2x_2}{dt^2} = -mg\frac{x_2}{l} + k(x_1 - x_2).$$

Letting $\alpha^2 = g/l$, $\beta^2 = k/m$ and assuming oscillatory solutions as $x_1 = \hat{x}_1 \exp(\tilde{\iota}\omega t)$, $x_2 = \hat{x}_2 \exp(\tilde{\iota}\omega t)$, and substituting them into the above equations, we obtain

$$(-\omega^2 + \alpha^2 + \beta^2)\hat{x}_1 - \beta^2\hat{x}_2 = 0,$$

$$-\beta^2\hat{x}_1 + (-\omega^2 + \alpha^2 + \beta^2)\hat{x}_2 = 0.$$

The necessary condition for $\hat{x}_1$ and $\hat{x}_2$ to have nontrivial solutions is

$$\begin{vmatrix} -\omega^2 + \alpha^2 + \beta^2 & -\beta^2 \\ -\beta^2 & -\omega^2 + \alpha^2 + \beta^2 \end{vmatrix} = 0.$$

Thus, the angular frequencies and the corresponding amplitudes are
$\omega = \sqrt{g/l}\, x_1 : x_2 = 1 : 1$,
$\omega = \sqrt{g/l + 2k/m}\, x_1 : x_2 = 1 : -1$.

Chapter 7

Answers to Chapter End Problems

1. 1.50×10^2 kg, The center of mass is at the position of 4.00 m from the end A.

2. The center of mass is 4.67×10^6 m from the center of the Earth to the center of the moon.

3. $X = 9.0$ cm, $Y = 4.0$ cm

4. On the symmetric axis and at $\dfrac{2a}{\pi}$ from the center of the circle.

5. On the symmetric axis and at $\dfrac{4}{3\pi}\dfrac{a^2 + ab + b^2}{a + b}$ from the center of the circles.

6. On the symmetric axis and at $\dfrac{3a}{8}$ from the center of the sphere.

7. On the symmetric axis and at $\dfrac{a}{2}$ from the center of the spherical shell.

8. On the symmetric axis and at $\dfrac{3}{8}\dfrac{(b+a)(b^2+a^2)}{b^2+ba+a^2}$ from the center of the spheres.

Chapter 8

Problem 1. $-1.04 \times 10^4\,\mathrm{N}$

Problem 2. The kinetic energy before the collision is
$$K = \frac{1}{2}m_1 v_1{}^2 + \frac{1}{2}m_2 v_2{}^2 .$$
The kinetic energy after the collision is

$$
\begin{aligned}
K &= \frac{1}{2}m_1 v_1'^2 + \frac{1}{2}m_2 v_2'^2 \\
&= \frac{m_1}{2(m_1+m_2)^2}\left\{(m_1-m_2)^2 v_1{}^2 + 4m_2{}^2 v_2{}^2 + 4m_2(m_1-m_2)v_1 v_2\right\} \\
&\quad + \frac{m_2}{2(m_1+m_2)^2}\left\{4m_1{}^2 v_1{}^2 + (m_2-m_1)^2 v_2{}^2 + 4m_1(m_2-m_1)v_1 v_2\right\} \\
&= \frac{1}{2}m_1 v_1{}^2 + \frac{1}{2}m_2 v_2{}^2 .
\end{aligned}
$$

Therefore, the kinetic energy is conserved before and after the collision.

Problem 3. (1) $\dfrac{1}{2}v$ (2) $\dfrac{1}{2}mv^2$ (3) $\dfrac{1}{4}mv^2$ (4) $\dfrac{1}{4}mv^2$

Answers to Chapter End Problems

1. (1) $\tau = 2.01\,\mathrm{s}$, $\alpha = -6.90\,\mathrm{m\,s^{-2}}$.

   (2) $F = ma = 1.50 \times 10^3 \times (-6.90) = -1.04 \times 10^4\,\mathrm{N}$

   (3) $0 - mu_0 = F\tau$
   $$F = -\frac{mu_0}{\tau} = -\frac{1.50 \times 10^3 \times 13.9}{2.01} = -1.04 \times 10^4\,\mathrm{N}$$

   (4) $0 - \dfrac{1}{2}mu_0{}^2 = Fs$, $F = -\dfrac{mu_0{}^2}{2s} = -1.04 \times 10^4\,\mathrm{N}$

2. (1) $v' = v/5$, positive x-direction

   (2) $\Delta K = 27mv^2/5$

3. (1) $v_1 = -2v$, $v_2 = v$

   (2) $e = 1$

4. $V_0 = \sqrt{\dfrac{gd}{\sin 2\theta}\left(1 + \dfrac{1}{e}\right)}$

5. (1) $v_1 = \dfrac{m - eM}{m + M}v$, $V_1 = \dfrac{m}{m + M}(1 + e)v$

$$(2) \quad v_2 = \frac{m + e^2 M}{m + M} v, \quad V_2 = \frac{m}{m + M}(1 - e^2)v$$

$$(3) \quad v_\infty = \frac{m}{m + M} v, \quad V_\infty = \frac{m}{m + M} v$$

6. We can obtain the answer integrating (8.50) with respect to λ from $\pi/2$ to $3\pi/2$,

$$N_{\text{forward}} = \frac{1000}{4} \int_0^{\pi/2} \sin\frac{\lambda}{2} d\lambda + \frac{1000}{4} \int_{3\pi/2}^{2\pi} \sin\frac{\lambda}{2} d\lambda$$

$$= \frac{1000}{2}\left[-\cos\frac{\lambda}{2}\right]_0^{\pi/2} + \frac{1000}{2}\left[-\cos\frac{\lambda}{2}\right]_{3\pi/2}^{2\pi} = 1000\left(1 - \frac{1}{\sqrt{2}}\right) = 293.$$

Thus, 293 particles are scattered forward.

7. We can obtain the answer integrating (8.57) with respect to λ from 0 to $\pi/2$,

$$N_{\text{forward}} = \frac{1000}{2} \int_0^{\pi/2} \sin\lambda d\lambda = \frac{1000}{2}\left[-\cos\lambda\right]_0^{\pi/2} = 500.$$

Thus, 500 particles are scattered forward.

Chapter 9

Problem 1. The force turning a steering wheel of a car, the force turning handlebars of a bicycle, the force turning the axis of a top, the force turning a screwdriver, the force turning a gimlet.

Problem 2. $23.2°$, $16.5°$

Answers to Chapter End Problems

1. The angular momentum of particle 2 after separating particle 2 is $2mlv$ and the angular momentum of the system is conserved before and after separating particle 2 so that
$3mlv = mlv' + 2mlv, \quad v' = v$.
The tangential speed of particle 1 does not change. The kinetic energy before separating particle 2 is
$$K = \frac{1}{2}I\omega^2 = \frac{3}{2}ml^2\left(\frac{v}{l}\right)^2 = \frac{3}{2}mv^2.$$
The kinetic energy after cutting off particle 2 is
$$K_1 = \frac{1}{2}I\omega^2 = \frac{1}{2}ml^2\left(\frac{v}{l}\right)^2 = \frac{1}{2}mv^2. \quad K_2 = \frac{1}{2}2mv^2 = mv^2,$$
$$K = K_1 + K_2 = \frac{3}{2}mv^2.$$
Therefore, the kinetic energy of the system is conserved before and after separating particle 2.

2. (1) The area density of the disk σ is
$$\sigma = \frac{M}{\pi a^2} .$$
The torque exerting on a ring of inner radius r and outer radius $r + \delta r$ is
$$\delta N = -\sigma 2\pi r^2 \delta r \mu g .$$
The total torque is obtained by integrating the above equation with respect to r from 0 to a,
$$N = \int_0^a dN = -\frac{2}{3}\mu M g a .$$

(2) The angular momentum of the ring of inner radius r and outer radius $r + \delta r$ is
$$\delta L = \sigma 2\pi r \delta r r^2 \omega = 2\pi \sigma r^3 \delta r \omega .$$
The angular momentum of the disk is
$$L = \int_0^a dL = \frac{1}{2} M a^2 \omega .$$

(3) $\dfrac{d\omega}{dt} = -\dfrac{4}{3}\dfrac{\mu g}{a}$

(4) $\omega = \omega_0 - \dfrac{4}{3}\dfrac{\mu g}{a} t, \qquad \tau = \dfrac{3}{4}\dfrac{a\omega_0}{\mu g}$

3. (1) $dN = -k\omega r^3 dr d\theta$

(2) $N = -\displaystyle\int_0^{2\pi}\int_0^a k\omega r^3 dr d\theta dr d\theta = -\frac{\pi}{2}k\omega a^4$

(3) $L = \displaystyle\int_0^{2\pi}\int_0^a \sigma\omega r^3 dr d\theta = \frac{1}{2}M a^2 \omega$

(4) $\dfrac{1}{2}M a^2 \dfrac{d\omega}{dt} = -\dfrac{\pi}{2}k a^4 \omega$

(5) $\omega = \omega_0 \exp\left(-\dfrac{\pi k a^2}{M}t\right)$

(6) $\tau = \dfrac{M}{\pi k a^2}$

4. (1) $\theta = \cos^{-1}(2a/l)^{1/3}$

(2) $\theta = 29.1°$

Chapter 10

Problem 1. $I = \dfrac{1}{12}M l^2$

Problem 2. $I = \dfrac{1}{3}M l^2$

Problem 3.   The moment of inertia of the piano wire is not negligible.

Answers to Chapter End Problems

1. $I = \dfrac{1}{2}M(a^2 + b^2)$

2. $I = \dfrac{1}{4}M(a^2 + b^2) + \dfrac{1}{12}Md^2$

3. $I = \dfrac{2}{3}Ma^2$

4. $I = \dfrac{2}{5}M\dfrac{b^4 + ab^3 + a^2b^2 + a^3b + a^4}{b^2 + ab + a^2}$

5. Letting the tangential velocity be v, and the angular velocity of the ball at point P be ω, we find

$$v = a\omega.$$

Taking the reference level of the potential energy at the level surface, we find from the mechanical energy conservation,

$$mg(a + 2b) = mg\{b + (a + b)\cos\theta\} + \frac{1}{2}mv^2 + \frac{1}{2}\frac{2}{5}Ma^2\omega^2. \quad (\text{D.7})$$

The centrifugal force and the normal component of the gravity exerting on the ball is balanced at point P, so that

$$m\frac{v^2}{a + b} = mg\cos\theta. \quad (\text{D.8})$$

where g is the magnitude of the acceleration due to gravity. From (D.7) and (D.8), we have

$$\theta = \cos^{-1}\frac{7}{10} = 54.0°; \, .$$

6. (1) $9.693 \times 10^{37} \text{ kg m}^2$
   (2) $8.136 \times 10^{37} \text{ kg m}^2$
   (3) $+19.1\%$

7. Taking the semimajor axis as the x-axis, the semiminor axis as the y-axis and the z-axis to obey the right-hand rule, we will calculate the moment of inertia about the z-axis,

$$\rho = \frac{M}{\pi abd},$$

$$I_{zz} = \int_0^d \int_{-b\sqrt{a^2-x^2}/a}^{b\sqrt{a^2-x^2}/a} \int_{-a}^a \rho(x^2 + y^2)\,dx\,dy\,dz$$

$$= \rho d \int_{-a}^a \left(x^2 [y]_{-b\sqrt{a^2-x^2}/a}^{b\sqrt{a^2-x^2}/a} + \frac{1}{3}[y^3]_{-b\sqrt{a^2-x^2}/a}^{b\sqrt{a^2-x^2}/a} \right) dx$$

$$= 2\rho d \frac{b}{a} \int_{-a}^a \left\{ x^2\sqrt{a^2 - x^2} + \frac{1}{3}\frac{b^2}{a^2}(a^2 - x^2)\sqrt{a^2 - x^2} \right\} dx.$$

Transforming an independent variable $x = a\cos\theta$ yields

$$\int_{-a}^{a} x^2\sqrt{a^2 - x^2}\,dx = -\int_{\pi}^{0} a^4\sin^2\theta\cos^2\theta\,d\theta$$

$$= a^4\int_{0}^{\pi}\frac{1}{4}\sin^2 2\theta\,d\theta = \frac{a^4}{4}\int_{0}^{\pi}\frac{1 - \cos 4\theta}{2}\,d\theta = \frac{\pi a^4}{8},$$

$$\int_{-a}^{a}\sqrt{a^2 - x^2} = -\int_{\pi}^{0} a^2\sin^2\theta\,d\theta = \frac{\pi a^2}{2}.$$

Therefore,

$$I_{zz} = 2\rho d\frac{b}{a}\left[\frac{\pi}{8}a^4 + \frac{b^2}{3a^2}\left\{\frac{\pi a^4}{2} - \frac{\pi}{8}a^4\right\}\right] = \frac{1}{4}M(a^2 + b^2).$$

8. $F_{\min} = \dfrac{\sqrt{h(2a - h)}}{a}Mg$

9. We will take the coordinate axis as the moving direction of the two bobs along the string. As shown in Fig. D.2, let the tension exerted by the string on bob 1 be S_1 and the tension exerted by the string on bob 2 be S_2. Due to Newton's third law, bob 1 exerts torque $S_1 a$ and bob 2 exerts torque $-S_2 a$ on the pulley. Letting the angular velocity of the pulley be ω and the falling speed of bob 1 be v, the angular momentum equation for the pulley, and the equation of motion for bob 1 and bob 2 are

$$I\frac{d\omega}{dt} = (S_1 - S_2)a, \quad m_1\frac{dv}{dt} = m_1 g - S_1, \quad m_2\frac{dv}{dt} = S_2 - m_2 g.$$

As there is no slip between the pulley and the string,

$$v = a\omega.$$

Eliminating S_1 and S_2 using the above four equations, we obtain

$$\left\{\frac{1}{2}Ma^2 + (m_1 + m_2)a^2\right\}\frac{dv}{dt} = a^2(m_1 - m_2)g.$$

We may find the particular solution applying the initial condition that $v = 0$ at $t = 0$,

$$v = \frac{(m_1 - m_2)g}{M/2 + (m_1 + m_2)}t.$$

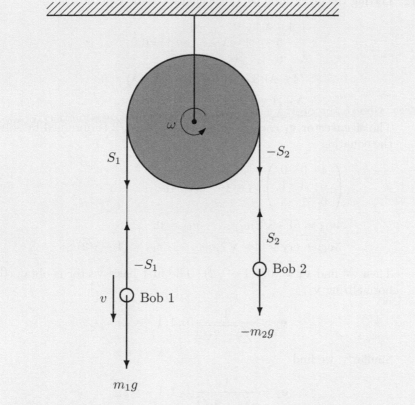

Figure D.2: Vertical motion of bobs hung from the fixed pulley.

10. Components of the inertia tensor are

$$I_{xx} = m[\{1^2 + 1^2 + (-1)^2\} + \{0^2 + 0^2 + 1^2\}] = 4m$$
$$I_{xy} = -m[1 \times 1 + (-1) \times 1 + 0 \times (-1)] = 0$$
$$I_{xz} = -m[1 \times 0 + (-1) \times 1 + 1 \times 0] = 0$$
$$I_{yx} = -m[1 \times 1 + (-1) \times 1 + 0 \times (-1)] = 0$$
$$I_{yy} = m[\{0^2 + 0^2 + 1^2\} + \{1^2 + (-1)^2 + 0^2\}] = 3m$$
$$I_{yz} = -m[1 \times 0 + 1 \times 0 + (-1) \times 1] = m$$
$$I_{zx} = -m[0 \times 1 + 0 \times (-1) + 1 \times 0 = 0$$
$$I_{zy} = -m[1 \times 0 + 1 \times 0 + (-1) \times 1] = m$$
$$I_{zz} = m[\{1^2 + (-1)^2 + 0^2\} + \{1^2 + 1^2 + (-1)^2\}] = 5m$$

$$\tilde{I} = m \begin{pmatrix} 4 & 0 & 0 \\ 0 & 3 & 1 \\ 0 & 1 & 5 \end{pmatrix}$$

11. Letting the eigenvalue be λ, the characteristic equation is

$$\begin{vmatrix} 4-\lambda & 0 & 0 \\ 0 & 3-\lambda & 1 \\ 0 & 1 & 5-\lambda \end{vmatrix} = 0,$$

$$(4-\lambda)(3-\lambda)(5-\lambda) - (4-\lambda) = 0,$$

$$\lambda = 4 - \sqrt{2}, 4, 4 + \sqrt{2}.$$

The eigenvector $\mathbf{v}_1$ corresponding to $\lambda = 4 - \sqrt{2}$ is obtained by solving the equation

$$\begin{pmatrix} 4 & 0 & 0 \\ 0 & 3 & 1 \\ 0 & 1 & 5 \end{pmatrix} \mathbf{v}_1 = (4 - \sqrt{2})\mathbf{v}_1,$$

$$4v_{11} = (4 - \sqrt{2})v_{11}. \qquad v_{11} = 0,$$

$$3v_{12} + v_{13} = (4 - \sqrt{2})v_{12}. \qquad v_{13} = (1 - \sqrt{2})v_{12}.$$

Then we find $\mathbf{v}_1 = (0, 1, 1 - \sqrt{2})$. The first basis vector is obtained by normalizing $\mathbf{v}_1$,

$$\mathbf{e}_1 = \frac{1}{\sqrt{4 - 2\sqrt{2}}}(0, 1, 1 - \sqrt{2}).$$

Similarly, we find

$$\mathbf{e}_2 = \frac{1}{\sqrt{4 + 2\sqrt{2}}}(0, 1, 1 + \sqrt{2}),$$

$$\mathbf{e}_3 = (1, 0, 0).$$

The orthogonalized inertial tensor is

$$\tilde{I} = m \begin{pmatrix} 4 - \sqrt{2} & 0 & 0 \\ 0 & 4 & 0 \\ 0 & 0 & 4 + \sqrt{2} \end{pmatrix}.$$

12. The density of the spheroid is

$$\rho = \frac{3M}{4\pi a^2 b}.$$

The moment of inertia of a circular disk of mass M and radius a is

$$I = \frac{1}{2}Ma^2.$$

Then the moment of inertia about the z-axis is

$$I_z = \int_{-b}^{b} \frac{1}{2}\rho\pi a^2 \left(1 - \frac{z^2}{b^2}\right) a^2 \left(1 - \frac{z^2}{b^2}\right) dz$$

$$= \frac{1}{2}\rho\pi\frac{a^4}{b^4}\int_{-b}^{b}(b^2-z^2)^2dz = \frac{1}{2}\rho\pi\frac{a^4}{b^4}b^4\int_{\pi}^{0}\sin^4\theta(-b\sin\theta)d\theta$$

$$= \frac{1}{2}\rho\pi a^4 b\int_{0}^{\pi}(1-\cos^2\theta)^2\sin\theta d\theta = \frac{2}{5}Ma^2.$$

The moment of inertia of an elliptic disk of mass M, the semimajor axis a and the semiminor axis b is

$$I = \frac{1}{4}M(a^2+b^2).$$

Then the moment of inertia about the y-axis is

$$I_y = \int_{-a}^{a}\frac{1}{4}\rho\pi\frac{b}{a}(a^2-y^2)\left\{(a^2-y^2)+\frac{b^2}{a^2}(a^2-y^2)\right\}dy$$

$$= \int_{-a}^{a}\frac{3M(a^2+b^2)}{16a^5}(a^4-2a^2y^2+y^4)dy = \frac{1}{5}M(a^2+b^2).$$

Similarly,

$$I_x = \frac{1}{5}M(a^2+b^2).$$

Chapter 11

Problem 1. When Mercury and the moon were born, it is thought that they had the liquid state inner structure. Tidal forces by the Sun and/or the Earth exerted on the inner structure to brake their rotation, so that their orbital periods and rotational periods synchronized.

Problem 2. Let the position of a planet be (x, y) in Fig. 11.11.

$$\overline{OF_1} = \overline{OA} - \overline{F_1A} = \frac{\eta}{1-\varepsilon^2} - \frac{\eta}{1+\varepsilon} = \frac{\varepsilon\eta}{1-\varepsilon^2} = c,$$

$x = r\cos\theta + c, \quad y = r\sin\theta, \quad r = \sqrt{(x-c)^2+y^2}.$

Substituting the above equations into (11.49), we obtain

$$\frac{x^2}{\eta^2/(1-\varepsilon^2)^2} + \frac{y^2}{\eta^2/(1-\varepsilon^2)} = 1,$$

where

$$a = \overline{OA} = \frac{\eta}{1+\varepsilon} + \frac{\eta\varepsilon}{1-\varepsilon^2} = \frac{\eta}{1-\varepsilon^2}, \quad b = \frac{\eta}{\sqrt{1-\varepsilon^2}}$$

holds, then we find

$$\frac{x^2}{a^2} + \frac{y^2}{b^2} = 1.$$

Problem 3.
Let's calculate one fourth area $(x > 0, \ y > 0)$ of an ellipse,
Putting $x = a\sin\theta$,

$$\frac{1}{4}S = ab\int_{0}^{\pi/2}\cos^2\theta d\theta = ab\int_{0}^{\pi/2}\frac{1+\cos(2\theta)}{2} = \frac{ab}{2}\left[\frac{\pi}{2}+\frac{1}{4}\sin(2\theta)\right]_{0}^{\pi/2}$$

$= \pi ab/4$

$S = \pi ab$

Answers to Chapter End Problems

1. 3.83×10^8 m

2. $F_z = -2\pi Gm\sigma \left(1 - \dfrac{z}{\sqrt{a^2 + z^2}}\right)$

3. $F_z = -2\pi Gm\sigma$

4. $\Phi = -G\dfrac{mM}{z}, \qquad F_z = -G\dfrac{mM}{z^2}$

5. $\Phi = -G\dfrac{mM}{a}, \qquad F_z = 0$

6. (1) $7.962 \times 10^{-5}\,\mathrm{s}^{-1}$

   (2) Taking the square root of (11.34) and substituting it into (11.37) yields

   $$r^{-1/2}dr = -\dfrac{2I}{m\sqrt{GM}}d\Omega.$$

   (3) 3.70×10^8 m

7. When the orbit is a circle, the universal gravitation and the orbital speed is constant. Letting the orbital radius be r, the orbital period be T and the orbital speed be v, we get

   $$v = \dfrac{2\pi r}{T}, \quad T^2 = \dfrac{4\pi^2 r^2}{v^2}.$$

   Letting the mass of the Sun be M, the mass of the planet be m and the gravitational constant be G, the equation of motion concerning the orbital motion of the planet becomes

   $$m\dfrac{v^2}{r} = G\dfrac{mM}{r^2}$$

   Using the above two equations, we get

   $$T^2 = 4\pi^2 r^2 \dfrac{r}{GM} = \dfrac{4\pi^2}{GM}r^3.$$

8. $\sqrt{(x-c)^2 + y^2} + \sqrt{(x+c)^2 + y^2} = 2a$

   $2(x^2 + y^2 + c^2) + 2\sqrt{(x^2 + y^2 + c^2)^2 - 4c^2 x^2} = 4a^2$

   $\dfrac{a^2 - c^2}{a^2}x^2 + y^2 = a^2 - c^2$

   $\dfrac{x^2}{a^2} + \dfrac{y^2}{b^2} = 1$ (because $b^2 = a^2 - c^2$)

9. Let the center of the Earth be O, the center of the satellite be O', one end of the satellite be A, the other end be B, the angle between the orbit and $\overline{AB}$ be θ, the distance between $\overline{AO}$ be r_A, and the distance between $\overline{BO}$ be r_B. We find using the Pythagorean theorem

$$r_B{}^2 = (r + l \sin \theta)^2 + (l \cos \theta)^2 = r^2 + l^2 + 2rl \sin \theta \,,$$
$$r_A{}^2 = (r - l \sin \theta)^2 + (l \cos \theta)^2 = r^2 + l^2 - 2rl \sin \theta \,.$$

The gravitational potential exerting on the satellite by the Earth is

$$\Phi = -GMm \left(\frac{1}{r_A} + \frac{1}{r_B} \right)$$
$$= -GMm \left\{ \left(r^2 + l^2 - 2rl \sin \theta \right)^{-1/2} + \left(r^2 + l^2 + 2rl \sin \theta \right)^{-1/2} \right\}$$
$$\cong -GMm \left\{ \left(1 + \frac{l^2}{r^2} - \frac{2l \sin \theta}{r} \right)^{-1/2} + \left(1 + \frac{l^2}{r^2} + \frac{2l \sin \theta}{r} \right)^{-1/2} \right\}$$
$$\cong -GMm \left\{ 1 + \frac{2l \sin \theta}{r} + \frac{1}{2l} \left(-\frac{3}{4} \right) \left(\frac{2l \sin \theta}{r} \right)^2 \right.$$
$$\left. + 1 - \frac{2l \sin \theta}{r} + \frac{1}{2l} \left(-\frac{3}{4} \right) \left(\frac{2l \sin \theta}{r} \right)^2 \right\}$$
$$= -GMm \left\{ 2 - \frac{3l^2}{2r^2} \sin^2 \theta \right\} \,.$$

The extreme values of Φ is obtained by differentiating the above equation and equating the resultant equation zero.

$$\frac{d\Phi}{d\theta} = GMm \frac{3l^2}{2r^2} \sin 2\theta = 0,$$
$$\theta = \frac{\pi}{2}.$$

Because $d^2\Phi(\pi/2)/d\theta^2 < 0$, Φ is minimum at $\pi/2$. Therefore, the satellite is stable when AB is vertical to the orbit.

10. $T = \frac{2\pi}{\omega} = 2\pi \sqrt{\dfrac{lr^2}{6GM}}$.

11. The gravitational force between a sphere of radius r, density ρ and an infinitesimal mass element $\rho r^2 \cos \theta d\theta d\phi dr$ at distance ξ is

$$\delta F = -\frac{G}{\xi^2} \frac{4}{3} \rho \pi r^3 \cdot \rho r^2 \cos \theta d\theta d\phi dr.$$

Then the work necessary for moving an infinitesimal volume element $r^2 \cos\theta d\theta d\phi dr$ from the spherical surface to ∞ is

$$\delta W = \int_r^\infty \frac{G}{\xi^2}\left(\frac{4}{3}\rho\pi r^3\right)\rho r^2 \cos\theta d\theta d\phi dr d\xi$$

$$= \frac{4}{3}G\rho^2 \pi r^4 \cos\theta d\theta d\phi dr.$$

The work necessary for moving all mass of the Earth to ∞ is obtained integrating δW through the whole sphere,

$$W = \frac{4}{3}G\rho^2\pi \int_0^a \int_0^{2\pi} \int_{-\pi/2}^{\pi/2} r^4 \cos\theta d\theta d\phi dr$$

$$= \frac{16}{15}G\rho^2\pi^2 a^5.$$

Therefore, the potential energy possessed by the Earth of uniform density ρ is,

$$U = -\frac{16}{15}G\rho^2\pi^2 a^5,$$

taking $U = 0$ at ∞. After the gravitational differentiation, the potential energy of a sphere of radius a_1 and the heavier density ρ_1 is

$$U_1 = -\frac{16}{15}G\pi^2\rho_1{}^2 a_1{}^5.$$

The work necessary for moving an infinitesimal volume element $r^2 \cos\theta d\theta d\phi dr$ from the spherical surface of radius r, inner radius a_1 and density ρ_2 to ∞ is

$$\delta W = \int_r^\infty \frac{G}{\xi^2}\frac{4}{3}\pi\left\{\rho_1 a_1{}^3 + \rho_2(r^3 - a_1{}^3)\right\}\rho_2 r^2 \cos\theta d\theta d\phi dr d\xi$$

$$= \frac{4}{3}G\pi\left\{(\rho_1 - \rho_2)\rho_2 a_1{}^3 r + \rho_2 r^4)\right\}\cos\theta d\theta d\phi dr.$$

The work necessary for moving all material at radius $\geq a_1$ to ∞ is

$$W_2 = \frac{4}{3}G\pi(\rho_1 - \rho_2)\rho_2 a_1{}^3 \int_{a_1}^a \int_0^{2\pi} \int_{-\pi/2}^{\pi/2} r \cos\theta d\theta d\phi dr$$

$$+ \frac{4}{3}G\pi\rho_2{}^2 \int_{a_1}^a \int_0^{2\pi} \int_{-\pi/2}^{\pi/2} r^4 \cos\theta d\theta d\phi dr$$

$$= \frac{4}{3}G\pi(\rho_1 - \rho_2)\rho_2 a_1{}^3 4\pi\left[\frac{1}{2}r^2\right]_{a_1}^a + \frac{4}{3}G\pi\rho_2{}^2 4\pi\left[\frac{1}{5}r^5\right]_{a_1}^a$$

$$= \frac{8}{3}G\pi^2(\rho_1 - \rho_2)\rho_2 a_1{}^3(a^2 - a_1{}^2) + \frac{16}{15}G\pi^2\rho_2{}^2(a^5 - a_1{}^5).$$

Then the potential energy possessed by the Earth after the gravitational differentiation is

$$U_1 + U_2 = -\frac{8}{3}G\pi^2(\rho_1 - \rho_2)\rho_2 a_1{}^3(a^2 - a_1{}^2) - \frac{16}{15}G\pi^2 \rho_2{}^2(a^5 - a_1{}^5)$$
$$- \frac{16}{15}G\pi^2 \rho_1{}^2 a_1{}^5.$$

Therefore, the potential energy released by the gravitational differentiation is

$$U - (U_1 + U_2) = -\frac{16}{15}G\pi^2\left\{(\rho^2 - \rho_2{}^2)a^5 - (\rho_1{}^2 - \rho_2{}^2)a_1{}^5\right\}$$
$$+ \frac{8}{3}G\pi^2(\rho_1 - \rho_2)\rho_2 a_1{}^3(a^2 - a_1{}^2).$$

12. $\Delta U = 2.365 \times 10^{31} \text{ [J]}$

13. $\Delta T = 6.11 \times 10^3 \text{ [K]}$

Chapter 12

Problem 1.

Integrating (12.47) from o to the atmospheric height H, we obtain

$$p(H) - p(0) = -\rho g H$$
$$H = \frac{p(0)}{\rho g} = \frac{1.01 \times 10^5}{1.00 \times 9.80} = 1.03 \times 10^4 \text{ [m]}$$

Problem 2.

$$\mathbf{\Omega} \times \mathbf{r} = \begin{vmatrix} \mathbf{e}_r & \mathbf{e}_\theta & \mathbf{e}_\phi \\ \Omega\sin\theta & \Omega\cos\theta & 0 \\ r & 0 & 0 \end{vmatrix} = \mathbf{e}_\phi(-r\Omega\cos\theta).$$

Using the rotation of a spherical coordinates in subsection B.2.3

$$\nabla \times (\mathbf{\Omega} \times \mathbf{r}) = \mathbf{e}_r \frac{1}{r\cos\theta}\frac{\partial}{\partial\theta}(-r\Omega\cos^2\theta) + \mathbf{e}_\theta \frac{1}{r}\left(\frac{\partial(r^2\Omega\cos\theta)}{\partial r}\right)$$
$$= 2\Omega(\mathbf{e}_r\sin\theta + \mathbf{e}_\theta\cos\theta) = 2\mathbf{\Omega}.$$

Answers to Chapter End Problems

1. Putting $\partial/\partial t = 0$ and $\Phi = T$ in (12.2), we get

$$\frac{dT}{dt} = v\frac{\partial T}{\partial y},$$
$$\frac{\partial T}{\partial y} = \frac{1}{v}\frac{dT}{dt} = \frac{6.00 \times 10^{-4}}{1.50 \times 10^2} = 4.00 \times 10^{-6} \text{ [K m}^{-1}\text{]}.$$

2. Suppose that the mass of a fluid is m, the volume at temperature T is V and the volume at temperature T_0 is V_0, we find

$$V = V_0\{1 + \alpha(T - T_0)\},$$
$$\rho = \frac{m}{V} = \frac{m}{V_0\{1 + \alpha(T - T_0)\}} \cong \rho_0\{1 - \alpha(T - T_0)\},$$

neglecting all terms of order $(T - T_0)^2$ and higher. Therefore,

$$\delta\rho\mathbf{g} = (\rho - \rho_0)\mathbf{g} = -\rho_0\alpha(T - T_0)\mathbf{g}.$$

3. The equation of state of ideal gas is

$$pV = mRT.$$

From the definition of the coefficient of thermal expansion,

$$\alpha = \frac{1}{V}\left(\frac{\partial V}{\partial T}\right)_p = \frac{1}{V}\frac{mR}{p} = \frac{1}{T}.$$

The buoyant force is obtained from (12.35),

$$\delta\rho\mathbf{g} = -\rho_0\alpha(T - T_0)\mathbf{g} = -\frac{T - T_0}{T_0}\rho_0\mathbf{g}.$$

4. The Coriolis parameter at the observatory is

$$f = 2\Omega\sin 30 = 7.27 \times 10^{-5}\,[\text{s}^{-1}].$$

Thus, the geostrophic wind is

$$u = -\frac{1}{f\rho_0}\frac{\partial p}{\partial y} = \frac{10 \times 10^2/1.0 \times 10^6}{7.27 \times 10^{-5} \times 1.20} = 11.5\,[\text{m s}^{-1}].$$

The velocity change is obtained using (12.50) and (12.51),

$$\delta u = f(v - v_g)\delta t = 7.27 \times 10^{-5} \times 2.5 \times 60^2 = 0.65\,[\text{m s}^{-1}].$$
$$\delta v = -f(u - u_g)\delta t = -7.27 \times 10^{-5} \times 1.5 \times 60^2 = -0.39\,[\text{m s}^{-1}].$$

The velocity components become $u = 13.7\,[\text{m s}^{-1}]$ and $u = 2.1\,[\text{m s}^{-1}]$.

5. The azimuthal velocity is $\mathbf{v} = \mathbf{\Omega} \times \mathbf{r}$, so that using vector identity and the divergence of cylindrical coordinates in the Appendix,

$$\nabla \times (\mathbf{\Omega} \times \mathbf{r}) = \mathbf{\Omega}(\nabla \cdot \mathbf{r}) - \mathbf{r}(\nabla \cdot \mathbf{\Omega}) - (\mathbf{\Omega} \cdot \nabla)\mathbf{r} + (\mathbf{r} \cdot \nabla)\mathbf{\Omega}$$
$$= \Omega\frac{\partial(r^2)}{r\partial r} = 2\Omega.$$

Chapter 13

Problem 1.

$$\Phi_x = \left(1 - \frac{1 + \exp\left(-\pi\right)}{2\pi}\right)\rho\delta_E u_g,$$

$$\Phi_y = \frac{1 + \exp\left(-\pi\right)}{2\pi}\rho\delta_E u_g.$$

Problem 2.

$$\Phi_r = 0,$$

$$\Phi_y = -\{1 + \exp\left(-\pi\right)\}\frac{\tau}{f}.$$

Problem 3. $3.2 \times 10^2\,\mathrm{s}$, $1.0 \times 10^4\,\mathrm{s}$

Problem 4. From (13.101), we find

$$\frac{\partial}{\partial x}T = \frac{\nu}{\alpha g}\nabla^4\psi,$$

Using (13.94), we will nondimensionalize temperature perturbation as

$$T = (T_1 - T_2)T^* = d\gamma T *,$$

Together with (13.105), we obtain

$$\frac{d\gamma}{d}\frac{\partial T^*}{\partial x *}T = \frac{\nu^2}{\alpha g d^4}\nabla^{*4}\psi^*,$$

Supposing wave structure of the wavenumber k in the x-direction and omitting asterisks, we get

$$T = \frac{\nu^2}{k\alpha\gamma g d^4}\left(D^2 - k^2\right)^2\Psi = \frac{1}{kG_r}\left(D^2 - k^2\right)^2\Psi.$$

Problem 5. The solution of the case of the both rigid boundaries and the odd mode satisfies the boundary conditions of a free surface at $z = 0$, so that we can assume a material boundary at $z = 0$. This is the situation of one rigid and one free boundary problem of fluid depth $d/2$. So, we can obtain the marginal Rayleigh number and the marginal wave length as follows:
$R_{ac} = 17610.048 \times (1/2)^4 = 1100.628$, $k_c = 5.365 \times (1/2) = 2.683$

Answers to Chapter End Problems

1. The pressure gradient force balances with the centrifugal force balance, so that

$$\frac{V^2}{r} = \frac{1}{\rho}\frac{\partial p}{\partial r},$$

$$r\omega^2 = \frac{1}{\rho}\frac{\partial\rho g H}{\partial r},$$

$$H = H_0 + \frac{\omega^2 r^2}{2g}.$$

2.

$$u = u_g \left[1 - \exp\left(-\gamma z\right) \cos \gamma z\right],$$
$$v = -u_g \exp\left(-\gamma z\right) \sin \gamma z,$$

where $\gamma = \sqrt{|f|/(2\nu)}$.

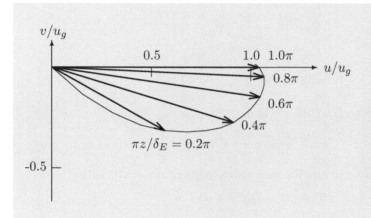

Figure D.3: Hodograph of the Ekman flow for the Southern Hemisphere. Velocity vectors are shown at normalized height $\pi z/\delta_E = 0.2\pi$, 0.4π, 0.6π, 0.8π, 1.0π.

3.

$$u = \frac{\tau}{2\rho_0 \gamma \nu} \left(\cos \gamma z + \sin \gamma z\right) \exp\left(\gamma z\right),$$
$$v = \frac{\tau}{2\rho_0 \gamma \nu} \left(\cos \gamma z - \sin \gamma z\right) \exp\left(\gamma z\right).$$

4.

$$u = V \exp\left(-\gamma z\right) \sin \gamma z = r\omega \exp\left(-\gamma z\right) \sin \gamma z,$$
$$v = V \left\{1 - \exp\left(-\gamma z\right) \cos \gamma z\right\} = r\omega \left\{1 - \exp\left(-\gamma z\right) \cos \gamma z\right\}.$$
$$w(\delta_E) = -\frac{\omega}{\gamma} \left\{1 + \exp\left(-\pi\right)\right\}.$$

5. In the inviscid layer, the pressure gradient force and centrifugal force exactly balance. But in the viscous boundary layer, the tangential velocity is reduced by the frictional force and the pressure gradient force overcomes the centrifugal force, so that tea leaves gather to the center of the teacup near the bottom.

6. $13.8\,\mathrm{m}$

7. Solving the general solutions (13.44) and (13.45) applying the boundary conditions that $u = 0$, $v = rw$, at $z = 0$ and $u = v = 0$, at $z = -\infty$, we obtain the following solution,

$$u = -rw \exp(\gamma z) \sin \gamma z \,,$$
$$v = rw \exp(\gamma z) \cos \gamma z \,.$$

Horizontal divergence is

$$\boldsymbol{\nabla}_{\mathrm{H}} \cdot \mathbf{v} = \frac{1}{r} \frac{\partial(ru)}{\partial r} + \frac{\partial v}{r\partial \theta} = -2\Omega \exp(\gamma z) \sin \gamma z.$$

Integrating the continuity equation from $z = -\delta_{\mathrm{E}}$ to $z = 0$,

$$\int_{-\delta_{\mathrm{E}}}^{0} \frac{\partial w}{\partial z} dz = -\int_{-\delta_{\mathrm{E}}}^{0} \boldsymbol{\nabla}_{\mathrm{H}} \cdot \mathbf{v} dz = -\frac{\omega}{\gamma}\left(1 + \exp(-\pi)\right),$$
$$w(0) - w(\delta_{\mathrm{E}}) = \frac{\omega}{\gamma}\left(1 + \exp(-\pi)\right),$$
$$w(\delta_{\mathrm{E}}) = \frac{\omega}{\gamma}\left(1 + \exp(-\pi)\right).$$

There is an upward motion $w = \Omega(1 + \exp(-pi))/\gamma$.

8. $4.47 \times 10^5\,\mathrm{s} = 5.18\,\mathrm{days}$

9. Putting $\bar{v} = r\Omega$ and substituting it into (13.163), we obtain

$$\frac{d^2\Omega}{dr^2} + \frac{3}{r}\frac{d\Omega}{dr} = 0.$$

Putting $\Omega' = d\Omega/dr$, the above equation becomes

$$\frac{d\Omega'}{\Omega'} = -\frac{3}{r}dr,$$
$$\Omega' = \frac{d\Omega}{dr} = C_1 r^{-3},$$
$$\Omega = C_2 - \frac{C_1}{2}r^{-2}\,,$$
$$\bar{v} = r\Omega = Ar + \frac{B}{r}.$$

Chapter 14

Answer of Problem 1.

$$\mathbf{c} \cdot \mathbf{c}_g = \frac{N^2 k}{(k^2 + m^2)^3}(km^2 - km^2) = 0.$$

We find that the phase velocity and the group velocity are orthogonal to each other.

Answers to Chapter End Problems

1. (1)

$$\rho = \frac{(28 \times 0.8 + 32 \times 0.2) \times 10^{-3}}{22.4 \times 10^{-3}} = 1.29\,[\mathrm{kgm}^{-3}].$$

(2) Letting the volume of the atmosphere of 1 mol be v_0, we find using the ideal gas equation

$$\frac{1.00 \times 10^3 \times 10^2 v_0}{288} = \frac{1.013 \times 10^3 \times 10^2 \times 22.4 \times 10^{-3}}{273},$$

$$v_0 = 23.9 \times 10^{-3}\,[\mathrm{m}^3],$$

$$\rho_0 = \frac{(28 \times 0.8 + 32 \times 0.2) \times 10^{-3}}{23.9 \times 10^{-3}} = 1.21\,[\mathrm{kgm}^{-3}].$$

(3) Letting the volume of the atmosphere of 1 mol be v', we find using the ideal gas equation

$$\frac{p_0 v_0}{T_0} = \frac{p' v'}{T'}.$$

$$v' = \frac{1.00 \times 10^3 \times 10^2}{8.5 \times 10^2 \times 10^2} \frac{279}{288} \times 23.9 \times 10^{-3} = 27.2 \times 10^{-3}\,[\mathrm{m}^3].$$

$$\rho' = \frac{28.8 \times 10^{-3}}{27.2 \times 10^{-3}} = 1.06\,[\mathrm{kgm}^{-3}].$$

(4)

$$N = \sqrt{\frac{\Delta \rho g}{\rho z}} = 2.94 \times 10^{-2}\,[\mathrm{s}^{-1}].$$

2.

$$c = \sqrt{gH} = \sqrt{9.80 \times 4.00 \times 10^3} = 1.98 \times 10^2\,\mathrm{ms}^{-1},$$

$$t = \frac{1.64 \times 10^7}{1.98 \times 10^2} = 8.28 \times 10^4\,\mathrm{s} = 23.0\,\mathrm{hr};\,.$$

3. Substituting h' into (14.20), we find

$$u' = \frac{gk}{\hat{\sigma}} h'.$$

Substituting h' into (14.22), we get

$$-\frac{\partial u'}{\partial x} = -\tilde{\imath}\frac{\hat{\sigma}}{H} h'.$$

For the wave propagating in the positive x-direction, u' and h' is just in phase, and the phase of horizontal convergence advances that of h' by

$\pi/2$. As a consequence, crests move to the positive x-direction and the wave propagates in the positive x-direction. For the wave propagating in the negative x-direction, u' and h' is π out of phase, and the phase of horizontal convergence lags that of h' by $\pi/2$. Then, crests move to the positive x-direction and the wave propagates in the negative x-direction.

4.

$$c = \sqrt{\frac{(1.03 - 1.00) \times 10^3}{1.03 \times 10^3}} \times 9.80 \times 10.0 = 1.69 \,[\text{ms}^{-1}].$$

5. Substituting (14.24) into (14.23), we get

$$(-\tilde{i}\sigma + \tilde{i}k\bar{u})^2 \hat{h} + gH(k^2 + l^2)\hat{h} = 0,$$
$$-(\sigma - k\bar{u})^2 + gH(k^2 + l^2) = 0,$$
$$\sigma - k\bar{u} = \pm\sqrt{gH(k^2 + l^2)}.$$

6. Differentiating (14.76) partially by k, we find

$$c_{gx} = \frac{\partial \hat{\sigma}}{\partial k}$$
$$= \pm(k^2 + l^2 + m^2)^{-1}[2kN^2\frac{1}{2}\left\{N^2(k^2 + l^2) + f^2m^2\right\}^{-1/2}$$
$$\quad - \left\{N^2(k^2 + l^2) + f^2m^2\right\}^{1/2} 2k\frac{1}{2}(k^2 + l^2 + m^2)^{-1/2}]$$
$$= \pm\frac{\left[kN^2(k^2 + l^2 + m^2) - k\left\{N^2(k^2 + l^2) + f^2m^2\right\}\right]}{(k^2 + l^2 + m^2)^{3/2}}$$
$$= \pm\frac{(N^2 - f^2)km^2}{(k^2 + l^2 + m^2)^{3/2}}.$$

The y and z-components of the group velocity c_{gy} and c_{gz} are similarly calculated.

7. The Coriolis parameter of the midlatitude is

$$f = \frac{2\pi}{24 \times 60 \times 60} \sin 45° = 1.03 \times 10^{-4} \,[\text{s}^{-1}].$$

From (14.86), we find

$$N^2k^2 = f^2m^2,$$
$$\frac{N}{L_x} = \frac{f}{L_z},$$
$$\theta = \tan^{-1}\frac{L_z}{L_x} = \tan^{-1}\frac{f}{N} = \tan^{-1}\frac{1.03 \times 10^{-4}}{1.20 \times 10^{-2}} = 0.49°.$$

Chapter 15

Problem 1.

Letting $l_d = \varepsilon^\alpha \nu^\beta$, we find

$$L = (L^2 T^{-3})^\alpha (L^2 T^{-1})^\beta = L^{2\alpha + 2\beta} T^{-(3\alpha + \beta)} .$$

Thus, we get

$$\begin{cases} 2(\alpha + \beta) = 1 , \\ 3\alpha + \beta = 0 . \end{cases}$$

Solving the above equations, we obtain $\alpha = -1/4, \beta = 3/4$.

Answers to Chapter End Problems

1. The pressure gradient in the atmosphere is successively produced by the thermal contrast of the Earth surface, while the oceanic pressure gradient if any is quickly canceled by gravity waves.

2.

$$|c_g|^2 = \frac{\beta^2}{\kappa^8} \{(k^2 - l^2)^2 + 4k^2 l^2\} = \frac{\beta^2}{\kappa^4}$$

$$|c_g| = \frac{\beta}{\kappa^2} = \left|\frac{\sigma}{k}\right| = |c_x|$$

3.

$$\frac{\partial \bar{u}}{\partial z} \frac{\partial \rho}{\partial x} = \Lambda \frac{\partial}{\partial x} \frac{-1}{g} \frac{\partial p}{\partial z} = -\frac{\Lambda}{g} \frac{\partial^2 p}{\partial z \partial x}$$

$$\frac{\partial \bar{\rho}}{\partial y} \frac{\partial v}{\partial z} = -\frac{1}{g} \frac{\partial}{\partial z} \frac{\partial \bar{p}}{\partial y} \frac{\partial}{\partial z} \left(\frac{1}{f \rho_0} \frac{\partial p}{\partial x} \right) = \frac{\Lambda}{g} \frac{\partial^2 p}{\partial z \partial x}$$

4. Differentiating (14.79) with respect to z , we get

$$\left(\frac{\partial}{\partial t} + \bar{u} \frac{\partial}{\partial x}\right) \frac{\partial \rho}{\partial z} + \frac{\partial \bar{u}}{\partial z} \frac{\partial \rho}{\partial x} + \frac{\partial v}{\partial z} \frac{\partial \bar{\rho}}{\partial y} + v \frac{\partial}{\partial y} \left(\frac{\partial \bar{\rho}}{\partial z}\right) + \frac{\partial w}{\partial z} \frac{\partial \bar{\rho}}{\partial z} = 0.$$

In the above equation

$$\frac{\partial \bar{u}}{\partial z} \frac{\partial \rho}{\partial x} = \frac{g}{f \rho_0} \frac{\partial \bar{\rho}}{\partial y} \frac{\partial \rho}{\partial x} ,$$

$$\frac{\partial v}{\partial z} \frac{\partial \bar{\rho}}{\partial y} = -\frac{g}{f \rho_0} \frac{\partial \rho}{\partial x} \frac{\partial \bar{\rho}}{\partial y},$$

$$\frac{\partial}{\partial y} \left(\frac{\partial \bar{\rho}}{\partial z}\right) = 0.$$

Using (14.86), the above equation becomes

$$\frac{\partial w}{\partial z} = -\frac{1}{\rho_0 N^2} \left(\frac{\partial}{\partial t} + \bar{u} \frac{\partial}{\partial x}\right) \frac{\partial^2 p}{\partial z^2}.$$

5. We can obtain the zonal velocity using (12.53)

$$u(d) = u(0) + \frac{du}{dz}d = \frac{\Delta\rho g d}{f\rho_0(b-a)},$$

$$R_\mathrm{o} = \frac{1}{f(b-a)}\frac{\Delta\rho g d}{f\rho_0(b-a)} = \frac{\Delta\rho g d}{4\Omega^2\rho_0(b-a)^2}.$$

Multiplying the above equation by four, we get

$$\Theta = 4R_\mathrm{o}.$$

Appendix E

Further Reading

1. Barger V. D. and Olsson M. G.: *Classical Mechanics–A Modern Perspective, 2nd edn.*, McGraw–Hill Inc., New York (1973).

2. Goldstein H., Poole Jr C. P. and Safko J. L.: *Classical Mechanics, 3rd edn.*, Pearson Education Inc., San Francisco (2001).

3. Greenspan H. P.: *The Theory of Rotating Fluids*, Cambridge University Press, Cambridge (1968).

4. Holton J. R.: *An Introduction to Dynamic Meteorology, 4th edn.*, Elsevier Academic Press, Burlington MA (2004).

5. Lindzen R. S.: *dynamics in atmospheric physics*, Cambridge University Press, Cambridge (1990).

6. McCall M. W.: *Classical Mechanics, 2nd edn.*, John Wiley and Sons Ltd., Hoboken NJ (2011).

7. Shimadu Y.: *Physics of the Earth's Interior*, Shokabo Company Ltd., Tokyo (1966), *in Japanese.*

8. Sommerfeld A.: *Mechanik der Deformierbaren Medien*, Becker and Erler, Leipzig (1945).

DOI: 10.1201/9781003310068-E

Appendix E

Further Reading

1.
2.
3.
4.
5.
6.
7.

Index